RIVER MECHANICS

RIVER MECHANICS

M. Selim Yalin

M.A.S.C.E., M.I.C.E., M.E.I.C., M.J.S.C.E.

Emeritus Professor of Civil Engineering,
Queen's University, Kingston, Canada

PERGAMON PRESS

OXFORD • NEW YORK • SEOUL • TOKYO

U.K. Pergamon Press Ltd, Headington Hill Hall,
Oxford OX3 0BW, England

U.S.A. Pergamon Press, Inc, 660 White Plains Road,
Tarrytown, New York 10591-5153, U.S.A.

OREA Pergamon Press Korea, KPO Box 315, Seoul 110-603,
Korea

JAPAN Pergamon Press Japan, Tsunashima Building Annex,
3-20-12 Yushima, Bunkyo-ku, Tokyo 113, Japan

First edition 1992

British Library Cataloguing in Publication Data
A catalogue record for this book is available from the British Library

Library of Congress Cataloging-in-Publication Data
A catalogue record for this book is available from the Library of Congress

ISBN 0 08 040190 2

Printed in Great Britain by BPCC Wheatons Ltd, Exeter

CONTENTS

3. Bed forms and friction factor

4. Regime channels

5. Meandering and braiding

PREFACE

This book concerns the processes and phenomena taking place in a river flowing in alluvium. It is intended primarily for graduate students and researchers in hydraulic engineering, geomorphology, physical geography and other disciplines dealing with the behaviour and evolution of rivers; however, it may have some appeal also to practising professionals.

Many fluvial processes have not yet been explained in a generally accepted manner, and therefore it would be only appropriate to discuss their mechanisms and formulations for the simplest cases – the introduction of possible but unnecessary complexities would only obscure the issue. Considering this, the present study of "ideal rivers" and associated processes is carried out for the minimum number of phenomenon-defining parameters: the flow rate is always constant, the alluvium cohesionless and homogeneous.

As is well known, at certain stages of sediment transport the initially flat flow boundaries become undulated: they become covered by bed forms (sand waves). After the discovery of turbulent bursts, it became evident that if antidunes, which are due to standing surface waves, are excluded, then all the remaining bed forms generated by a rectilinear flow are caused by turbulence (as has been anticipated for a long time by many researchers). More specifically, dunes and bars are caused by the sequences of vertical and horizontal bursts respectively, ripples being due to the viscous structures of the flow at the bed. Chapter 2 deals exclusively with the interpretation of turbulence in the light of the recent advances in that field: the emphasis is on turbulent structures and their scales. Chapter 3 concerns the bed forms themselves.

The current prevailing view is that an alluvial stream evolves because one of its energy-related properties tends to acquire its extreme value: the evolution terminates (the regime state is achieved) when the extreme value is reached. The present text adheres to this view (Chapter 4). In fact, the regime trend is used not only to determine the width, depth and slope of an alluvial stream, but also to explain, and whenever possible to formulate, its behaviour in plan. Questions such as, "Will an initially straight stream remain as it is, or will it braid or meander; and in the latter case, to what extent?" are answered in this book with the aid of the regime concept. In Chapter 5 it is shown that the answer to such questions depends largely on how the initial or valley slope of a stream compares with its regime slope.

Very often in fluvial hydraulics one deals with experimental curves whose equations cannot be deduced theoretically. And yet, for computational purposes, it would be desirable to represent an experimental point pattern by a best-fit equation. Considering this, certain "computing equations" (in short, "comp-eqs.") are suggested in Chapters 1, 3 and 4 – these equations have no claim other than that their graphs pass through the midst of the respective data-point patterns.

Letter symbols, particularly the conventional ones, are often used in the present text without definition. To compensate for this, a List of Symbols is included at the beginning of the book.

This text contains many ideas of my own which have not been published before. Thus, for imperfections in any statement, formula or diagram presented without reference, I alone am responsible.

It is my pleasant duty to express my thanks to Dr. D. W. Bacon, former Dean of the Faculty of Applied Science, Queen's University, for his generous financial and moral support.

I am also indebted to Prof. E. A. Walker and Dr. K. R. Hall for reviewing the English of my original text.

My deepest gratitude goes to my graduate student A. M. Ferreira da Silva, M.Sc. (LNEC-Lisbon), for her competent and enthusiastic help in the preparation of the scientific content of this book.

M. S. Yalin

Kingston, Ontario

LIST OF RELEVANT SYMBOLS

1. General

g	acceleration due to gravity
t	time
x	direction of rectilinear flow
y	direction vertically perpendicular to x
z	direction horizontally perpendicular to x
y_b	elevation of bed surface
f_A	dimensional function determining a quantity A
Φ_A, ϕ_A, Ψ_A, ψ_A	dimensionless functions determining a quantity A
$\approx$	approximately equal to, comparable with
$\sim$	proportional to (proportionality factor may not be constant)
∇	nabla-operator ("del")

Subscripts:

av	signifies average value of a quantity
cr	signifies the value corresponding to the initiation of sediment transport (to the "critical stage")
R	signifies regime value of a quantity

2. Physical Properties of Fluid and Granular Material

ρ	density of fluid
$\gamma = \rho g$	specific weight of fluid
ν	kinematic viscosity
ρ_s	grain density
γ_s	specific weight of grains in fluid
D	typical grain size (usually D_{50})
w	terminal (settling) velocity of grains

3. Flow

B	flow width at the free surface
$\overline{B}$	flow width at the bed
B_c	width of the two-dimensional flow region

h	flow depth
S	slope
Q	flow rate
q	flow rate per unit width (specific flow rate)
v	average flow velocity
U_i	instantaneous local flow velocity in the direction i ($i = x, y, z$)
u_i'	fluctuating component of U_i
u_i	time average component of U_i
U and u	abbreviations for U_x and u_x respectively
u_{max}	maximum value of u (at the free surface, if B/h is large)
u_b	typical value of u at the bed
B_s	roughness function
$\kappa \approx 0.4$	Von Karman constant
τ	shear stress
τ_v, τ_t	viscous and turbulent components of τ
τ_0	shear stress interacting between the flow and the bed surface
$(\tau_0)_f$, $(\tau_0)_\Delta$	pure friction and form-drag components of τ_0
$v_* = \sqrt{\tau_0/\rho}$	shear velocity
$c = v/v_*$	dimensionless Chézy friction factor
c_f, c_Δ	pure friction and form-drag components of c
$\bar{c}$	flat bed value of c
k_s	granular roughness of bed surface
k_s'	granular roughness of banks
K_s	total roughness of the bed (covered by bed forms)

4. Mechanics of Sediment Transport

ϵ	the value of y separating bed-load and suspended-load regions
q_{sb}	specific volumetric bed-load rate (within ϵ)
q_{ss}	specific volumetric suspended-load rate (within $(h - \epsilon)$)
$q_s = q_{sb} + q_{ss}$	total specific volumetric transport rate past the bed
$q_s' = q_{sb}' + q_{ss}'$	total specific volumetric transport rate past the banks
Q_s	volumetric transport rate (through the whole cross-section of the flow)
p_s	volumetric deposition rate per unit area (from an external source)
C	local volumetric concentration of suspended particles
C_ϵ	the value of C at $y = \epsilon$
V_b	vertical displacement velocity of a point of the bed surface
U_b	migration velocity of bed forms
Λ	developed bed form length
Δ	developed bed form height
$\delta = \Delta/\Lambda$	developed bed form steepness

Λ_i, Δ_i, δ_i signify the quantities above corresponding to the bed form i ($i = a$ (alternate bars); $= b$ (bars); $= d$ (dunes); $= g$ (antidunes); $= r$ (ripples))

Λ_t	bed form length at a time t
Δ_t	bed form height at a time t
T_i	duration of development of a bed form i

5. Turbulence

i) Vertical turbulence:

L	burst length
L'	burst width
T	burst period
e	burst-forming eddy
E	macroturbulent eddy (e at $t = T$)
$\mathcal{E}$	largest turbulent eddy (of "pre-burst-era")
l	eddy size (diameter)

ii) Horizontal turbulence:

Subscript H: marks the horizontal counterparts of the quantities above

ϵ	thickness of the "free" (not rubbing the bed) eddy e_H
ϵ_{max}	thickness of the "free" E_H (of the free e_H at $t = T_H$)
N	number of burst-rows (not to be confused with dimensionless specific flow rate N)
$\omega \sim (B/h)/c^2$	dimensionless variable determining N
$\tau_H \sim (\partial u/\partial z)^n$	horizontal shear stress

6. Meandering and Braiding

Channel-fitted cylindrical coordinates:

l	direction along channel center line in plan; $l = 0$ at inflection point
r	radial direction
y	vertical direction ($y = 0$ at the flow bed)

Natural coordinates of a streamline:

s	flow direction
n	direction normal to s

x	general direction of flow (in a sinuous channel)

z	direction horizontally perpendicular to x
Λ_m	meander wave length
Δ_m	meander amplitude
L_m	meander length (along l)
$\sigma = L_m / \Lambda_m$	sinuosity
R	radius of channel curvature in plan
θ	deflection angle at a section l (Fig. 5.2)
θ_0	deflection angle at $l = 0$
$J_0(\theta_0)$	Bessel function of first kind and zero-th order
S_v	valley slope
ω	deviation angle (of streamlines from curvilinear parallelism)
U_m	migration velocity of a meandering stream (along x)
V_m	expansion velocity of meander loops (along r)
R_m, R_b	meandering and braiding regions in the $(B/h;\ h/D)$-plane
k	number of consecutive "splits" of a braiding stream

Subscripts:

a	signifies value of a quantity at the apex section (of a sinuous channel)
1	signifies quantities at the free surface
2	signifies quantities near the bed

7. Dimensionless Combinations

$Fr = \dfrac{v^2}{gh}$	Froude number
$Re = \dfrac{vh}{\nu}$	flow Reynolds number
$Re_* = \dfrac{v_* k_s}{\nu}$	roughness Reynolds number
$X = \dfrac{v_* D}{\nu}$	grain size Reynolds number
$Y = \dfrac{\rho v_*^2}{\gamma_s D}$	mobility number
$Z = \dfrac{h}{D}$	dimensionless flow depth
$W = \dfrac{\rho_s}{\rho}$	density ratio
$\xi^3 = \dfrac{X^2}{Y}$	material number
$\xi^3 = \dfrac{\gamma_s D^3}{\rho \nu^2}$	

$\eta = \dfrac{Y}{Y_{cr}}$ relative flow intensity

$\Phi = \dfrac{\rho^{1/2} q_s}{\gamma_s^{1/2} D^{3/2}}$ Einstein's dimensionless transport rate

$N = \dfrac{q}{D v_{*cr}}$ dimensionless specific flow rate (not to be confused with number N of burst-rows)

CHAPTER 1

FUNDAMENTALS

1.1 Width-to-Depth Ratio of a River

Following the example of Ref. [11], we begin this text by pointing out that a natural river is, as a rule, a very wide object: the width-to-depth ratio B/h of a large alluvial stream (lower reaches of a river) is usually a three-digit number. The (undistorted) cross-section of a natural river is thus more likely to be as in Fig. 1.1a (where $B/h \approx 100$), than as in Figs. 1.1b and c, which are the typical textbook sketches. The portrayal of a river with the aid of vertically exaggerated cross-sections as in Figs. 1.1b and c, during the theoretical considerations and laboratory experiments, led to the emergence

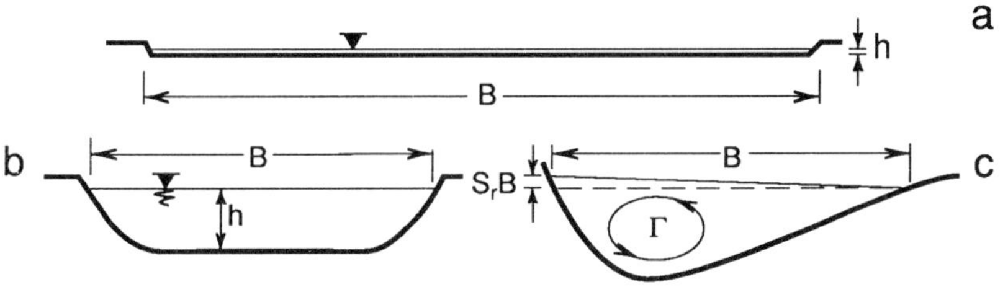

Fig. 1.1

of the correspondingly exaggerated notions and formulations. Thus the expression of the hydraulic radius is often unnecessarily encumbered, the importance of bank friction is overestimated, the role of the cross-sectional circulation (Γ) and/or of the transverse free surface slope S_r on the formation of meanders is overstressed, etc. With the increment of B/h the cross-sectional circulation tends to lose its meaning, while S_r inevitably approaches zero − yet it is exactly the rivers having large B/h which exhibit prominent meandering.

The mathematical treatment of a phenomenon always requires the removal of "natural arbitrariness", the replacement of natural conditions by their idealized counterparts − and river mechanics is no exception. Accordingly, we will assume that the cross-section of a straight stretch of an *ideal river* has the form of a symmetrical trapezoid as shown in Fig. 1.2: the bed $\overline{b_2 b_2'}$ is horizontal, the banks $b_1 b_2$ and $b_1' b_2'$ are curvilinear. At b_1 and and b_1' the banks are inclined by the angle of repose ϕ (of the cohesionless alluvium); at b_2 and b_2' they are tangent to the bed. If B/h is sufficiently large,

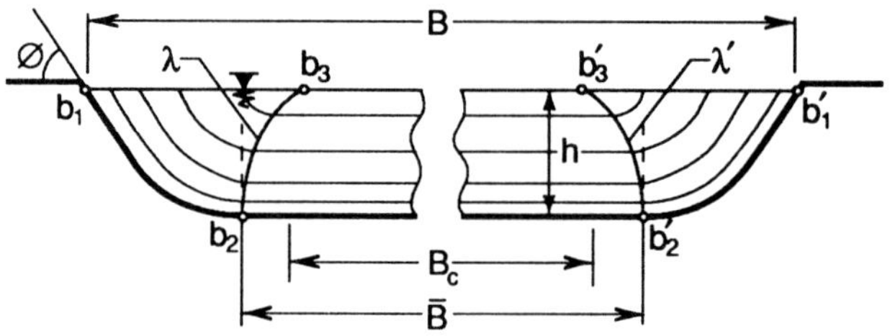

Fig. 1.2

then the cross-section possesses a *central region* where the flow can be regarded and treated as two-dimensional. Let λ and λ' be the lines formed by the points where the curvature of the isovels practically vanishes.[1] The width B_c of the central region must thus be identified with the distance $\overline{b_3 b_3'}$, which is somewhat smaller than the bed width $\overline{b_2 b_2'} = \overline{B}$. According to [36],

$$B_c = B\left(1 - 2m\frac{h}{B}\right) \tag{1.1}$$

where m varies depending on the cross-section geometry and boundary roughness.[2] In the present book we will be dealing exclusively with channels corresponding to large values of B/h (larger than ≈ 10, say). For such "wide" channels, the distinction between various widths ($B, \overline{B}, B_c$) becomes inconsequential. Thus, when needed, a channel flow can be specified by a single width (as in the case of regime channels studied in Chapter 4, Part I), the width B_c can be replaced by $\overline{B}$ (Chapter 4, part II), the total Q and Q_s can be identified with those of the two-dimensional flow past the bed, etc.[3]

It is fortunate that most of the natural alluvial streams are wide, for this makes it possible to treat them on the simplest two-dimensional basis (which is, in fact, the only reliable basis at present).

---

[1] The flow boundary $b_1 b_2 b_2' b_1'$ is itself an isovel ($u = 0$); hence λ and λ' are normal to the bed at b_2 and b_2'.

[2] The functional relation determining m will be discussed at the end of 2.3.5.

[3] The last mentioned identification is also enhanced by the fact that the specific rates q and q_s progressively decrease along $b_2 b_1$ (q vanishes at b_1, q_s before b_1 is reached).

1.2 Dimensional Methods

1.2.1 *Characteristic parameters*

Dimensional methods are particularly useful in the study of those phenomena and processes whose physical mechanism is not quite known and whose definition involves a large number of quantities. The fluvial processes obviously belong to this category; hence, extensive use of the dimensional methods is made in the present text.

A physical phenomenon of a specified geometry is defined by a limited number (n) of independent quantities[4]

$$a_1, a_2, a_3, \dots, a_n, \tag{1.2}$$

which are referred to as the *characteristic parameters* (of that phenomenon). Any quantitative property A of a phenomenon must be a certain function of its characteristic parameters a_i ($i = 1, 2, \dots, n$):

$$A = f_A(a_1, a_2, a_3, \dots, a_n) \tag{1.3}$$

(where $A \neq a_i$).

1- Characteristic parameters a_i are *not* the variables of a phenomenon (although they may appear as such in (1.3)): they are merely the "ingredients" needed to form the actual (dimensionless) variables. The "task" of the parameters a_i is to define (or determine) a phenomenon, and therefore any quantity which contributes to its definition – constant or variable, positive or negative, dimensional or dimensionless – must be included in the set (1.2).[5] The numerical value of each characteristic parameter a_i remains constant in the course of a given experiment, and the fact that a_i can be a variable quantity means that its numerical value may vary from one experiment to another.

If A is a quantity which varies as a function of position (x_k) and/or time (t), then

$$A = F_A(a_1, a_2, a_3, \dots, a_n, x_k, t)$$

which, for $x_k = const_k$ and $t = const_t$, reduces into (1.3). In other words, if A varies in space and/or time, then (1.3) is to be interpreted as the expression of A which corresponds to a specified location and/or time.

2- Characteristic parameters a_i are not required to be of any particular physical nature: what matters is their number (n) and independence

---

4 n quantities are "independent" if none of them is expressible as a function of (some of) the remaining $n - 1$ quantities.

5 It is thus clear that it is misleading to refer to the characteristic parameters a_i as the "dimensional variables" – as is often done in the literature. More information on characteristic parameters can be found in [36], [37], [32].

(physical laws have more a mathematical than physical basis). For example, a steady state flow in a circular pipe is defined by five independent parameters: they can be $[Q, D, k_s, \rho, \mu]$, $[Q, v, (k_s/D), \rho, \nu]$, $[u_{max}, v_*, D, \rho, k_s]$, ... , and so on.

1.2.2 *Dimensionless variables and functions*

A relation such as (1.3) reflects a law of the physical world which is supposed to exist independently of the human mind. But if so, then the values supplied by (1.3) should not depend on the activity of our mind. Yet if A is a dimensional quantity, then its numerical value given by (1.3) will obviously vary depending on the units *we* choose to evaluate a_i. Hence the relation (1.3), as it stands, cannot be the proper (or the ultimate) form of the expression of a natural law. The proper form must be *dimensionless* – for only the values of the dimensionless quantities remain the same in all systems of units.

The dimensionless version of the relation (1.3) can be expressed as

$$\Pi_A = \phi_A(X_1, X_2, X_3, \dots, X_N), \tag{1.4}$$

where $N = n - 3$ quantities X_j are the *dimensionless variables* of the phenomenon and Π_A is the *dimensionless counterpart* of A. According to the "π-theorem" of the theory of dimensions,[6] X_j and Π_A are determined, with the aid of a_i and A, as

$$X_j = a_1^{x_j} a_2^{y_j} a_3^{z_j} a_{j+3}^{m_j} \qquad (j = 1, 2, \dots, N) \tag{1.5}$$

and

$$\Pi_A = a_1^{x_A} a_2^{y_A} a_3^{z_A} A. \tag{1.6}$$

The "repeaters" a_1, a_2 and a_3 in the power products above can be *any* three parameters having *independent dimensions*.[7] The exponents m_j can be selected at random: the exponents x_j, y_j and z_j must be determined (depending on the selected m_j) so that the power products X_j become dimensionless. Similarly, the exponents x_A, y_A and z_A must be determined (depending on the dimension of A) so that Π_A is dimensionless. If A is dimensionless, then $x_A = y_A = z_A = 0$, and Π_A is equal to A itself; if one of a_{j+3} is dimensionless, then $X_j = a_{j+3}$.

The following three points are relevant:

1- The form of the function ϕ_A in (1.4) is not specified. Hence from (1.4) it does not follow, and it is in fact false, that all N dimensionless variables X_j must necessarily be present in the expression of every dimensionless

---

6 The derivation of the π-theorem is given e.g. in [22], [14], [5].

7 Three quantities a_1, a_2, a_3 have independent dimensions if none of them has the dimension expressible by the dimensions of the remaining two (more on the topic in [22], [36], [37]).

property Π_A. The n parameters a_i and thus the N dimensionless variables X_j, which are necessary and sufficient for the expression of *all* properties (A and Π_A) of a phenomenon, should be regarded merely as sufficient as far as the expression of any single property is concerned. Thus, one can have e.g. $\Pi_{A_1} = \phi_{A_1}(X_3, X_4, X_{N-1})$, $\Pi_{A_2} = \phi_{A_2}(X_1)$, $\Pi_{A_3} = const$, ... , etc. (alongside $\Pi_{A_4} = \phi_{A_4}(X_1, X_2, X_3, \ldots, X_N)$, say). The analogous is valid for f_A in (1.3).

2- The relations (1.3) and (1.4) correspond to a phenomenon of a specified geometry. This means that if the geometry of a phenomenon varies, then the form of the functions f_A and ϕ_A must be expected to vary as well (for all A and Π_A). One can say that the form of a function f_A or ϕ_A is itself a function of the geometry. (Recall, for example, that the drag coefficients of a sphere and of a cylinder are determined by the different functions (curves) of the same dimensionless variable, viz $X = DU_\infty/\nu$).

3- Note from (1.5) that, in general (i.e. when x_j, y_j, $z_j \neq 0$), the repeaters a_1, a_2 and a_3 appear in the expression of every dimensionless variable, yet each of the remaining parameters appears in the expression of only one (a_4 in X_1, a_5 in X_2, ... , etc.). Hence, it is said that the influence of the parameter a_4 (on Π_A) is reflected by the dimensionless variable X_1, the influence of a_5 by X_2, ... , etc. None of the dimensionless variables X_j reflects the influence of a repeater; and if the consideration of the influence of a_1, say, by a special variable is desired, then a_1 should not be selected as one of the repeaters.

1.3 Two-Phase Motion and its Dimensionless Variables

1.3.1 *Basic formulation*

The simultaneous motion of liquid and solid phases (of the transporting flow and the transported sediment) constitutes a mechanical totality which can be referred to as the *two-phase motion* (or the transport phenomenon) [36]. If the geometry of granular material – determined by the shape of grains and of the dimensionless grain-size distribution curve – is specified, then the stationary and uniform (or quasi-uniform)[8] two-dimensional two-phase motion (in the central region B_c of a "wide" river) can be determined [36], [37], [38] by the following seven characteristic parameters:

$$\rho,\ \nu,\ \rho_s,\ D,\ h,\ v_*,\ \gamma_s \tag{1.7}$$

---

[8] The flow is quasi-uniform if its non-uniformities (along a distance Λ, say) are uniformly distributed in the flow direction x, that is, if any of its properties A varies as a periodic function: $A = f(x, y, z) = f(x + k\Lambda, y, z)$, where k is an integer (see [36], Chapter 1).

(see List of Symbols).[9] Identifying the repeaters a_1, a_2 and a_3 with ρ, D and v_*, one determines, on the basis of (1.5), the dimensionless variables

$$\left.\begin{aligned} X_1 &= X = \frac{v_* D}{\nu} \\ X_2 &= Y = \frac{\rho v_*^2}{\gamma_s D} \\ X_3 &= Z = \frac{h}{D} \\ X_4 &= W = \frac{\rho_s}{\rho} \,. \end{aligned}\right\} \qquad (1.8)$$

The *grain size Reynolds number* X reflects the influence of ν; the *mobility number* Y (which is the most important variable of sediment transport) takes into account the role of γ_s; the variables Z and W reflect the roles of h and ρ_s, respectively. The grain mass, and thus ρ_s, is of importance only with regard to the (accelerated) motion of an individual grain [36], [37], [38]. In the present book we will be concerned only with the grain motion *en masse*, and therefore any dimensionless property Π_A of the two-phase motion will be treated as a function of (to the most) X, Y and Z only:

$$\Pi_A = \rho^{x_A} D^{y_A} v_*^{z_A} A = \phi_A(X, Y, Z) \,. \qquad (1.9)$$

1.3.2 *Equivalent forms*

It is not necessary to express a property Π_A literally in terms of X ($= X_1$), Y ($= X_2$) and Z ($= X_3$): any X_j can be replaced by a quantity which is known to be a function of that X_j and of one (or both) of the remaining two variables. To illustrate this point, consider the initiation of sediment transport (the "critical stage" of a mobile bed) which, as is well known, is determined by a certain relation between the critical values of X and Y:

$$Y_{cr} = \Phi(X_{cr}) \,. \qquad (1.10)$$

The experimental curve (*Shields' curve*) implying this relation is shown in Fig. 1.3. Note, on the other hand, that the ratio X_{cr}^2/Y_{cr} does not depend on v_{*cr}, and therefore it is equal to the ratio X^2/Y corresponding to any stage of the two-phase motion:

---

[9] The consideration of $v_* = \sqrt{gSh}$ and h in the set of characteristic parameters (1.7) is equivalent to that of gS and h. This means that in the case of a uniform or quasi-uniform two-phase motion (specified by (1.7)) the parameters g and S co-operate in the form of a single parameter gS. This parameter is the gravity component which generates the flow and thus the whole phenomenon [36], [37]. If the flow is quasi-uniform, then the variable (within Λ) properties may be determined, among others, by g and S separately, but their averaged (along x) values are determined by the product gS.

$$\frac{X_{cr}^2}{Y_{cr}} = \frac{X^2}{Y} = \frac{\gamma_s D^3}{\rho \nu^2} = \xi^3. \qquad (1.11)$$

	SOURCE		MATERIALS		SOURCE		MATERIALS
⊙	Casey, H.J.	[1]	Sand + water	◇	Shields, A.	[14]	Amber + water
◉	Gilbert, G.K.	[2]	" "	◆	" "	"	lignite + water
+	Grass, A.J.	[3]	" "	◈	" "	"	granite + water
□	Iwagaki, Y.	[4]	" "	◇	" "	"	Barite + water
■	Karahan, E.	[6]	" "	✦	U.S.W.E.S.	[15]	sand + water
◇	Kramer, H.	[7]	" "	◖	Vanoni, V.A.	[16]	" "
▽	Mantz, P.A.	[8]	" "	◕	" "	"	glass beads + water
⊖	Meyer-Peter, E.} Müller, R.	[9]	" "	▼	White, C.M.	[17]	sand + oil
◄	Neill, C.R.	[11]	" "	○	White, S.J.	[18]	sand + water
◁	" "	"	glass beads + water	●	" "	"	sand + oil
				★	Yalin, M.S.} Karahan, E.	[35]	sand + glycerole
				☆	" "	"	sand + water

(Ref. numbers are those used in [35])

$Y_{cr} = 0.1 \, X_{cr}^{0.3}$ — Laminar

- ? — Turbulent

$$10 \qquad 1 \qquad 10^{-1} \qquad 10^{-2}$$

Y_{cr}

$$10^{-1} \qquad 1 \qquad 10 \qquad 10^2 \qquad 10^3$$

$X_{cr} \longrightarrow$

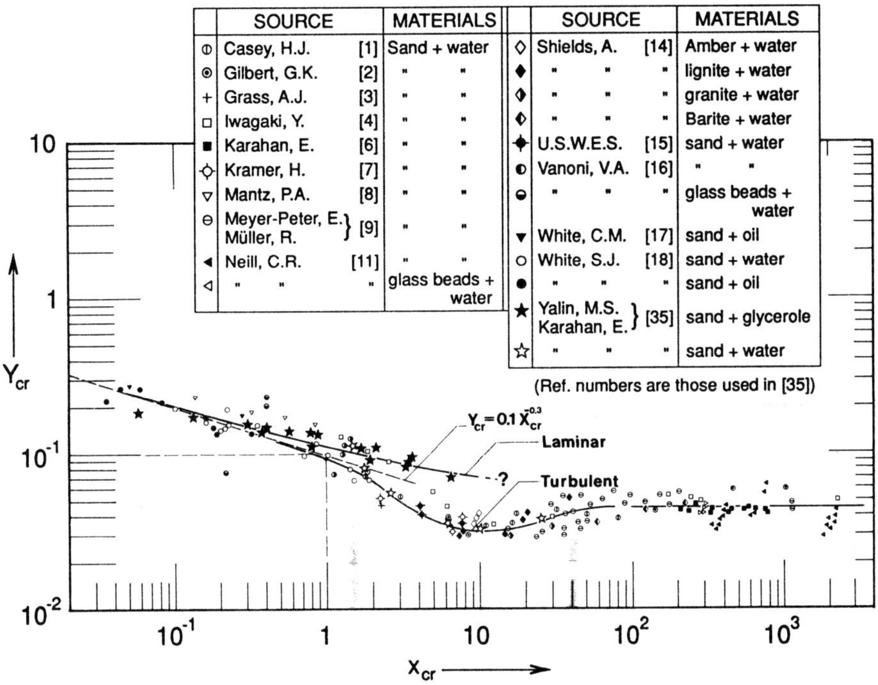

Fig. 1.3 (from Ref. [35])

Substituting $X_{cr} = \sqrt{\xi^3 Y_{cr}}$ in (1.10), one obtains

$$Y_{cr} = \Phi(\sqrt{\xi^3 Y_{cr}}) \quad \text{i.e.} \quad Y_{cr} = \Psi(\xi). \qquad (1.12)$$

The *modified Shields' curve* (1.12) (which can be unambiguously determined from the original Shields' curve (1.10) (see [36])) contains υ_{*cr} in Y_{cr} only; hence, it can supply υ_{*cr} without "trial-and-error".

Using $X = \sqrt{\xi^3 Y}$ in (1.9), one determines

$$\Pi_A = \phi_A(\sqrt{\xi^3 Y}, Y, Z) = \phi_{A_1}(\xi, Y, Z), \qquad (1.13)$$

which indicates that any Π_A can equally well be treated as a function of ξ, Y and Z; or, to that matter, as a function (ϕ_{A_2}) of X, ξ and Z.

Dividing the Y-number by Y_{cr}, we introduce its normalized counterpart

$$\eta = \frac{Y}{Y_{cr}} = \frac{Y}{\Psi(\xi)}, \qquad (1.14)$$

which is referred to as the *relative flow intensity*. Substituting $Y = \eta \Psi(\xi)$ in (1.13), one obtains

$$\Pi_A = \phi_{A_1}(\xi, \eta\Psi(\xi), Z) = \phi_{A_3}(\xi, \eta, Z) \tag{1.15}$$

which indicates that any Π_A can be expressed also as a function of ξ, η and Z. Further equivalent forms are also possible.

1.3.3 *(B/h)-ratio as the dimensionless variable*

It was tacitly assumed throughout the considerations above that the flow-specifying parameters h and v_* (or, which is the same, h and S) can be assigned any physically possible values. Yet, in the case of a natural river, h and S emerge as a result of the channel formation process, and therefore h and S (and also B) are themselves certain functions of the "channel forming" Q (and of the parameters reflecting the physical nature of the liquid and solid phases involved).[10]

Although the flow width B determines the extent of the central region B_c, it does not affect the (two-dimensional) transport phenomenon within B_c directly; hence, B has not been included in the list of parameters (1.7). The possible indirect influence of B can be explained as follows. If the ratio B/h is smaller than a certain value which depends on $Z = h/D$ – in short if $B/h < \psi_\epsilon(Z)$ – then (as will be explained in Chapter 3) the bed is either flat or it is covered by the bed forms (viz dunes, antidunes and/or ripples) whose length and steepness do not depend on B/h. In this case, B/h has no influence on the two-dimensional transport phenomenon within B_c at all. If, on the other hand, $B/h > \psi_\epsilon(Z)$, then the bed may be covered by the bed forms (viz alternate or multiple bars) whose size and configuration *do* depend on B/h. In this case, B/h *has* an (indirect) influence on the phenomenon in B_c: it (directly) affects the geometry of bed forms which (directly) affects the two-phase motion.

In Chapter 3 it will be shown that with the increment of B/h from $\psi_\epsilon(Z)$ onwards, the linear scale of bars progressively changes from B to h so as to become equal to h in the limit $(B/h) \to \infty$. Correspondingly, the influence of B/h on the phenomenon in B_c must also progressively decrease, so as to vanish completely when $(B/h) \to \infty$; i.e. when the flow becomes truly two-dimensional.

1.4 Mechanical Properties of Flow

1.4.1 *Shear stress and velocity distributions*

Consider a two-dimensional turbulent open-channel flow in the central region B_c. We assume that this flow is stationary and uniform: the bed is flat, and $k_s \ll h$.

---

[10] The determination of these functions is one of the topics of Chapter 4.

The time average shear stress distribution along y (Fig. 1.4a) is given by the linear relation

$$\tau = \tau_0\left(1 - \frac{y}{h}\right) \tag{1.16}$$

where

$$\tau_0 = \rho v_*^2 = \gamma S h. \tag{1.17}$$

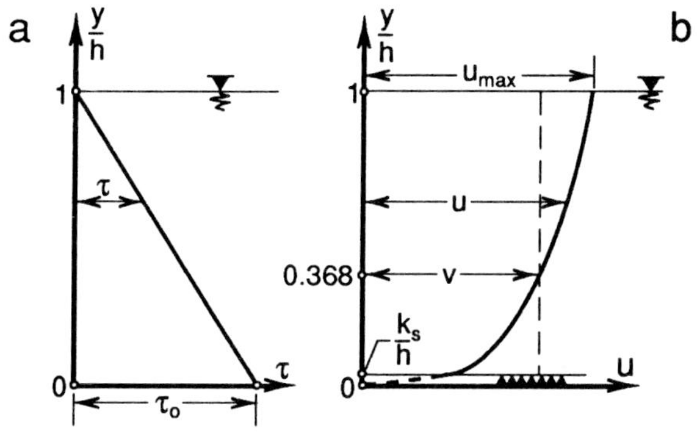

Fig. 1.4

For the time average velocities u (Fig. 1.4b), we have

$$\frac{u}{v_*} = \frac{1}{\kappa}\ln\frac{y}{k_s} + B_s\,, \tag{1.18}$$

which can be expressed also as

$$\frac{u}{v_*} = \frac{1}{\kappa}\ln\left(A_s\frac{y}{k_s}\right). \tag{1.19}$$

The values of A_s and B_s are interrelated by

$$A_s = e^{\kappa B_s}\,. \tag{1.20}$$

If the fluid is "clear" (no sediment in suspension), then[11] $\kappa \approx 0.4$, the *roughness function* $B_s = \phi_{B_s}(Re_*)$ (where $Re_* = v_* k_s/\nu$) being given by the experimental curve in Fig. 1.5. Note that

$$\text{if } Re_* > \approx 70, \text{ then } B_s = 8.5 \text{ (rough turbulent regime)} \tag{1.21}$$

and

$$\text{if } Re_* < \approx 5, \text{ then } B_s = \frac{1}{\kappa}\ln Re_* + 5.5 \text{ (hydro-smooth regime)}. \tag{1.22}$$

---

[11] The graph representing the *variation* of the Von Karman "constant" κ is given e.g. in Fig. 5.23 of Ref. [36]. An extensive account of the influence of sediment concentration on the distribution of flow velocities can be found in the recent work, Ref. [15].

The straight lines S_1 and S_2 in Fig. 1.5 are the graphs of (1.21) and (1.22) respectively. If $\approx 5 < Re_* < \approx 70$ (transitional regime), then the value of B_s must be taken from Fig. 1.5.[12]

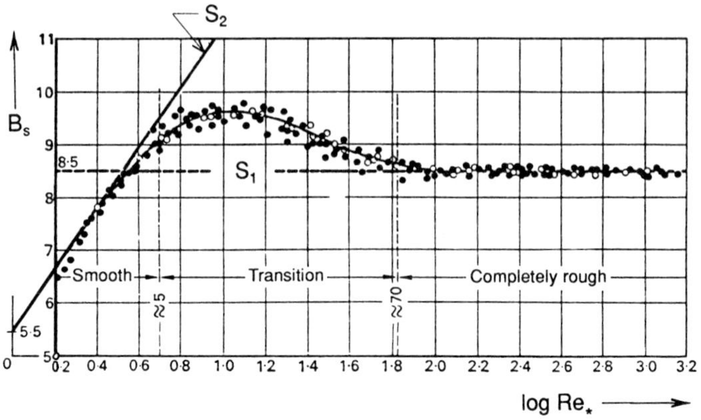

Fig. 1.5 (after Ref. [21])

The relation (1.18) (and thus (1.19)) can be used, for all practical purposes, throughout the range $y_{min} \leq y \leq h$ (see e.g. [21], [36]), where y_{min} is either k_s or $11.6\,\nu/\upsilon_*$, whichever is the larger.[13] Accordingly, the maximum velocity at $y = h$ can be given by

$$\frac{u_{max}}{\upsilon_*} = \frac{1}{\kappa}\ln\frac{h}{k_s} + B_s = \frac{1}{\kappa}\ln\left(A_s\frac{h}{k_s}\right). \tag{1.23}$$

Using the definition of the average flow velocity, viz the expression

$$(h - y_{min})\,\upsilon = \int_{y_{min}}^{h} u\,dy\,, \tag{1.24}$$

in conjunction with (1.18) (or (1.19)) and (1.23), and considering that $y_{min} \ll h$, one determines

$$\frac{\upsilon}{\upsilon_*} = \frac{u_{max}}{\upsilon_*} - 2.5 \tag{1.25}$$

(where $2.5 \approx 1/\kappa$).

---

[12] The roughness function $B_s = \phi_{B_s}(Re_*)$ can be characterized by the comp-eq.

$$B_s = 8.5 + [2.5\ln(Re_*) - 3]e^{-0.127[\ln(Re_*)]^2}.$$

The solid curve in Fig. 1.5 is the graph of this equation.

[13] If $y_{min} = 11.6\,\nu/\upsilon_* > k_s$, then the u-distribution of the viscous flow within the layer $k_s < y < (11.6\,\nu/\upsilon_*)$ is given by the linear form $u/\upsilon_* = (\upsilon_*/\nu)y$ (see [36]).

From (1.25), (1.23) and (1.18), it follows that v is equal to that u which is at the dimensionless level $y/h = e^{-1} \approx 0.368$ (Fig. 1.4b). Hence, the average flow velocity v can be expressed also (using $\kappa = 0.4$) as

$$\frac{v}{v_*} = 2.5 \ln\left(b_s \frac{h}{k_s}\right) \tag{1.26}$$

(where $b_s = e^{\kappa B_s - 1}$).

If the flow is rough turbulent, then $\kappa B_s - 1 \approx (0.4)(8.5) - 1 = 2.4$ (see (1.21)), and (1.26) yields

$$\frac{v}{v_*} = 2.5 \ln\left(11 \frac{h}{k_s}\right). \tag{1.27}$$

This relation is often approximated by the power form

$$\frac{v}{v_*} \approx 7.66\left(\frac{h}{k_s}\right)^{1/6}. \tag{1.28}$$

If the flow is hydraulically smooth, then $\kappa B_s - 1 \approx \ln Re_* + 1.2$ (see (1.22)), and (1.26) gives

$$\frac{v}{v_*} = 2.5 \ln\left(3.32 \frac{v_* h}{\nu}\right). \tag{1.29}$$

(A more detailed derivation of the expressions above can be found e.g. in [13], [17], [21], [36]).

1.4.2 *Friction factor*

i- *Flat bed surface (having granular roughness k_s)*

Friction factor is a quantity which relates the flow velocity v to the shear velocity v_*. The ratio

$$c = \frac{v}{v_*}, \tag{1.30}$$

referred to as the *dimensionless Chézy coefficient*, is the simplest (and most sound) friction factor. Substituting $v_* = \sqrt{\tau_0/\rho} = \sqrt{gSh}$ in (1.30), one obtains immediately the proper[14] version of the Chézy resistance equation for two-dimensional flows:

---

[14] The "improper" feature of the original Chézy formula $v = C\sqrt{SR}$ (where $R = h$, if the flow is two-dimensional) is due to its dimensional coefficient C (which has the disadvantage of having different numerical values in different systems of units). Moreover, the original Chézy formula does not indicate how v is determined by g which generates it: clearly $C = c\sqrt{g}$, and thus $v \sim \sqrt{g}$. The analogous is valid for the Manning formula $v = (k/n)R^{2/3}S^{1/2}$ where the total coefficient (k/n) is dimensional (though n is dimensionless). Comparing Manning's formula with $v = c\sqrt{gSR}$ (which is the generalized version of (1.31)), one determines the interrelation between (k/n) and c, viz $(k/n) = c(\sqrt{g}/R^{1/6})$. The

$$v = c\sqrt{gSh}\,. \tag{1.31}$$

Observe that the relations (1.26) to (1.29) are, in effect, the expressions of the friction factor c. A stationary and uniform two-dimensional open-channel flow past a flat bed having granular roughness k_s is completely determined by the characteristic parameters[15]

$$\rho,\ \nu,\ k_s,\ h,\ v_*\,, \tag{1.32}$$

and thus by the dimensionless variables $Re_* = v_* k_s/\nu$ and h/k_s. Hence c, which is a dimensionless property (Π_v) of this flow, must be a certain function of Re_* and h/k_s:

$$c = \bar{\Phi}_c(Re_*\,,\ h/k_s)\,. \tag{1.33}$$

Note that each of the relations (1.26) to (1.29) is determined indeed by Re_* and/or h/k_s only (for $B_s = \phi_{B_s}(Re_*)$). Consider the limiting cases. If the flow is rough turbulent, then it does not depend on ν ($\sim Re_*^{-1}$) any longer and c becomes a function of h/k_s alone (Eqs. (1.27) and (1.28)). If the flow is hydraulically smooth, then it does not depend on k_s, and c becomes a function of the product $(h/k_s)Re_* = v_* h/\nu$ which does not contain k_s (Eq. (1.29)).

ii- *Undulated bed surface (having granular roughness k_s)*

1- Suppose now that the bed is undulated; i.e. that it is covered by a series of triangular bed forms as shown in Fig. 1.6a. We assume that these bed

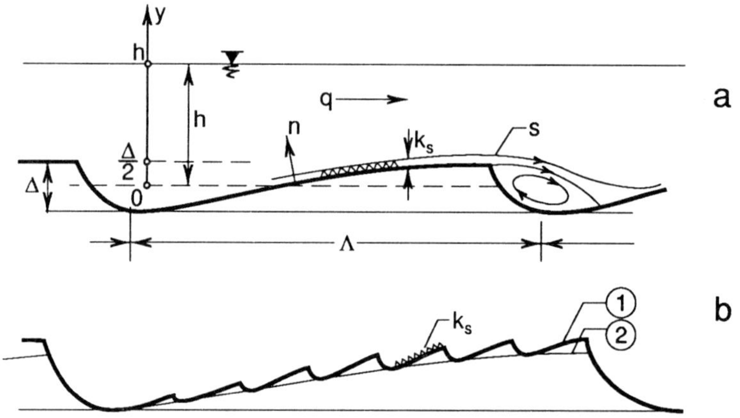

Fig. 1.6

dimensionless Darcy-Weisbach friction factor f does not involve any "hidden parameters" such as g or R, and it is related to c by $f = 8/c^2$.

[15] Note that (1.32) is but the subset of (1.7) where the grains are "immobilized" by removing ρ_s and γ_s, while D is converted into the "yard-stick" (k_s) measuring the size of the granular bed roughness.

forms extend in the direction z, perpendicular to the drawing, indefinitely (two-dimensional bed forms): their length and height are Λ and Δ. Clearly, Λ and Δ are the additional characteristic parameters, and (1.33) becomes generalized into

$$c = \Phi_c(Re_*,\ h/k_s,\ \Delta/\Lambda,\ \Lambda/h)\,.\qquad (1.34)$$

[16]

In Refs. [39] and [6] the following c-relation has been derived (independently, and by using different methods):

$$\frac{1}{c^2} = \frac{1}{c_f^2} + \frac{1}{2}\left(\frac{\Delta}{\Lambda}\right)^2 \frac{\Lambda}{h}\,. \qquad (1.35)$$

Here c_f is a function of (only) Re_* and/or h/k_s, and therefore (1.35) is consistent with (1.34).[17] The validity of (1.35) has been verified by field and laboratory data (Ref. [34]).[18]

In some cases, more than one mode of bed forms can be present simultaneously; for example, we can have ripples superimposed on dunes, as depicted schematically in Fig. 1.6b. Since energy losses are additive, the losses due to the bed forms 1 and 2 can be added [30], [31], [33], and the relation (1.35) can be generalized into

$$\frac{1}{c^2} = \frac{1}{c_f^2} + \frac{1}{2}\sum_{i=1}^{2}\left(\frac{\Delta}{\Lambda}\right)_i^2 \frac{\Lambda_i}{h}\,. \qquad (1.36)$$

Sometimes this formula is expressed as

$$\frac{1}{c^2} = A_f + A_1 + A_2 \qquad (1.37)$$

or

---

[16] If k_s is still a repeater (as in the previous paragraph), then the π-theorem supplies Λ/k_s and Δ/k_s (in addition to Re_* and h/k_s). However, since the consideration of Λ/k_s, Δ/k_s and h/k_s is obviously equivalent to that of $(\Delta/k_s)/(\Lambda/k_s)$, $(\Lambda/k_s)/(h/k_s)$ and h/k_s, the replacement of Λ/k_s and Δ/k_s by the physically more meaningful Δ/Λ and Λ/h is permissible.

[17] The combination $(\Delta/\Lambda)^2\Lambda/h$ was used also in Ref. [24].

[18] The relation (1.35) gives c^{-2} as a linear function of the dimensionless complex $[(\Delta/\Lambda)^2\Lambda/h] = (\Delta/\Lambda)(\Delta/h)$. However, if Δ/Λ is finite and Δ/h is "small" (ripples (only) on the bed of a deep flow), then the bed forms (ripples) manifest themselves as an ordinary bed roughness, with $K_s \sim \Delta$ ($\ll h$). Clearly, for such cases c should turn into the logarithmic function of the complex $[(\Delta/\Lambda)^2\Lambda/h]^{-1}$ (as suggested first in Ref. [24]); and experiment shows that this is indeed so (see e.g. [34]). It has been found in [34] that the transition from (1.35) to the logarithmic form takes place for $[(\Delta/\Lambda)^2\Lambda/h] < \approx 10^{-2}$.

$$\frac{1}{c^2} = \frac{1}{c_f^2} + \frac{1}{c_\Delta^2}, \tag{1.38}$$

where

$$\frac{1}{c_f^2} = A_f \quad \text{and} \quad \frac{1}{c_\Delta^2} = A_1 + A_2 = \frac{1}{2}\sum_{i=1}^{2}\left(\frac{\Delta}{\Lambda}\right)_i^2 \frac{\Lambda_i}{h} \tag{1.39}$$

are the *pure friction* and *form-drag* components of $1/c^2$ respectively.

2- Let $\bar{c}$ and $\bar{v}$ be the values of c and v which correspond to a flat bed. Observe, e.g. from (1.36), that if the bed is flat ($(\Delta/\Lambda)_i = 0$), then

$$c_f = c \; (= \bar{c}), \tag{1.40}$$

where $\bar{c}$ is given by the relations (1.26) to (1.29). At the present state of knowledge it would be reasonable to evaluate c_f by using the expressions of $\bar{c}$, viz (1.26) to (1.29), even if the bed is *not* flat. This method of evaluation of c_f is commonly used today, and it will be used in the present book.

If $(\Delta/\Lambda)_i > 0$, then (as is clear from (1.36)) $c < c_f \; (= \bar{c})$. Hence if c is substituted in (1.26) to (1.28), then these expressions will no longer be satisfied by k_s: they will be satisfied by a different length, K_s say, which is referred to as the *total bed roughness*. Clearly

$$K_s \geq k_s, \tag{1.41}$$

where $K_s = k_s$ when $(\Delta/\Lambda)_i = 0$.

From (1.36) and (1.40), one determines the following relation for the ratio c/c_f which will be denoted by λ_c:

$$\lambda_c = \frac{c}{c_f} = \left[1 + \frac{\bar{c}^2}{2}\sum_{i=1}^{2}\left(\frac{\Delta}{\Lambda}\right)_i^2 \frac{\Lambda_i}{h}\right]^{-1/2}. \tag{1.42}$$

Here $\bar{c}$ can be evaluated (with the aid of (1.26) to (1.29)) by using the space average flow depth h (Fig. 1.6a). The relations (1.36) to (1.42) can be used for the case of a mobile bed if k_s, Δ_i and Λ_i are known. We will deal with the mobile bed value of k_s in Section 1.6; the expressions of Δ_i and Λ_i will be determined in Chapter 3.

3- Consider now the shear stresses. Multiplying both sides of (1.38) with ρv^2 and taking into account that $v^2 = c^2\tau_0/\rho$, one obtains

$$\tau_0 = (\tau_0)_f + (\tau_0)_\Delta, \tag{1.43}$$

where

$$(\tau_0)_f = \frac{\rho v^2}{c_f^2} \quad \text{and} \quad (\tau_0)_\Delta = \frac{\rho v^2}{c_\Delta^2} \tag{1.44}$$

are the *pure friction* and *form-drag* components of the *total shear stress* τ_0 ($= \gamma h S$) respectively. From (1.44) it is clear that

$$\frac{(\tau_0)_\Delta}{(\tau_0)_f} = \left(\frac{c_f}{c_\Delta}\right)^2. \tag{1.45}$$

Using (1.45), (1.38) and (1.43), one determines[19] the following relation:

$$(\tau_0)_f = \tau_0\left(\frac{c}{c_f}\right)^2 = \tau_0 \lambda_c^2 \tag{1.46}$$

(where $\lambda_c = c/c_f$ is given by (1.42)), which is useful in determining the transport rate past an undulated bed (next section).

The friction factor relation (1.35) has been derived (in both [39] and [6]) from two-dimensional considerations. Thus the more regular the bed forms are, the better is the agreement of (1.35) (and (1.36)) with the measurements. There is no reason to expect that the dimensionless complex $(1/2)[(\Delta/\Lambda)^2\Lambda/h]$ should still remain as it stands if the bed forms are no longer regular (if they are three-dimensional). And from comparison with the data it follows that in the latter case the complex mentioned should be somewhat modified (Section 3.6).

1.5 Sediment Transport

1.5.1 *Two modes of sediment transport, transport rates*

If $Y > Y_{cr}$ (or $\eta > 1$), then the grains forming the uppermost layer [36] of the movable bed are in motion – flow is transporting the sediment. The detachment of grains is due to $(\tau_0)_f$;[20] their downstream motion due to the local flow velocities u. As has already been mentioned, the motion of sediment is usually accompanied by the wave-like deformation of the bed surface. If η is within $1 < \eta < \approx 10$, say, then most of the grains are transported in the neighbourhood of the bed: and the grain motion is by "jumps" (paths P_b in Fig. 1.7). This (first) mode of transport is referred to as the *bed-load*. The thickness ϵ of the bed-load layer can be identified with the height δ_b of the deterministic grain paths P_b. With the increment of η, some of the moving grains "diffuse", due to turbulence, into the body of flow and form the *suspended-load*. The grains forming this (second) mode of transport move along the probabilistic grain paths P_s. Suspended-load is not a "substitute for", but an "addition to" the bed-load (whenever suspended-

---

[19] Express (1.38) and (1.43) as $(c_f/c)^2 = 1 + (c_f/c_\Delta)^2$ and $\tau_0/(\tau_0)_f = 1 + (\tau_0)_\Delta/(\tau_0)_f$ respectively, and consider (1.45).

[20] The detachment of a grain from the uppermost layer is, of course, due to the hydrodynamic force $\mathbf{F}$ acting on it [36]. However, the magnitude of $\mathbf{F}$ (and thus of its components $\mathbf{F}_y$ and $\mathbf{F}_x$) is directly proportional to $(\tau_0)_f$:
$F_y \sim F_x \sim F = \phi(X_f)\,\rho D^2 (v_*)_f^2 = [\phi(X_f) D^2](\tau_0)_f$, where $X_f = (v_*)_f D/\nu$.

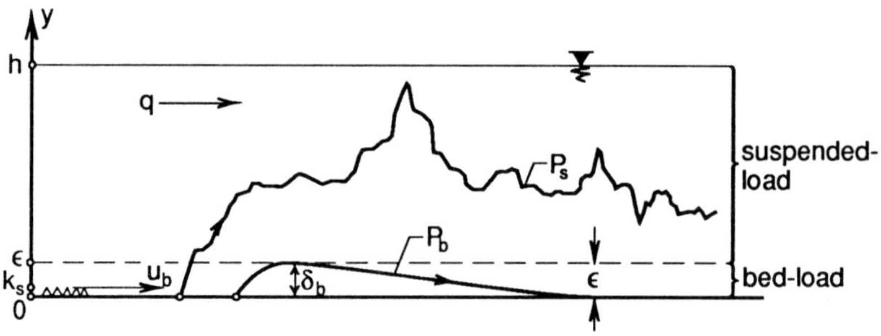

Fig. 1.7

load is present, bed-load is also present) – although the percentage of grains transported in suspension progressively increases with the increment of η.

The specific (volumetric) *transport rate* q_s can thus be given as the sum of the *bed-load rate* q_{sb} and the *suspended-load rate* q_{ss}:

$$q_s = q_{sb} + q_{ss} , \tag{1.47}$$

where the dimension of any q_{si} is

$$[q_{si}] = [\text{length}]\,[\text{velocity}] . \tag{1.48}$$

The most favourable dimensionless form of a q_{si} is the Einstein's combination

$$\Phi_i = \left(\frac{\rho}{\gamma_s}\right)^{1/2} \frac{q_{si}}{D^{3/2}} . \tag{1.49}$$

This combination does not involve flow parameters (v_* and/or h), and therefore, for a given granular material and fluid, it remains related to q_{si} by the same (constant) proportion for all flow stages.

From the definition of the transport rate (as the volume of solids passing through a flow section per unit time), it follows that the sum (1.47) implies

$$\int_0^h U_s C\,dy = \int_0^\epsilon U_s C\,dy + \int_\epsilon^h U_s C\,dy , \tag{1.50}$$

where C is the $(x;t)$-average dimensionless volumetric concentration of solids at a level y, and U_s is their $(x;t)$-average migration velocity.[21]

The present state of knowledge does not permit an exact evaluation of the integral form (1.50), and therefore the existing transport formulae (which have a theoretical origin) usually stem from some less rigorous premises. Most

---

[21] Since the bed-load grains are moving within $k_s < y < \epsilon$, the lower limit of the integrals in (1.50) should be k_s rather than 0. However, we will continue to use 0, for in the present context k_s loses its meaning when the bed is no longer flat. Moreover, the grain motion within $k_s < y < \epsilon$ is the same as that within $0 < y < \epsilon$.

of the prominent transport formulae are of comparable accuracy. These formulae will not be reviewed here, as they have already been extensively discussed in many previous works (see e.g. [8], [9], [19], [36]).

1.5.2 *Bed-load*

Consider the first term on the right-hand side of (1.50). This term, which implies q_{sb}, can be expressed (by the average value theorem) as

$$q_{sb} = \int_0^{\epsilon} U_s C\, dy = \frac{U_{s\epsilon}}{\gamma_s}\left(\gamma_s \int_0^{\epsilon} C\, dy\right), \tag{1.51}$$

where $U_{s\epsilon}$ can be viewed as the migration velocity of the "grain cloud" within $0 < y < \epsilon$ (where $\epsilon \ll h$).

i- *Flat bed* ($(\tau_0)_f/\tau_0 = \lambda_c^2 = 1$)

Since $U_{s\epsilon}$ is generated by the flow velocities u, it must be proportional to a typical flow velocity of the bed-load layer. The most appropriate typical u in the layer mentioned is $u = u_b$ at the roughness level $y = k_s$ (Fig. 1.7). Thus, one can write

$$U_{s\epsilon} = \alpha_1 u_b \quad \text{(where } u_b = v_* B_s\text{)}. \tag{1.52}$$

The proportionality factor α_1 in this expression is one of the dimensionless properties of the two-phase motion and therefore, in general, it must be expected to be a certain function of (some of) the dimensionless variables ξ, η and Z, say (if the version (1.15) is adopted). The bed-load does not depend on $h \sim Z$ (see e.g. [36]): for a given granular material and fluid, it is determined by $\tau_0 \sim \eta$ alone. Hence, α_1 must be a function (to the most) of ξ and η:

$$\alpha_1 = \phi_1(\xi, \eta). \tag{1.53}$$

The expression in brackets on the right-hand side of (1.51) is the weight $G_{s\epsilon}$ of the grain cloud moving within $0 < y < \epsilon$ over each unit area of the bed surface. Since the detachment of grains is due to the shear action, $G_{s\epsilon}$ must be an increasing function of τ_0 or, to be more exact, of $[\tau_0 - (\tau_0)_{cr}]$ (for $G_{s\epsilon} \equiv 0$ when $[\tau_0 - (\tau_0)_{cr}] < 0$). But the dimensions of $G_{s\epsilon}$ and $[\tau_0 - (\tau_0)_{cr}]$ are identical, and therefore $G_{s\epsilon}$ must increase in proportion to $[\tau_0 - (\tau_0)_{cr}]$. i.e.

$$G_{s\epsilon} = \alpha_2 [\tau_0 - (\tau_0)_{cr}], \tag{1.54}$$

where, as in the case of α_1,

$$\alpha_2 = \phi_2(\xi, \eta). \tag{1.55}$$

Substituting (1.52) and (1.54) in (1.51), one determines

$$q_{sb} = (\alpha_1 \alpha_2) u_b [\tau_0 - (\tau_0)_{cr}] / \gamma_s , \tag{1.56}$$

which is completely analogous to Bagnold's [1] bed-load formula

$$q_{sb} = \beta u_b [\tau_0 - (\tau_0)_{cr}] / \gamma_s . \tag{1.57}$$

From Ref. [1], it follows that $\beta = \alpha_1 \alpha_2$ is, in fact, a function of ξ only. The graph of the function $\beta = \phi_\beta(\xi)$ is given in Ref. [1] (and it is reproduced also in Fig. 5.5 of Ref. [36]).[22] Note from this graph that the function $\beta = \phi_\beta(\xi)$ tends to be a constant ($\beta \approx 0.5$) if $Re_* > \approx 70$, i.e. if the flow is rough turbulent.

Bagnold's relation (1.57) is one of the simplest among the prominent bed-load formulae. And yet it is exactly this relation which naturally emerges from the fundamental form (1.51). Considering this, in the present text q_{sb} will invariably be interpreted with the aid of (1.57). Using (1.49), one can express (1.57) in the dimensionless form as

$$\Phi_b = (B_s \beta) Y^{1/2} (Y - Y_{cr}) \quad \text{or} \quad \Phi_b = [(B_s \beta) Y_{cr}^{3/2}] \eta^{1/2} (\eta - 1) , \tag{1.58}$$

where $(B_s \beta) = (8.5)(0.5) = 4.25$ and $Y_{cr} \approx 0.05$, if the flow is rough turbulent.

ii- *Undulated bed* ($(\tau_0)_f / \tau_0 = \lambda_c^2 < 1$)

Let $(u_b)_\Delta$ be the flow velocity at the level $y = 0$, which is the average level of an undulated bed (Fig 1.6a). We identify $(u_b)_\Delta$ with the average value of u along the streamline s at the grain level $n = k_s$ (which can be regarded as the lower boundary of the main flow).[23]

From (1.31) it is clear that if the flat bed becomes undulated, and thus if $\bar{c}$ changes to c, $\tau_0 = \gamma S h$ remaining constant, then $\bar{v}$ changes to v so that $v / \bar{v} = c / \bar{c} = \lambda_c$. Clearly, this reduction of the average flow velocity $\bar{v}$ must be accompanied by the reduction of all flow velocities u – including the reduction of u_b into $(u_b)_\Delta$. Since the reduction of various u cannot differ from each other significantly, one can approximate the reduction of u_b by that of $\bar{v}$:

$$\frac{v}{\bar{v}} = \lambda_c \approx \frac{(u_b)_\Delta}{u_b} . \tag{1.59}$$

It follows that the the undulated-bed version of an originally flat-bed q_{sb}-formula can be obtained by applying the following rule: "Multiply each $\tau_0 \sim Y$ with λ_c^2 and each velocity with λ_c; leave $(\tau_0)_{cr}$ and/or v_{*cr} as they are" (for $(\tau_0)_{cr}$ and/or v_{*cr} "do not belong" to the given (undulated) stage). For example, the undulated-bed counterpart of Bagnold's formula (1.57) can thus be expressed as

---

[22] In Fig. 5.5 of Ref. [36], $e_b / \tan \psi_0$ means β.

[23] This identification ensures the continuity in the transition $\lim(\Delta \to 0)\, (u_b)_\Delta = u_b$.

$$q_{sb} \approx \beta (u_b \lambda_c)[(\tau_0 \lambda_c^2) - (\tau_0)_{cr}] / \gamma_s . \tag{1.60}$$

(Observe that if $(\tau_0 \lambda_c^2) \gg (\tau_0)_{cr}$, then the reduction of q_{sb} is by the factor $\approx \lambda_c^3$).

1.5.3 *Suspended-load*

Consider now the second term on the right of (1.50). This term, which implies q_{ss}, can be expressed as

$$q_{ss} = \int_\epsilon^h U_s C \, dy = \int_\epsilon^h u \, C \, dy, \tag{1.61}$$

for we assume that the suspended particles are conveyed downstream with the local flow velocities u:

$$U_s = u . \tag{1.62}$$

Unlike the bed-load layer ϵ, the suspended-load region $(h - \epsilon)$ is, in general, not "thin": and the C-distribution along it, which varies markedly with the relative flow intensity η, cannot be standardized. Hence q_{ss} can be determined only by the factual integration of (1.61).

There is a general agreement that the C-distribution along y can be expressed best by the Rouse-Einstein form

$$\frac{C}{C_\epsilon} = \left[\frac{\epsilon}{y} \frac{h - y}{h - \epsilon} \right]^m \tag{1.63}$$

(with $m = 2.5 w / v_*$),

and it is only the determination of ϵ and C_ϵ which is still under investigation. For the distribution of u along y the relations presented in 1.4.1 can be used, as long as the suspended-load is not too heavy.[24]

i- *Flat bed*

In this case, the lower limit $(y = \epsilon)$ of the suspended-load region can be identified with the thickness of the bed-load layer, i.e. with the jump height δ_b of the bed-load grains. Note that the bed-load rate can be expressed also in the form

$$q_{sb} = U_{s\epsilon} (C_\epsilon)_{av} \epsilon . \tag{1.64}$$

---

[24] The influence of C on the u-distribution increases continuously with the increment of C, and therefore the answer to "How heavy is *too heavy*?" depends largely on the degree of accuracy required from the intended computations of u. The generalized u-distributions which take into account the presence of suspended-load can be found e.g. in Refs. [4], [10], [16], [18] (see also [3]): most of them yield the clear fluid relations in 1.4.1 in the limit $C \to 0$.

Here $U_{s\epsilon}$ is proportional to u_b (see (1.52)), while the average concentration $(C_\epsilon)_{av}$ of the grain cloud moving within $0 < y < \epsilon$ is proportional to C_ϵ at $y = \epsilon$. Considering this and equating (1.64) with (1.57), one determines

$$\epsilon C_\epsilon \sim [\tau_0 - (\tau_0)_{cr}]/\gamma_s \quad \text{i.e.} \quad \frac{\epsilon}{D} C_\epsilon \sim (Y - Y_{cr}), \tag{1.65}$$

and consequently

$$\frac{\epsilon}{D} C_\epsilon \sim \Psi(\xi)(\eta - 1).$$

It is not likely that $U_{s\epsilon}$ and $(C_\epsilon)_{av}$ are related to u_b and C_ϵ by constant proportionalities: the ratios $U_{s\epsilon}/u_b$ and $(C_\epsilon)_{av}/C_\epsilon$ must be expected to vary with $(\eta - 1)$. Consequently, the relation above means, in fact,

$$\frac{\epsilon}{D} C_\epsilon = \Psi(\xi)\phi(\eta - 1). \tag{1.66}$$

It follows that ϵ/D and C_ϵ are interdependent: the product of their expressions must satisfy (1.66).

Adopting for the jump height δ_b the expression of L.C. Van Rijn [25], [26] with $\lambda_c = 1$, one can express ϵ ($= \delta_b$) as

$$\frac{\epsilon}{D} = 0.3\,\xi^{0.7}(\eta - 1)^{0.5}. \tag{1.67}$$

Furthermore, L.C. Van Rijn gives

$$\frac{\epsilon}{D} C_\epsilon = \frac{0.015}{\xi^{0.3}}(\eta\lambda_c^2 - 1)^{1.5} \tag{1.68}$$

where $\lambda_c = 1$ in the case of a flat bed. Observe that (1.68) is in line with (1.66). Hence if the bed is flat, then ϵ and C_ϵ can be determined from (1.67) and (1.68) respectively. Substituting (1.67) in (1.68), and taking into account $\lambda_c = 1$, one obtains $C_\epsilon = 0.05\xi^{-1}(\eta - 1)$, which is realistic as long as η is not too large. (The rate of change $\partial C_\epsilon/\partial\eta$ should decrease with the increment of η).

ii- *Undulated bed*

In this case, we usually have $\Delta \gg \delta_b \sim D$ and it would be only reasonable (following Refs. [25], [26]) to identify the suspended-load region with the region above the bed form crests, i.e. with $(\Delta/2) < y < h$ (Fig. 1.6a). Hence if the bed is undulated, then

$$\epsilon = \Delta/2, \tag{1.69}$$

C_ϵ still being given by (1.68) (with $\lambda_c < 1$). The average bed level $y = 0$ is also the y-origin of the C- and u-distributions. The latter distribution must be evaluated by K_s ($\gg k_s$).[25]

[Some of the recent expressions for C_ϵ and ϵ are presented in Refs. [3] and [7]. Most of these expressions were determined on the basis of experiments performed within certain ranges, and therefore they may not remain valid if they are "pushed to the limits". For example, C_ϵ cannot increase indefinitely with the increasing values of η (or v_*/w), and yet in many works C_ϵ is given as a continuously increasing function of η. The C_ϵ-expressions of Refs. [23] and [7], which are of the form

$$C_\epsilon = \frac{\alpha x}{1 + \beta x}, \tag{1.70}$$

are free from this particular imperfection,[26] as they possess (correctly) the asymptotic property, $\lim(x \to \infty) C_\epsilon = \alpha/\beta = const.$ On the other hand, the form proposed in [23] does not take into account the influence of ξ, while that in [7] reflects the influence of both ξ and $\lambda_c(v_*/w)$ by a single combination (x) – no explanation is offered in [7] for such a "similarity collapse". The consideration of $x \sim v_*^5$ in [7], instead of $(\eta\lambda_c^2 - 1)$, is hardly an improvement; for it implies that C_ϵ is different from zero for any $v_* > 0$. Although the large exponent of v_* (viz 5) ensures that the growth of C_ϵ with v_* (from zero onwards) is very feeble for small v_*, nonetheless the form in [7] implies that the suspended-load originates *before* the bed-load (which requires $v_* > v_{*cr}$ for its existence)].

1.6 Granular Skin Roughness

From laboratory experiments carried out with turbulent flows past a flat (scraped) movable bed with stationary grains ($Y < Y_{cr}$) [12], [20], [2], it follows that the value of the granular skin roughness k_s can be given by

$$k_s \approx 2D, \tag{1.71}$$

25 Knowing Λ_i, Δ_i, h and k_s, one can determine K_s e.g. as follows: Determine $\bar{c}$ ($= v/v_*$) from (1.26). Then compute λ_c (from (1.42)) and thus $c = \bar{c}\lambda_c$. Finally, solve K_s from (1.27) or (1.28) where v/v_* is to be identified with c and k_s with K_s.

26 In Ref. [23], $x = (\eta\lambda_c^2 - 1)$; in Ref. [7], $x = [\lambda_c(v_*/w)\,\xi^{0.9}]^5$ (Eqs. (38) and (43) in [7]). The quantity R_p of Ref. [7] implies $\xi^{3/2}$. Using $n = 0.6$, proposed in [7], in R_p^n (Eq. (37) in [7]) one determines $R_p^n = \xi^{(0.6)(3/2)} = \xi^{0.9}$ which appears in the expression of x.

where D can be identified with D_{50}.[27]

At the early stages of sediment transport the percentage of grains (of the uppermost layer) which are in motion is small, and therefore (1.71) should still be valid up to a certain $Y\,(> Y_{cr})$, even if the grains are moving. This "certain Y" can be inferred from the experiments of K.C. Wilson [27], [28], [29]. When conducting his sediment transport experiments in pressurized conduits, K.C. Wilson was able to work with Y-values that are by an order of magnitude larger than those commonly encountered in comparable open-channels. The values of k_s/D computed (via $c = v/v_*$) from his experiments are plotted versus Y in Fig. 1.8. Note from this graph that k_s/D tends to increase in proportion to Y (as $k_s/D \approx 5Y$) when Y is large, and it tends to become ≈ 2 (as to justify (1.71)) when Y is small. (D was identified with D_{50} of the reasonably uniform materials used).

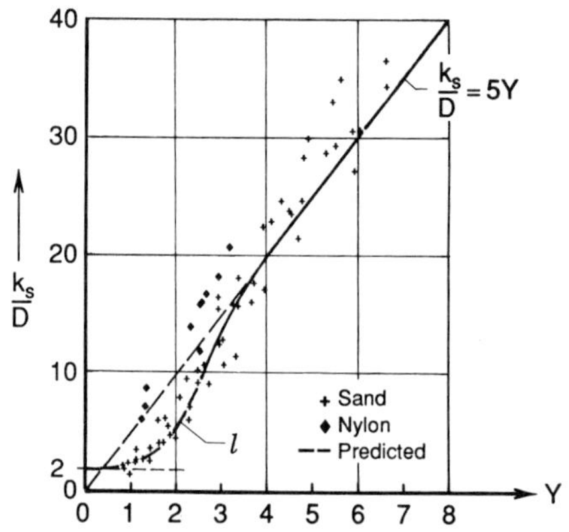

Fig. 1.8 (after Ref. [28])

From Fig. 1.8, one also infers that (1.71) can be used up to $Y \approx 1$ (i.e up to $\eta \approx 20$), say. If $Y > \approx 1$, then k_s/D can be taken from Fig. 1.8. The curve l in Fig. 1.8 is the graph of the following mathematical fit[28]

$$\left.\begin{aligned} &\frac{k_s}{D} = 2 \quad (\text{if } Y < 1) \\ &\frac{k_s}{D} = 5Y + (Y-4)^2\,[0.043\,Y^3 - 0.289\,Y^2 - 0.203\,Y + 0.125] \\ &\qquad\qquad\qquad\qquad\qquad\qquad\qquad\qquad (\text{if } 1 < Y < 4) \\ &\frac{k_s}{D} = 5Y \quad (\text{if } Y > 4)\,. \end{aligned}\right\} \qquad (1.72)$$

---

[27] The claim that D must necessarily be identified with D_{90} is unjustifiable, for the ratio $(D_{90} - D_{50})/D_{90}$ of the (reasonably uniform) materials subjected to experiments was much smaller than the relative scatter of experimental points used to determine (1.71).

[28] The curve l and the second equation in (1.72) are due to the author.

1.7 Transport Continuity Equation

In physics, an active surface usually either absorbs or emits something (light, heat, electrons, etc.): a mobile bed surface does both at the same time. During the transport of sediment some particles are landing on an area (δA) of the bed surface, while some others are detaching from it (l_i and d_i in Fig. 1.9a). And, in fact, the bed level is in equilibrium only when the number

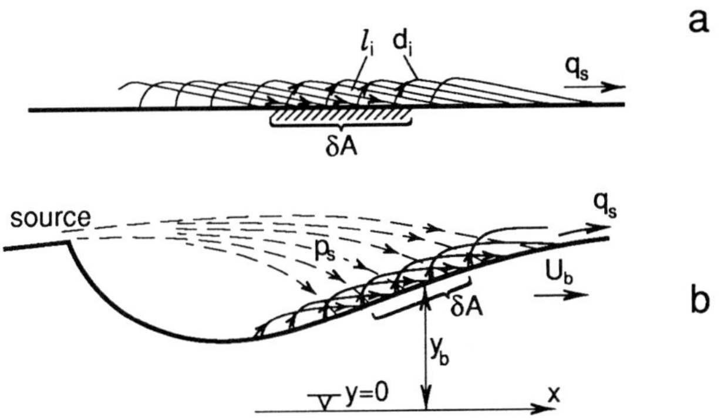

Fig. 1.9

of incoming and outgoing particles (per unit area and per unit time) are equal. In some cases, the particles landing on an area δA may not be only those which form the transport rate q_s: the particles from an external source (Fig. 1.9b) may also be landing on δA. The volumetric deposition rate of the "external" particles (per unit area and per unit time) will be denoted by p_s. Note that the dimension of p_s, viz

$$[p_s] = \frac{[\text{volume}]}{[\text{area}]\,[\text{time}]} = [\text{velocity}]\,, \tag{1.73}$$

is different from the dimension of q_s (see (1.48)). We will assume in the derivations below that q_s consists of the bed-load only ($q_s = q_{sb}$).

1.7.1 *Two-dimensional conditions*

Let us suppose that the area δA is situated indeed on the upstream slope of a bed form, as shown in Fig. 1.9b. The bed form itself is moving (migrating) in the flow direction x, its steepness increasing at the same time. Let U_b be the migration velocity of the bed form and $V_b = dy_b/dt$ the growth-speed of its ordinates.[29] In this case the area δA (which is assumed

[29] The growth-speed $V_b = dy_b/dt$ is thus the same as the substantive derivative DA/Dt of continuum mechanics, for V_b implies the time rate of growth of the ordinate y_b as it moves (along x) with the bed form.

to be "small" in comparison to the bed form length Λ) is displaced upwards per unit time by the amount

$$- U_b \frac{\partial y_b}{\partial x} + V_b \,. \tag{1.74}$$

This upward displacement of δA is due to the motion of sediment, and thus it must be equal to

$$- \frac{\partial q_s}{\partial x} + p_s \,. \tag{1.75}$$

Hence,

$$\frac{\partial q_s}{\partial x} + \frac{dy_b}{dt} = U_b \frac{\partial y_b}{\partial x} + p_s \tag{1.76}$$

which is the transport continuity equation.[30]

1.7.2 *Three-dimensional conditions*

i- In this case, the bed-load is a field vector ($\mathbf{q}_s = \mathbf{q}_s(x, z, t)$) and (1.76) becomes

$$\nabla \mathbf{q}_s + \frac{dy_b}{dt} = \frac{\partial q_{sx}}{\partial x} + \frac{\partial q_{sz}}{\partial z} + \frac{dy_b}{dt} = U_b \frac{\partial y_b}{\partial x} + p_s \,, \tag{1.77}$$

which corresponds to a migrating bed form determined by the ordinates $y_b = f_y(x, z, t)$.

ii- Consider now the beginning of a bed deformation experiment ($U_b = 0$; $p_s = 0$): at $t = 0$ the bed surface is flat ($y_b = f_y(x, z, 0) = 0$). We assume that the (bed-load transporting) initial flow is, in general, convective in plan; i.e. it is converging or diverging (because of the plan shape of its curvilinear banks, say). Clearly, in this case (1.77) reduces into

$$\nabla \mathbf{q}_s + \frac{\partial y_b}{\partial t} = 0 \,, \tag{1.78}$$

for $dy_b/dt = \partial y_b/\partial t$ when $U_b = 0$ (as can be inferred from footnote 30). The direction of bed-load at a location (point $P(x, z)$) on the bed plane is the

---

30 Since

$$dy_b = \frac{\partial y_b}{\partial x} dx + \frac{\partial y_b}{\partial t} dt \quad \text{and thus} \quad \frac{dy_b}{dt} = \frac{\partial y_b}{\partial x} U_b + \frac{\partial y_b}{\partial t} \; (= V_b),$$

the relation (1.76) can be expressed also as

$$\frac{\partial q_s}{\partial x} + \frac{\partial y_b}{\partial t} = p_s \,,$$

which reduces into the familiar Exner-Polya equation if $p_s = 0$.

same as the direction of flow at the bed there. Considering this, and using Bagnold's formula (1.57), one can express the bed-load vector as

$$\mathbf{q}_s = \mathbf{u}_b \bar{\phi}(\eta) \tag{1.79}$$

where

$$\mathbf{u}_b = \mathbf{u}_b(x, z, t) \quad \text{and} \quad \bar{\phi}(\eta) = \beta D Y_{cr}(\eta - 1)\,. \tag{1.80}$$

We focus our attention on the conditions at $t = 0$, or "just after" $t = 0$ (when the possible bed deformation (y_b) can still be regarded as infinitesimal).

Since η is proportional to τ_0, while τ_0 (in a given experiment) is uniquely determined by the magnitude u_b of $\mathbf{u}_b$ (i.e. since $\eta \sim \tau_0 \sim u_b^2$), the function $\bar{\phi}(\eta)$ can be taken as a function of u_b: as $\phi(u_b)$, say. Consequently, the consideration of (1.79) can be replaced by that of

$$\mathbf{q}_s = \mathbf{u}_b \phi(u_b)\,. \tag{1.81}$$

Now $\nabla \mathbf{u}_b = 0$ (continuity of fluid motion), and therefore

$$\nabla \mathbf{q}_s = \nabla[\mathbf{u}_b \phi(u_b)] = (\nabla \mathbf{u}_b)\phi(u_b) + \mathbf{u}_b \nabla \phi(u_b) \tag{1.82}$$

i.e.

$$\nabla \mathbf{q}_s = \mathbf{u}_b \nabla \phi(u_b) \tag{1.83}$$

which yields, in conjunction with (1.78),

$$\frac{\partial y_b}{\partial t} = -\,\mathbf{u}_b \nabla \phi(u_b)\,. \tag{1.84}$$

This relation indicates that if the flow is absolutely uniform (the same $u_b = const$ for any $P(x, z)$ on the bed plane), then the (flat) bed cannot undergo any deformation ($\nabla \phi(u_b) = 0$, and thus $V_b = \partial y_b / \partial t = 0$). If, however, the flow is convective ($u_b \neq const$), then the initial bed must necessarily deform with the passage of time ($\nabla \phi(u_b) \neq 0$, and thus $V_b = \partial y_b / \partial t \neq 0$).

If s is the natural longitudinal coordinate of a streamline, and $\mathbf{i}_s$ is the unit vector along it, then

$$\nabla \phi(u_b) = \mathbf{i}_s \frac{\partial \phi(u_b)}{\partial s}\,. \tag{1.85}$$

Using (1.85), one can express (1.84) as

$$\frac{\partial y_b}{\partial t} = -\,u_b \frac{\partial \phi(u_b)}{\partial s} \tag{1.86}$$

or as

$$\frac{\partial y_b}{\partial t} = -\,\alpha\, u_b \frac{\partial u_b}{\partial s}\,, \tag{1.87}$$

where $\alpha = \partial \phi(u_b)/\partial u_b$ is always positive and finite [for $\bar{\phi}(\eta)$ is an increasing function of η (> 1) (see (1.80)), and η is an increasing function of u_b (viz

$\eta \sim u_b^2)$]. The relation (1.87) indicates even more explicitly than (1.84) that the bed of a convective flow *must* deform (Chapter 5).[31]

1.8 Data-Plots

Usually, it is not the shortage of data but their interpretation and utilization which form the concern. Consider the dimensionless function

$$\Pi = \Phi(X_1, X_2, \ldots, X_n) . \tag{1.88}$$

If $n = 1, 2, 3$, etc., then (1.88) implies respectively a curve, a family of curves, a family of the family of curves, etc. Hence, no matter what the number n might be, the experimental determination of the function (1.88) means, in effect, the experimental determination of a curve or curves, each being a function of one variable:

$$\Pi_j = \phi(X_j) . \tag{1.89}$$

Each particular river-region or experimental flume can provide only a limited range, ΔX_j say, of the variable X_j. Moreover, the scatter in fluvial hydraulics plots is rather gross, and its average magnitude often is comparable with such a ΔX_j. But this means that it is hardly likely that a point-set obtained from measurements in a single flume or a river-region can reveal the (whole) curve (1.89) sought. (And in some cases the information supplied by an individual point-set can even be misleading. Take e.g. the plot shown in Fig. 2.8a. Here the set of "white" circles suggests that $\Pi_j = \upsilon T/h$ strongly decreases with $X_j = \upsilon h/\nu$ (line l). Yet, from the total point pattern it is clear that Π_j is practically independent of X_j).

From the aforementioned, it should be clear that a reliable way of determining (or verifying) an experimental relation in fluvial hydraulics should be by using *all* the field and laboratory data available to the researcher. The fact that the small- and large-scale streams exhibit different behaviour is not because they are determined by different variables, but because their (same) variables have different numerical values, X_j'' and X_j' say. The larger the difference $X_j' - X_j''$, the better for the discovery of the relation (1.89) (and the worse for the physical modelling (which rests on $X_j' = X_j''$) of the phenomenon it represents).

---

[31] If q_s does not consist of the bed-load only ($q_{ss} \neq 0$), then the left-hand sides of (1.76) and (1.77) have also the term

$$\frac{\partial}{\partial t}\int_0^h C dy = h\frac{\partial C_{av}}{\partial t} ,$$

implying the time rate of increment of the volume of suspended solids over the unit bed area.

References

1. Bagnold, R.A.: *The flow of cohesionless grains in fluids.* Philosophical Trans. Roy. Soc. London, A, 249, No. 964, Dec. 1956.
2. Bishop, C.T.: *On the time-growth of dunes.* M.Sc. Thesis, Dept. of Civil Engrg., Queen's Univ., Kingston, Canada, 1977.
3. Cioffi, F., Gallerano, F.: *Velocity and concentration profiles of solid particles in a channel with movable and erodible bed.* J. Hydr. Res., Vol. 29, No. 3, 1991.
4. Coleman, N.L.: *Velocity profiles with suspended sediment.* J. Hydr. Res., Vol. 19, No. 3, 1981.
5. Comolet, R.: *Introduction a l'analyse dimensionelle et aux problems de similitude en Mechanique des Fluides.* Masson et Cie, Paris, 1958.
6. Engelund, F.: *Hydraulic resistence of alluvial streams.* J. Hydr. Div., ASCE, Vol. 92, No. HY2, March 1966.
7. Garcia, M., Parker, G.: *Entrainment of bed sediment into suspension.* J. Hydr. Engrg., Vol. 117, No. 4, April 1991.
8. Garde, R.J., Raju, K.G.R.: *Mechanics of sediment transportation and alluvial stream problems.* Wiley Eastern, New Dehli, 1978.
9. Graf, W.H.: *Hydraulics of sediment transport.* McGraw-Hill Book Co., Inc., New York, 1971.
10. Itakura, T., Kishi, T.: *Open channel flow with suspended sediments.* J. Hydr. Div., ASCE, Vol. 106, No. HY8, Aug. 1980.
11. Jansen, Ph. et al.: *Principles of river engineering.* Pitman, London, 1979.
12. Kamphuis, J.W.: *Determination of sand roughness for fixed beds.* J. Hydr. Res., Vol. 12, No. 2, 1974.
13. Landau, L.D., Lifshitz, E.M: *Fluid Mechanics.* Pergamon Press, Oxford, 1986.
14. Langhaar, H.L.: *Dimensional analysis and theory of models.* John Wiley and Sons, New York-London (5th edition), 1962.
15. Lee, H.Y., Yu, W.S.: *Velocity profile of sediment-laden open channel flow.* Int. J. Sediment Research, IRTCES, Vol. 6, No. 1, March 1991.
16. Marchi, E.: *Sulle resistenze nelle correnti con transporto solido e sui problemi di modellazione degli alvei nei confronti della formazione dei meandri.* XIX Convegno di Idraulica e Construzioni Idrauliche, Pavia, 1984.
17. Monin, A.S., Yaglom, A.M.: *Statistical Fluid Mechanics: mechanics of turbulence.* Vol. 1 and 2, The MIT Press, 1975.
18. Parker, G., Coleman, N.L.: *Simple model of sediment laden flows.* J. Hydr. Engrg., ASCE, Vol. 112, No. 5, May 1986.
19. Raudkivi, A.J.: *Loose boundary hydraulics.* Pergamon Press, Oxford (2nd edition), 1976.
20. Riedel, H.P.: *Direct measurement of bed shear stress under waves.* Ph.D. Thesis, Dept. of Civil Engrg., Queen's Univ., Kingston, Canada, 1972.
21. Schlichting, H.: *Boundary layer theory.* McGraw-Hill Book Co. Inc., Verlag G. Braun (6th edition), 1968.
22. Sedov, L.I.: *Similarity and dimensional methods in Mechanics.* Academic Press Inc., New York, 1960.
23. Smith, J.D., McLean, S.R.: *Spatially averaged flow over a wavy surface.* J. Geophys. Res., 82, 12, 1977.
24. Vanoni, V.A., Li-San Hwang: *Relation between bed forms and friction in streams.* J. Hydr. Div., ASCE, Vol. 93, No. HY3, May 1967.
25. Van Rijn, L.C.: *Sediment transport.* Delft Hydraulic Laboratory, Publication No. 334, Feb. 1985.
26. Van Rijn, L.C.: *Sediment transport, part II: suspended load transport.* J. Hydr. Engrg., ASCE, Vol. 110, No. 11, Nov. 1984.
27. Wilson, K.C., Nnadi, F.N.: *Behaviour of mobile beds at high shear stress.* Proc. XXII Coastal Eng. Conf., Vol. 3, Chapter 192, July 2-6, Delft, 1990.

28. Wilson, K.C.: *Mobile-bed friction at high shear stress.* J. Hydr. Engrg., ASCE, Vol. 115, No. 6, June 1989.
29. Wilson, K.C.: *Frictional behaviour of sheet flow.* Prog. Rep. 67, Inst. Hydrodyn. and Hydr. Engrg., Tech. Univ. Denmark, June 1988.
30. Yalin, M.S., Kibbee, M.S.: *Physical modelling of sediment transporting flows.* in "Movable Bed Physical Models", H.W. Shen ed., NATO ASI Series, Kluwer Acad. Publishers, London, 1990.
31. Yalin, M.S., Silva, A.M.F.: *Physical modelling of self forming alluvial channels.* National Conf. on Hydr. Engrg., ASCE, July 30-Aug. 3, San Diego, 1990.
32. Yalin, M.S.: *Fundamentals of hydraulic physical modelling.* In "Recent Advances in Hydraulic Physical Modelling", R. Martins ed., NATO ASI Series, Kluwer Acad. Publishers, London, 1989.
33. Yalin, M.S., MacDonald, N.J.: *Determination of a physical river model with mobile bed.* Proc. XXII Congress IAHR, Lausanne, 1987.
34. Yalin, M.S., Lai, G.: *On the form drag caused by sand waves.* Proc. JSCE, No. 363/II-4, Nov. 1985.
35. Yalin, M.S., Karahan, E.: *Inception of sediment transport.* J. Hydr. Div., ASCE, Vol. 105, No. HY11, Nov. 1979.
36. Yalin, M.S.: *Mechanics of sediment transport.* Pergamon Press, Oxford, 1977.
37. Yalin, M.S.: *Theory of hydraulic models.* MacMillan, London, 1971.
38. Yalin, M.S.: *Similarity in sediment transport by currents.* Hydraulics Research Paper No. 6, Hydraulics Research Station, London, 1965.
39. Yalin, M.S.: *On the average velocity of flow over a mobile bed.* La Houille Blanche, No. 1, Jan/Feb. 1964.

CHAPTER 2

TURBULENCE

2.1 Bursting Processes

2.1.1 *General description*

Much of the recent research on turbulence is devoted to the study of bursting processes and coherent structures. So far, however, agreement appears to have been reached only with regard to the description of their general pattern: their detailed mechanism is still under investigation. Some authors [49], [63], [30] focus their attention on the transverse vortex which occurs during the development of a burst; others [51], [27], [5] emphasize the role of the viscous sublayer and various "structures" associated with it. However, in "his excellent and very extensive study" [9], A.J. Grass [17] has demonstrated, experimentally, that bursts occur irrespective of whether or not the viscous influence at the flow boundaries is present, i.e. that they occur in any regime of turbulent shear flow (see also [40], [12]). Considering this, bursts are explained in this chapter without any reference to the regime of turbulent flow. These (schematical) explanations rest on experimental observations and measurements reported by various sources (especially [54], [49] and [30]).

Turbulence is due to a multitude of *eddies* (fluid volumes with a sense of rotation) having various sizes l. These eddies are superimposed on each other and, ultimately, on the mean flow (see e.g. [35], [32]). The size of the smallest, or *microturbulent*, eddies is assumed to be $l_{min} \approx \nu/\upsilon_*$ (Kolmogorov scale), the size of the largest, or *macroturbulent*,[1] eddies being nearly equal to the flow thickness h: $l_{max} \approx h$.[2] Hence,

---

[1] The "large-scale turbulence" is formed by eddies whose size l scales with the flow thickness h. Among these eddies, only those whose size is (nearly) equal to h will be termed in this text as macroturbulent eddies. (In the literature "macroturbulence" and "large-scale turbulence" often are used as synonyms of each other).

[2] This book deals exclusively with the flows having a free surface. Hence in the following, unless stated otherwise, the turbulent shear flow will invariably be interpreted as an open-channel flow (of the thickness (depth) h).

$$\approx \frac{\nu}{\upsilon_*} \leq l \leq \approx h. \tag{2.1}$$

The fluctuating velocities of all eddies are of comparable magnitude, and just because of this their typical times, or periods, are different: the larger an eddy, the larger is its period. Eddies are not permanent – they are "born", they "live" and they "die".

"An important part in any turbulent flow is played by the largest eddies, whose size is of the order of dimension (h) of the region in which the flow takes place" [32]. Most of the "pre-burst-era" treatises on turbulence begin with a statement such as "first the largest eddies occur". Then they proceed to describe (adequately) the process of disintegration of these eddies: the associated energy cascade, the mechanism of energy dissipation, etc. (see e.g. [32], [35], [55]). Nothing is said, however, in these works, which form the "eddy-cascade theory", as to how the largest eddies come into being in the first place. And it is only after the discovery of bursting processes that some information on this score has been gained. Indeed, recent research indicates that the largest eddies do not originate in their full size $\approx h$. Rather, they are generated near the flow boundaries (that is, "near the bed" in the case of an open-channel flow), with a size much smaller than h: the nearer the "birth place" of an eddy to the bed, the smaller is its initial size in comparison to h. Immediately after its birth, the prospective macroturbulent eddy, e_1 say, is ejected upwards and, as it moves away from the bed, it is conveyed downstream by the flow. During this motion, the upward-velocity of e_1 progressively decreases, whereas its size l increases (due to coalescence and engulfment). Eventually, the size of the eddy e_1 becomes nearly equal to h: i.e. it becomes a macroturbulent eddy (E_1). However, soon after that, the (large and unstable) eddy E_1 hits the bed and is destroyed (*break-up phase*). Owing to reasons which will be clarifed later, the destruction of the eddy E_1 prompts the generation of a new (second) eddy e_2. Having passed through the same stages, this second eddy causes, in turn, the generation of the third eddy e_3, and so on. *Bursting process*, or simply *burst*, is the term which stands for the above outlined sequence of events, which occurs between the births of two consecutive eddies (e_i and e_{i+1}).[3] A burst-forming eddy e_i and the flow complex around it constitute a *coherent structure*, which evolves as it is conveyed downstream in the course of a burst cycle. It should be emphatically pointed out that the eddy e_i as well as the coherent structure around it are three-dimensional: their (continually increasing during the burst cycle) extent in the spanwise direction z is limited – no matter how large the width-to-depth ratio (B/h) of the flow might be.[4]

---

[3] More refined definitions of bursts can be found e.g. in [2], [4], [12].

[4] Although bursts are distributed randomly in space and time, the burst itself signifies a certain sequence of events. Owing to this reason, turbulence is not regarded today as a completely random phenomenon as has been assumed previously. (See the classical ex-

The current knowledge in the field suggests that a complete picture of turbulence and its energy structure must be expected to emerge from the complementary study of both:

> Bursting processes (which convey information on the development of the largest eddies E and thus on the generation of turbulence-energy),

and

> Eddy-cascade processes (which convey information on the disintegration of the largest eddies E and thus on the decay and dissipation of turbulence-energy).

2.1.2 *Sequence of events during a burst cycle*

i- Let us now take a closer look at a burst cycle. Recent research ([49], [50], [38], etc.) gives reason to postulate that a turbulent shear flow contains in its body a series of large-scale *high-speed* and *low-speed* regions, R_h and R_l say, which extend almost throughout the flow depth h, and which slowly deform as they are conveyed downstream [49]. From Refs. [12] and [38], one infers that these regions are likely to be distributed as shown in the highly idealized Fig. 2.1 (which is in line with Fig. 7 in Ref. [37]). Here, the large-scale lengths L and L' are of several flow depths, the transition from R_l to R_h, in

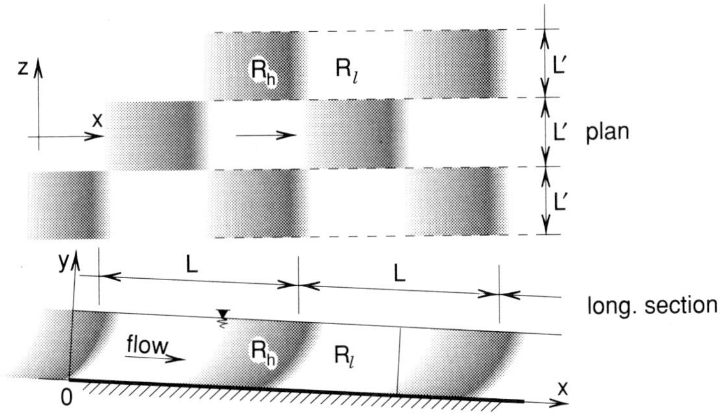

Fig. 2.1

the flow direction, being more gradual than from R_h to R_l [49] (as implied by the shading).

Fig. 2.2a shows, schematically, the velocity distribution diagrams (along y) at a fixed location $(x; z)$. During the passage of R_l, we have (for any y) $U_l < u$; during the passage of R_h, $U_h > u$. [The areas confined by the U_h- and U_l-diagrams are not equal, and the continuity of flow rate is satisfied by

pressions of turbulence e.g. in [35], [32], [55]; their updated versions in [19], [20], [42], [44], [45]).

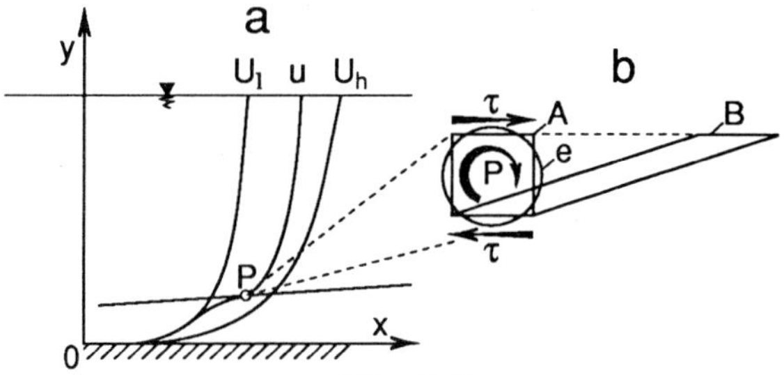

Fig. 2.2

the chessboard-like arrangement of R_h and R_l in plan]. The transition from a U_l-diagram to a U_h-diagram occurs by means of the inflectional U-diagrams ([3], [49], [16], etc.), the largest values of the instantaneous shear stress τ being at the inflection points P (where the instantaneous $\partial U/\partial y$ is the largest): the points P are mostly in the lower part of the flow. Consider a fluid element A subjected to the action of the shear stress τ at P (Fig. 2.2b). If the value of τ is sufficiently large then, instead of deforming into a highly strained parallelogram B, the overstressed element A "chooses" to roll-up into an eddy e [49], [24] (Helmholtz instability). Thus, the stress in the fluid is relieved and a burst-forming eddy is born. The subsequent "biography" of e is depicted in Figs. 2.3 and 2.4. The events associated with the eddy e are not affected, in principle, by its extent in the spanwise direction and therefore, for the sake of simplicity, they are described below in the two-dimensional manner.

Although the eddy e is formed by both low- and high-speed fluids (Fig. 2.3a), it travels along a trajectory s in the low-speed fluid region [49], its size continually increasing in the process (Fig. 2.3b). The displacement of e along s means the displacement of a fluid mass in that direction. This necessarily generates a circulatory motion e' in the low-speed region; the high-speed fluid overtakes the eddy e through the gap between this eddy and the free surface (Fig. 2.3b). Observe that the (encircled) part of the low-speed fluid forms an *ejection* (from m to e).[5] The part of R_l to the left of ejection contains the circulatory motion e'' induced by it.[6] The continual upward displacement of e and the consequent decrement of the gap between e and the free surface, on the one hand weakens the ejection flow, while, on the other, makes the passage of the overtaking high-speed fluid (through the gap) more and more difficult. At a certain stage, the "obstruction" (under e) due to the ejection flow and the circulatory motion e'' will be so weak, while the impediment

---

5 If the flow contacting the bed is viscous ($Re_* \ll 70$), then some fluid of this flow, at m, may be lifted by ejection.

6 The interface between e'' and R_h separates fluids moving in opposite directions. Hence, it is a high-shear zone which is likely to be the source of the "typical eddies" (of Falco [14]), which are "riding" on the boundary between R_h and R_l. See the sketch of these eddies in Fig. 31 of Ref. [38].

due to the reduction of the gap (above e) will be so strong, that the high-speed fluid will suddenly change its configuration so as to overtake the eddy e by passing *under* it. This suddenly arisen and convectively accelerated flow

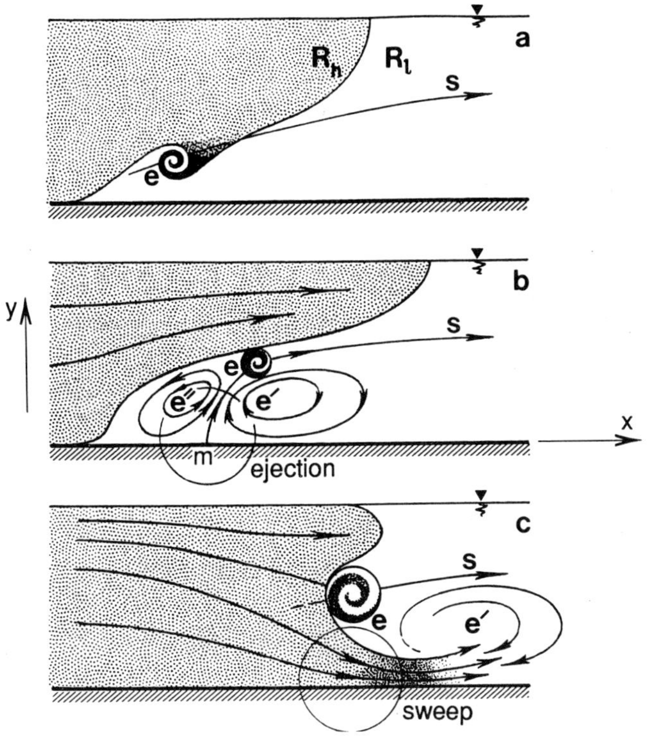

Fig. 2.3

of the high-speed fluid is referred to as *sweep* (it "sweeps" downstream the weakened ejection and all that is around it) (Fig. 2.3c). Having passed the eddy e, the sweep flow widens (decelerates) and interacts with the opposing circulatory motion e'. Thus the flow around e is neutralized. After this stage, the ever-enlarging eddy e is conveyed downstream in the body of an almost neutral flow (which overtakes the eddy only through the gap at the free surface [54] (Fig. 2.4a)). Eventually, the eddy e acquires a size nearly the

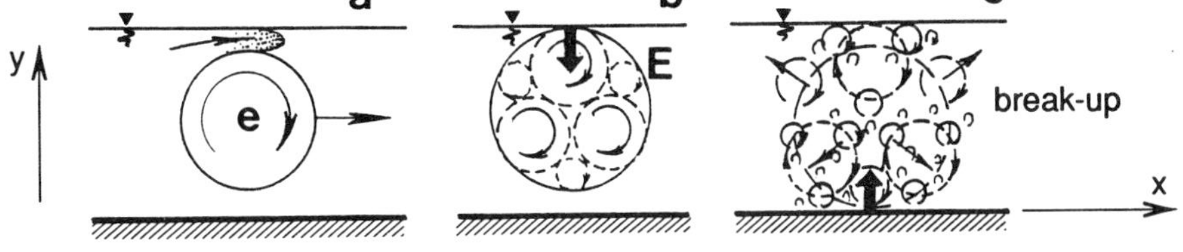

Fig. 2.4

same as the flow depth h: it becomes a macroturbulent eddy E. Owing to its proximity to the free surface, the eddy E interacts with it (Fig. 2.4b). As a result of this interaction, which is vividly described in [54] (see also [22]), the eddy E almost loses its coherence, and it is deflected toward the bed. The impact of E, or of what is left of it, with the bed completes its disintegration (break-up phase in Fig. 2.4c).[7]

According to the eddy-cascade theory ([32], [35]), a macroturbulent eddy E first disintegrates into a set of smaller eddies which, in turn, disintegrate into even smaller ones, and so on. This process of successive disintegration continues until the size of eddies is reduced to the microturbulent scale $l_{\min} \approx \nu/v_*$. The kinetic energy of E is transmitted by the "cascade" of eddy generations, almost without change, to the (last) level of microturbulence, where it is dissipated due to viscous friction. The disintegration of E means the termination of stress-relief in the fluid at that location. This prompts the initiation of a new stress-build-up there [23], [24], which manifests itself by the formation of the new high- and low-speed fluid regions R_h and R_l in the location previously occupied by E. The emergence of R_h and R_l leads, in turn, to the generation of a new eddy e, and the cycle depicted in Figs. 2.3 and 2.4 starts all over again.

The occurrence of bursts in all parts of the flow means the incessant conversion of the energy of the mean flow into the kinetic turbulence-energy, which is eventually dissipated by microturbulence. This is how turbulence is "going on". The energy of the mean flow, which is lost for the generation of bursts and thus for the maintainance of turbulence, is replenished by the continual loss of potential energy of the fluid mass forming open-channel flow (by the continual displacement of this mass *down*-stream). The potential energy first becomes converted into the strain energy concentrating (intermittently) at the locations of inflection points P (Fig. 2.2). With the rolling-up of the burst-forming eddies e, this strain energy turns into the kinetic energy of eddies and of the accompanying phenomena (e', e'', ejection, sweep, etc.): 70 to 80% of the kinetic turbulence-energy is generated at the early stages of a burst cycle [64].[8] It follows that the generation of turbulence manifests itself in the form of an organized (traceable) growth of eddies; its dissipation, by the disorganized spread of the "debris" [12] produced by the break-up's. Following A.A. Townsend, one can say that turbu-

---

[7] From these explanations, one should not infer that, for the occurrence of break-up's, a turbulent flow must necessarily have a free surface. Break-up's are present in boundary layer flows as well as in the flows in closed conduits. The existence of flow boundaries alone (onto which a continually growing eddy e will eventually impinge) is sufficient to ensure the occurrence of break-up's, and thus of bursts having a finite length.

[8] The quantitative formulation and evaluation of turbulence-energy production by bursts can be found in [27], [34], [41].

lence consists of (an organized) "universal" motion and (a disorganized) "irrelevant" motion.[9] As a result of this, "the large coherent structures ... are concealed in a tremendous clatter of noise" [12].

ii- Only some "sections" of coherent structures can be observed by means of contemporary methods, and not their three-dimensional totality (more on the topic in Ref. [16]). Owing to this reason, the three-dimensional form of coherent structures and their evolution in the course of a burst cycle is, for the present, a topic to theorize upon. None of the conceptual schemes produced to date for the "three-dimensional mechanics" of bursts (see e.g. [8], [15], [38], [43]) enjoys general acceptance. Thus, whether the eddy e is in fact the (longitudinal) section of a cigar-like three-dimensional eddy having its axis in the z-direction, or of a curvilinear "hairpin"-vortex, as shown in Fig. 2.5, etc., remains a matter of academic speculation.

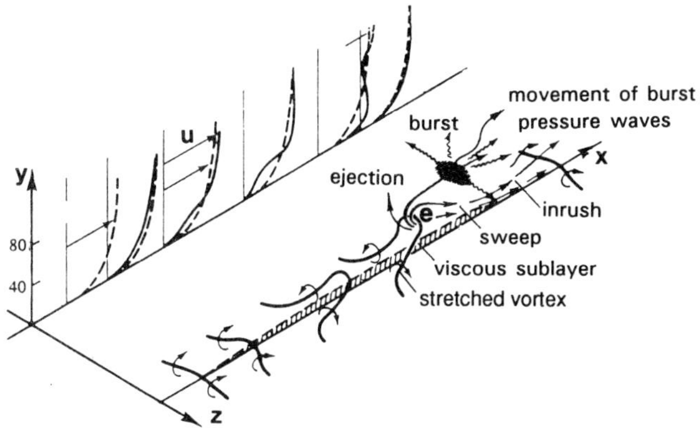

Fig. 2.5 (from Ref. [21])

Consider e.g. the scheme in Fig. 2.5. It implies that both ejections and sweeps originate from the same circulatory fluid motion around the stretched vortex, that the sweep overtakes the vortex (and thus e) by passing over it, and that the path of the sweep does not cross the path of ejection. Yet, from experimental observations it follows (see e.g. [49]) that ejections and sweeps originate from different fluid zones (R_l and R_h), that the sweep overtakes e by passing under it, and that it "sweeps the weakened ejection downstream" by moving across its path.

---

[9] Some authors maintain that the energy dissipation is, in fact, the *only* role microturbulence plays in the physics of turbulence [12], [34].

2.2 Turbulence Scales

2.2.1 *Burst length*

i- From the content of the preceding section it should be clear that the large-scale length L in Fig. 2.1 can be identified with that (average) distance in the flow direction x which is needed for the realization of a burst cycle, the large-scale length L' being the (average) maximum extent of the coherent structure in the spanwise direction z. In the following, L and L' will be referred to as *burst length* and *burst width* respectively.

The growth rates of turbulent eddies in the flow direction (x) exhibit a remarkable similarity. Consider, for example, the photographs in Fig. 2.6. They show the growth of eddies in three different turbulent flows: Fig. 2.6a – open-channel flow of water; Fig. 2.6b – mixing layer of two liquids; Fig. 2.6c – circular jet flow of a gas. Yet, in all these cases the size l of eddies appears to increase along x in almost the same manner, which can be approximated by the simple proportionality

$$l \approx \frac{1}{6} x \tag{2.2}$$

(where 1/6 is $\approx \tan\theta$).

Hence, it would be only natural to expect that the size (l) of a burst-forming eddy e grows, as it moves along its trajectory s (Fig. 2.7), also as implied by (2.2), i.e. as

$$l \approx \frac{1}{6} s. \tag{2.3}$$

Since the trajectory s is flat, we can write

$$l = l_{max} = h \quad \text{when} \quad s \approx x = L \tag{2.4}$$

which yields, in conjunction with (2.3),

$$L \approx 6h. \tag{2.5}$$

This relation is in line with experimental measurements which indicate that the burst length L is approximately equal to six flow depths h [29], [30], [31]. Assuming, with an accuracy sufficient for all practical purposes, that the burst-forming eddy e is conveyed downstream with the average flow velocity v, one can identify the average period T of a burst cycle with the ratio L/v, and one can express (2.5) as

$$\frac{L}{h} = \frac{vT}{h} \approx 6. \tag{2.6}$$

The measurement of the average burst period T at a location of flow is by far easier than that of the burst length L, and therefore in the current research the value of L is usually determined by measuring T (i.e. as L =

Fig. 2.6

$= \upsilon T$). Most of the measurements of T were conducted for boundary-layer flows, where u_∞ and δ figurate instead of υ and h (see e.g. [52], [7]). However, the values of $u_\infty T/\delta$ and $\upsilon T/h$ are comparable, and "now it appears to be fairly well established that T scales with outer variables,[10] and the generally accepted number is $u_\infty T/\delta \approx 6$" (Cantwell 1981 [9]).

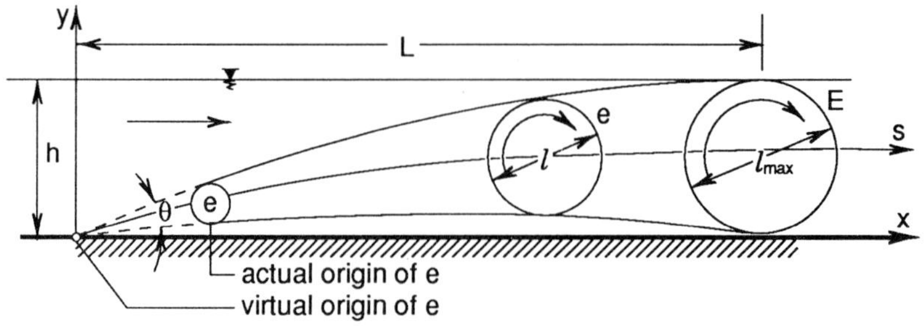

Fig. 2.7

ii- It would be reasonable to expect that the "generally accepted" value $\upsilon T/h = L/h \approx 6$ should nonetheless be affected by the relative roughness k_s/h and/or the Reynolds number $Re_* = \upsilon_* k_s/\nu$, at least when they are sufficiently small; and some attempts to reveal the influence of these dimensionless variables have already been made. Consider, for example, Figs. 2.8a and b which show the measured values of $\upsilon T/h = L/h$ plotted[11] versus the flow Reynolds number[12] $\upsilon h/\nu$. Fig. 2.8a contains laboratory data of boundary-layer flows past a smooth bed. Fig. 2.8b contains field data corresponding to various types of bed roughness; it includes the results of measurements carried out in the Missouri river and various canals (see [25] for more information on the data). No influence of $\upsilon h/\nu$ or k_s/h is detectable from these graphs: the gross scatter of experimental points around $L/h \approx 5$ to 6 is random. However, it would be imprudent to conclude on the basis of Figs. 2.8 alone that L/h is not affected by k_s/h and/or $\upsilon_* k_s/\nu$: first, because the influence of these variables may be only feeble (and thus undetectable owing to the gross scatter); and second, because the data plotted may not correspond to their sufficiently small values.

---

[10] "outer variables" are u_∞ and δ; "inner variables" υ_*, ν and k_s (see e.g. [52], [1]).

[11] Figs. 2.8a and b are adopted from Jackson 1976 [25]. The original coordinates (of Figs. 3 and 13 in Ref. [25]) are altered to correspond to the present text by identifying δ and u_∞ with h and υ respectively. The momentum thickness θ was evaluated as $(7/72)\,\delta \approx$ $\approx 0.097h \approx 0.1h$. The Eulerian integral time scale T_E was taken (in accordance with Ref. [25]) as $0.36T$.

[12] Since $\upsilon h/\nu = c(h/k_s)\,Re_*$ where $c = \bar{\Phi}_c(Re_*, h/k_s)$ (see (1.33)), the consideration of $\upsilon h/\nu$ and k_s/h is equivalent to that of $Re_* = \upsilon_* k_s/\nu$ and k_s/h.

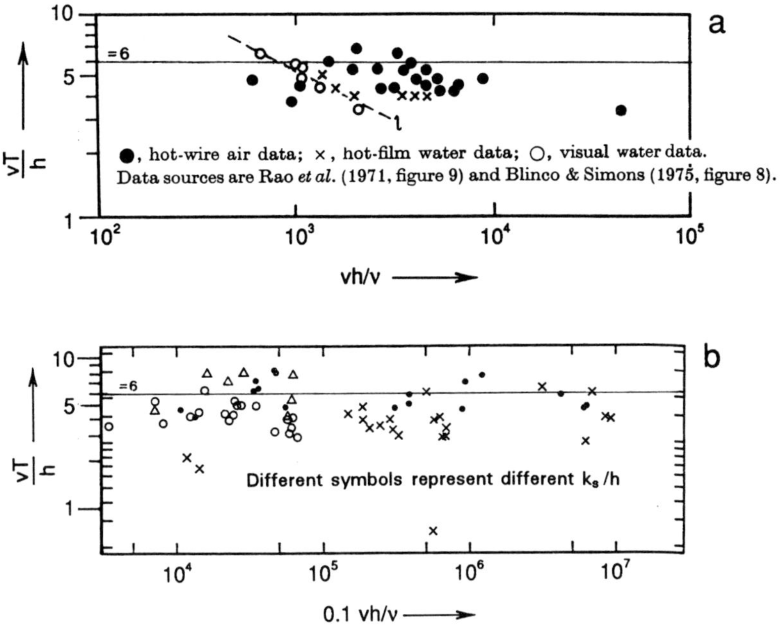

Fig. 2.8 (after Ref. [25])

2.2.2 *Burst length and friction factor*

A systematic study of bursts was carried out in laboratory flumes by A.B. Klaven and Z.D. Kopaliani [29] (see also [18]), and from their work it follows that L/h *is* affected by Re_* and/or k_s/h. According to Ref. [29], the influence of Re_* and/or k_s/h (on L/h) is by means of the friction factor c: the ratio L/h is a function of c which is a function of Re_* and k_s/h. (It may be noted here that in most of the practical cases the flow is rough turbulent and c varies with k_s/h only). The following relation is proposed in Ref. [29]:

$$\frac{L}{h} \sim c^{2/3} . \qquad (2.7)$$

Although the consideration of L/h as a function of c appears to be sound, it is doubtful that this function should have a power form (as implied by (2.7)). Indeed, since c always *varies* with k_s/h and/or Re_*, a power function of c can never yield $L/h \approx 6 = const$. To put it differently, if L/h is a function of c, then it must be an asymptotic function (which is capable of becoming $L/h \approx const$ for sufficiently large c).

The following is an attempt to derive an asymptotic variation of L/h with c.

Consider the time-growth rate dl/dt of an eddy moving along its path s. Owing to dimensional reasons, the "velocity" dl/dt must necessarily be proportional to a typical flow velocity, $\overline{W}$ say, the proportionality factor being a function of the dimensionless position $\zeta = y/h$

$$\frac{dl}{dt} = \phi(\zeta)\overline{W} . \qquad (2.8)$$

One would expect that $\overline{W}$ is either the local flow velocity u (Fig. 2.9), or its increment Δu

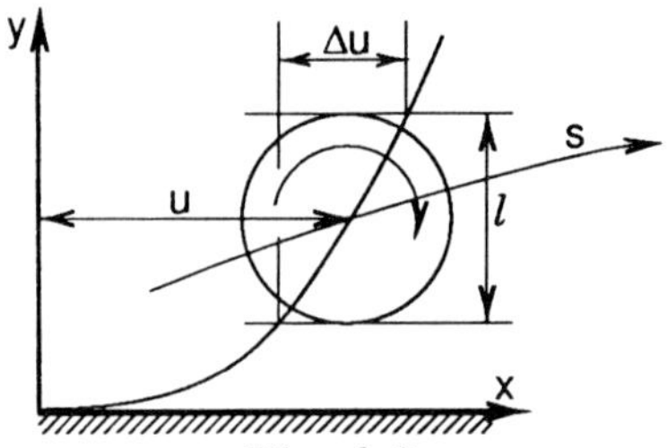

Fig. 2.9

(across the eddy size l), or their linear combination. Taking the latter, viz

$$\overline{W} = \alpha u + \beta \Delta u, \tag{2.9}$$

and considering

$$\frac{dl}{dt} = \frac{dl}{dx}\frac{dx}{dt} = \frac{dl}{dx}u, \tag{2.10}$$

one obtains from (2.8)

$$\frac{dl}{dx} = \phi(\zeta)\left[\alpha + \beta\frac{\Delta u}{u}\right]. \tag{2.11}$$

Adopting roughly $\alpha\phi(\zeta) \approx 1/6$, identifying u with v, and considering that $\Delta u/u = (du/dy)l/u = (v_*/\kappa y)l/u \approx v_*/u$ can be approximated by $v_*/v = 1/c$, one obtains from (2.11) the following expression of $L/h \approx (dl/dx)^{-1}$

$$\frac{L}{h} \approx 6\left[1 + \frac{\beta}{\alpha}\frac{1}{c}\right]^{-1}, \tag{2.12}$$

where the value of (β/α) is not known at present. Observe that (2.12) is a non-decreasing function of c which becomes $L/h \approx 6$ for sufficiently large c.[13]

2.2.3 "*Lesser bursts*"

As has already been mentioned, most of the turbulence-energy (in fact 70 to 80% of it) is generated at the early stages of a burst cycle [64]; the rest of it is produced by the roll-up's of some additional eddies, e_i say, at the later stages.

Since each eddy e_i maintains its coherence for a finite interval of time T_i, it travels during that time (in the flow direction) a certain distance L_i, which is referred to as the *coherence length* (of that eddy). It follows that the bursts of the length $L \approx 6h$ and the period $T \approx L/v$ studied so far are, in fact, only the largest bursts of a turbulent flow. In addition to them, a turbulent flow contains also a multitude of "lesser bursts" of the lengths and the periods which satisfy

$$L_i < L \approx 6h \quad ; \quad T_i < T. \tag{2.13}$$

In the present book, we will deal mainly with the largest, or usual, bursts of the length $L \approx 6h$; and therefore, unless stated otherwise, we will continue to use the term "burst" to designate them only.

---

[13] Note that it is not l but dl/dx which is affected by c via u.

2.2.4 *Burst width*

The cross-sectional development of a burst, in the convected $(y; z)$-plane, cannot be observed or measured directly: the current experimental methods are not suitable for that. Hence the present knowledge on the cross-sectional development of a burst is based on deductions and inferences. However, one aspect of this development appears to be certain: the coherent structure grows by means of counter-rotating and expanding circulatory motions, as shown schematically (by the phases 1 to 4) in Fig. 2.10.

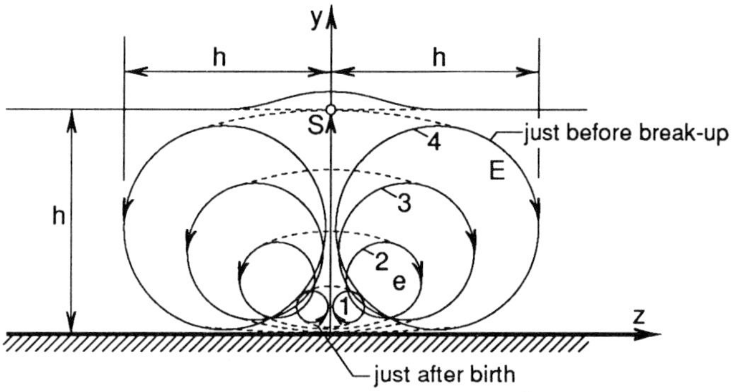

Fig. 2.10

If the pressure at the stagnation point S, where the upward-directed flow impinges the free surface, is sufficiently high, then the free surface may swell (Fig. 2.10). This local and temporary swelling, caused by the upward-directed (local and temporary) burst-currents, is referred to as a *boil*. (The conditions enhancing the occurrence of boils will be considered in Chapter 3). It has been found by R. Kinoshita [28], with the aid of aerial photo-surveys carried out in natural rivers, and also by J.M. Coleman [11] that the average spacing between boils, in the z-direction, is approximately $2h$ (see also [38], [39]). The fact that a boil is but an "outcome" of a burst, and consequently that the location of a boil specifies the location of the center line of a burst, has been demonstrated, on the basis of experimental data, by R.G. Jackson [25]. Hence, it would be only reasonable to identify the large-scale length L' (in Fig. 2.1) with $\approx 2h$:

$$L' \approx 2h \quad \text{and thus} \quad L' \approx \frac{1}{3} L . \tag{2.14}$$

From recent studies on cellular secondary currents ([36], [39], [46], [47], [48], [59]), it follows that the width of two adjacent counter-rotating cells of these currents is equal to $2h$, i.e. that it is equal to the (independently determined) burst width L'. The vertical sequences in Fig. 2.11 show schematically the consecutive phases (1 to 4) of cross-sectional development of bursts. These sequences, which are shifted because of the chessboard-like plan arrangement in Fig. 2.1, can be interpreted either as those along the flow direction x (at an instant t), or as those in time t (at a section x). Observe that it is not only the burst- and the secondary current-"channels" (I and II) themselves, but also the sense of rotation of their fluids which are in coincidence. This suggests that bursts and cellular secondary currents (which can occur only in turbulent flows) are likely to be interrelated. In fact, the

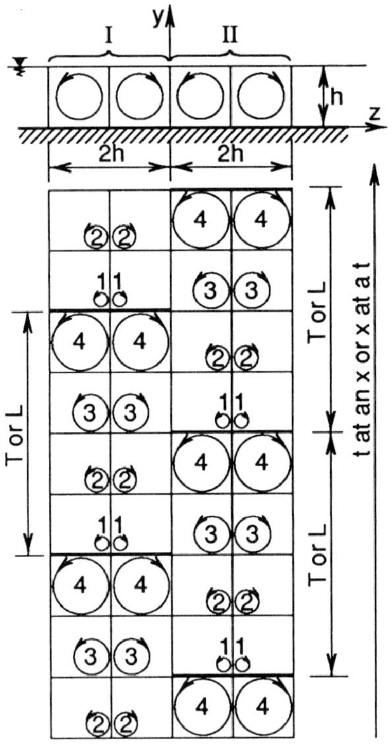

Fig. 2.11

secondary currents appear as if they were some regular counterparts (perhaps space-time averages) of the burst-induced currents. The possibility of an interrelation between bursts and cellular secondary currents is not mentioned in the works referred to above; only in Ref. [28] is a hint on such a possibility given.

2.2.5 *Viscous structures at the flow bed*

Consider the flow in the vicinity of the bed. Recent research indicates that the influence of turbulence can be detected right down to the bed surface. There is no (classical) viscous sublayer δ_L in the clear-cut sense depicted in Fig. 2.12a, where τ_t and τ_ν are turbulent and viscous components of the total shear stress $\tau = \tau_t + \tau_\nu$: the influence of ν vanishes along y in a continuous manner as shown schematically in Figs. 2.12b and c. Considering this, the explanations below are given without appealing to the viscous sublayer concept.

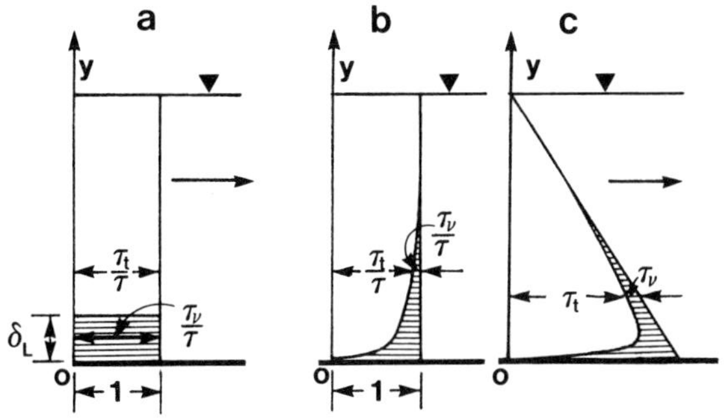

Fig. 2.12

The viscous flow in the neighbourhood of the bed is determined by the parameters ρ, ν, v_*, k_s, and therefore the dimensionless counterpart Π_A of any quantity A related to this flow is determined by

$$\Pi_A = \phi_A(Re_*) \qquad (Re_* = v_* k_s/\nu)\,. \qquad (2.15)$$

In particular, if A is a "length", λ say, then

$$\frac{\lambda}{k_s} = \phi_\lambda(Re_*)\,. \qquad (2.16)$$

If $Re_* < \approx 5$ (hydraulically smooth regime), then the flow contacting the bed surface is completely viscous and its properties can no longer be dependent on k_s. Hence when $Re_* < \approx 5$, then k_s must vanish from the expression (2.16), and this can be achieved only if $\phi_\lambda(Re_*)$ is of the form

$$\phi_\lambda(Re_*) = \frac{(const)}{Re_*}\,. \qquad (2.17)$$

Using (2.17) in (2.16), one obtains

$$\lambda = (const)\,\frac{\nu}{v_*}\,. \qquad (2.18)$$

It follows that if the flow contacting the bed is completely viscous, then any of its lengths λ is "measured" in terms of the viscous length unit ν/v_* only.

Experiment shows that turbulence disturbs the viscous flow at the bed as to produce (in the $(x\,;\,z)$-plane corresponding to a level $y\,(>k_s)$, however small) a series of adjacent high- and low-speed zones ("spots") which vary slowly with the passage of time. A sample of these quasi-periodic (along x and z) zones is shown in Fig. 2.13. Here they are depicted by the u-$const$ contours (at $y = 15\nu/v_*$) corresponding to three consecutive instants. In this graph, x^+ and z^+ imply xv_*/ν and zv_*/ν respectively. It has been found [9], [6], [5], [3] that the average length of each zone in the x-direction is comparable with $\approx 1000\,\nu/v_*$, the average longitudinal period, or the average "wave length", being thus

$$\lambda_x \approx 2000\,\frac{\nu}{v_*}\,. \qquad (2.19)$$

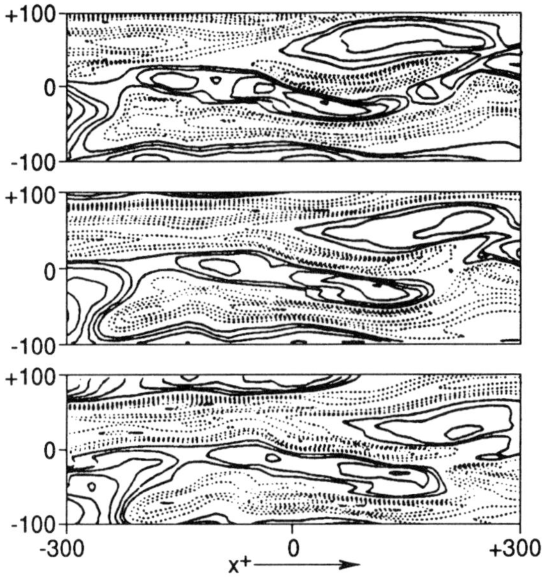

Fig. 2.13
(from Ref. [26]; see also [3], [33], [53], [58])

Clearly, the disturbance of the velocity field must be accompanied by the corresponding disturbance of streamlines. Hence, the streamlines of the viscous flow at the bed are undulated – also with the average wave length λ_x (Fig. 2.14). The above described configurations will be referred to hence-

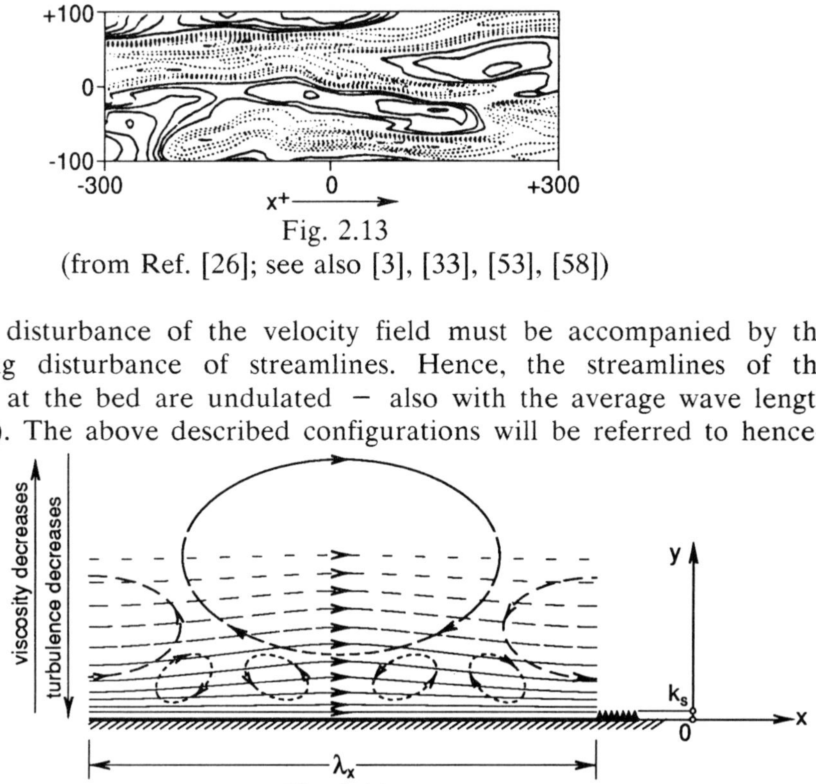

Fig. 2.14

forth as *viscous flow structures*, or *undulations* (at the bed of a turbulent flow). It should thus be clear that the viscous flow at the boundaries of a hydraulically smooth turbulent flow is not the same as that of a laminar flow: it is disturbed right down to $y = k_s$.

No information on λ_x, that is, on the form of its function $\phi_\lambda(Re_*)$ is available if the influence of k_s is no longer negligible (transitional regime: $\approx 5 < Re_* < \approx 70$). If $Re_* > \approx 70$ (rough regime), then there is no viscous flow at the bed.

2.2.6 *Sequence of bursts and its consequences*

Consider Fig. 2.15a which shows the (highly idealized) sequence of the largest bursts $\beta_1, \beta_2, \ldots$, along x. Here 1 and 4 symbolize the states of the burst-forming eddy e just after its birth at $t = 0$, and just before its disintegration at $t = T$. We assume that the burst sequence $\beta_1, \beta_2, \ldots$, is *regular,* i.e. that it consists of identical bursts "fired" (from their virtual origins O) in concert with the same time interval T. The variation of conditions in the lateral direction z has no bearing on the point intended to make, and therefore the burst sequence in Fig. 2.15 will be treated as two-dimensional.[14]

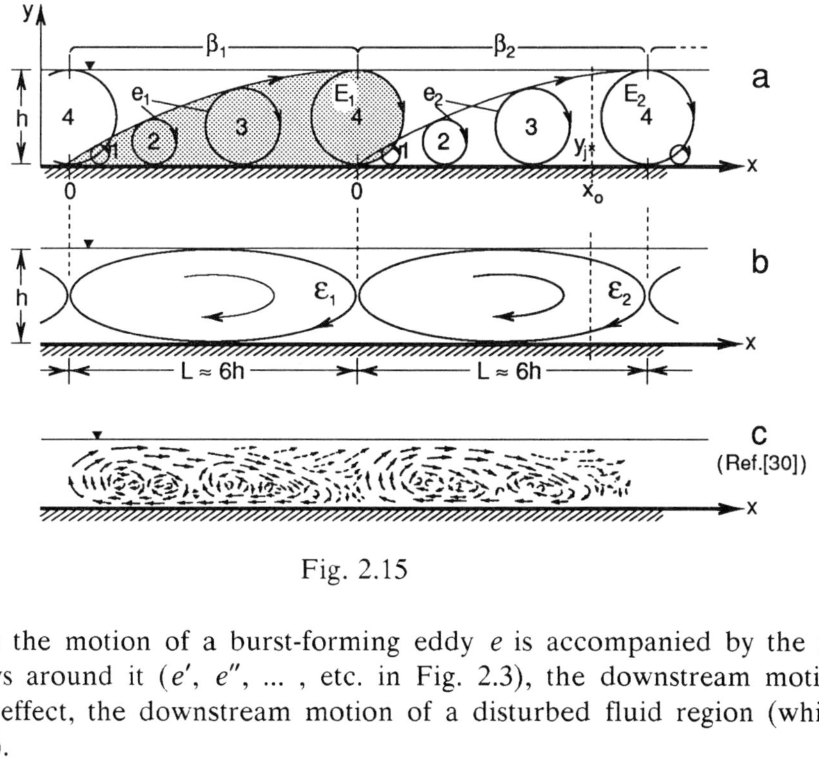

Fig. 2.15

Since the motion of a burst-forming eddy e is accompanied by the induced flows around it (e', e'', ... , etc. in Fig. 2.3), the downstream motion of e is, in effect, the downstream motion of a disturbed fluid region (which includes e).

Let $P(x_0, y_0)$ be a fixed point at the flow bed, and y_j ($j = 1, 2, \ldots, n$) the multitude of points on the vertical line passing through $P(x_0, y_0)$ ($y_j \in [0; h]$; Fig. 2.15). Suppose now that a "pre-burst-era" observer measures the time-variation of flow velocities U_i ($i = x, y, z$) at the points y_j and obtains $3n$ diagrams, which cumulatively imply

$$U_{ij} = F_i(t, y_j) = f_{ij}(t) \,. \tag{2.20}$$

---

[14] The burst sequence is conceivable, for, as has been mentioned earlier, the disintegration of E_i prompts the "birth" of e_{i+1}: the commencement of the sequence can be ensured by a "local discontinuity" (more on the topic in Chapter 3).

He will soon notice that these different $3n$ diagrams have one common property − they are all repetitive with the period T:

$$U_{ij} = f_{ij}(t) = f_{ij}(t + KT) \qquad (K \text{ integer}).$$

The observer, who is not aware of the existence of bursts, will justifiably attribute this repetitiveness, which is detectable for all points y_j, to the passage of the adjacent "largest eddies" $\mathcal{E}_1, \mathcal{E}_2, \ldots$ (Fig. 2.15b) whose dimensions along y and x are $\approx h$ and $\approx 6h$ ($= vT$) respectively.

This observer-story is a symbolic summary of what has really happened in the past. Considering this, a largest eddy ($\mathcal{E}$) referred to in a pre-burst-era publication will be interpreted in the following merely as an "envelope" containing a burst − provided that the linear dimensions of that eddy are $\approx h$ and $\approx 6h$, and its period is $T \approx 6h/v$.

Since the eddies e and $\mathcal{E}$ are conveyed downstream with the same speed ($\approx v$), and since the distance between the consecutive eddies $e_1, e_2, \ldots$, is the burst length L (which is also the longitudinal extent of $\mathcal{E}$), the "moving envelope" $\mathcal{E}$ contains only one eddy e at a time. The eddy e contained in $\mathcal{E}$ for the duration T evolves during that duration (the states $1, 2, \ldots$ in Fig. 2.15a). Consequently, the configuration of fluid in $\mathcal{E}$ evolves also as a function of $t \in [0\,; T]$. Fig. 2.15c (due to A.B. Klaven [30]) shows, in a convective frame of reference, the fluid configuration of two adjacent $\mathcal{E}$'s at an instant. Fig. 2.15c is from a frame of a cine-film taken by a camera moving downstream with the velocity v [31].

The largest turbulent eddies $\mathcal{E}$ (of the area $\approx hL$) should not be confused with the macroturbulent eddies E (of the area $\sim h^2$), which are merely e at $t = T$.

The area "swept" by the burst-forming eddy e in the course of a burst cycle (shaded area in Fig. 2.15a) will be referred to as *burst module*.

2.3 Horizontal Turbulence

2.3.1 *General description*

The turbulence dealt with so far can be referred to as *vertical turbulence*, for it is formed by the eddies whose axes of rotation are perpendicular to the vertical $(x; y)$-planes (to the longitudinal sections of flow). Yet an open-channel flow has turbulent eddies also in plan (see Fig. 2.16),[15] and one can say (following Yokosi [66]) that these eddies, whose axes are perpendicular to the (nearly) horizontal $(x; z)$-plane, form *horizontal turbulence*.[16]

---

[15] The photograph in Fig. 2.16 shows the (horizontal) turbulent eddies at the free surface of a flume flow (see also [62]).

[16] The concept of horizontal turbulence has already been used in some monographs on fluvial hydraulics (see e.g. [18] and [56]).

The characteristics of horizontal turbulence are marked in the present text by the subscript H.

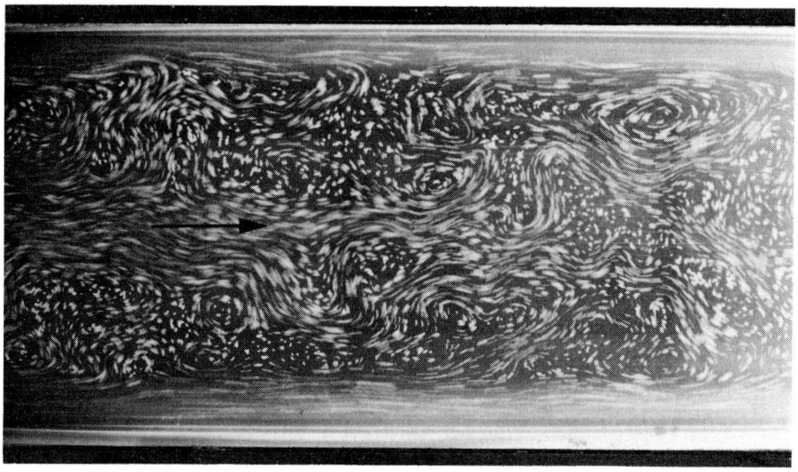

Fig. 2.16 (from Ref. [61])

To the author's knowledge, the bursts of horizontal turbulence have not yet been studied for their own sake in open-channels. On the other hand, the related observations and measurements indicate that horizontal turbulence too has its coherent structures and bursts, and that they are analogous to those of the vertical turbulence. The difference appears to be mainly due to the length-scale: all "lengths" of the large-scale vertical turbulence are proportional to the flow depth h; those of the large-scale horizontal turbulence, to the flow width B.[17] Horizontal bursts are produced by the horizontal burst-forming eddies e_H, which are likely to originate at the banks (where the horizontal shear stresses τ_H are the largest), and which subsequently move away from them as they are conveyed downstream. During their motion, the (horizontal) size l_H of the eddies e_H increases. If the (B/h)-ratio of the flow is not too large, then the eddies e_H may grow until they acquire their full macroturbulent size $(l_H)_{\max} = B$ − then, they break-up. The size $(l_H)_{\max} = B$ is acquired by an eddy e_H after it has travelled the distance $L_H \approx \alpha B$, which is the length of horizontal bursts. Owing to the universality of the divergence angle θ (see 2.2.1), the value of α must be expected to be comparable, or even equal, to ≈ 6. [The photograph in Fig. 2.17 shows the plan view of a coherent structure in an open-channel flow: the overall shape of this horizontal structure is similar to the burst modules shown in earlier illustrations. See also the experimentally determined patterns of horizontal turbulence in Ref. [60]]. From the measurements carried out by S. Yokosi [66] in the Uji River (Japan) and laboratory flumes, it follows that a turbulent open-channel flow

[17] No difference can be expected because of the (vertical or horizontal) orientation of the axes of turbulent eddies, for the structure of turbulence does not depend on gravity.

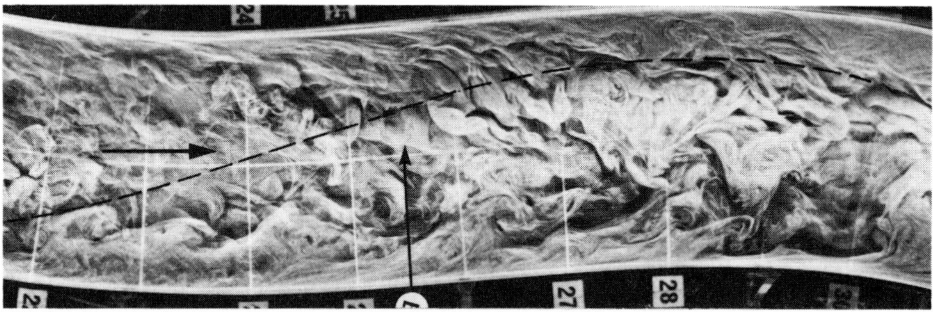

Fig. 2.17
(courtesy of Prof. H. Ohnari, Tokuyama College of Technology)

(having $B \gg h$) contains in plan a sequence of adjacent largest eddies which manifest themselves in the form of (comparatively thin) horizontal "disks". The thickness of these disks is equal to the flow depth h, their plan dimensions along z and x being $\approx B$ and αB respectively (where α is comparable with ≈ 6). Adopting *mutatis mutandi* the content of 2.2.6 for a horizontal "flow ribbon" (and thus replacing h, L, T, ... etc. by B, L_H, T_H, ... etc.), one realizes that the "largest eddies" reported in the pre-burst-era work of S. Yokosi [66] are but the eddies $\mathcal{E}_H$ (horizontal counterparts of $\mathcal{E}$), containing horizontal bursts (of the average length $L_H = \alpha B$ and the period $T_H = L_H / v$). We go now to the determination of $\alpha = L_H / B$.

2.3.2 *Length of horizontal bursts*

i- If the oscillogram of the time-variation (due to turbulence) of a velocity U_i is averaged over the consecutive equal time intervals Δt, then the "smoothed" oscillogram will contain only those U_i-fluctuations whose period is larger than Δt. Thus, by selecting a sufficiently large Δt, one can reveal those longest periods, or lowest frequencies, of U_i-fluctuations, which are caused by the passage of the largest eddies $\mathcal{E}_H$.

Fig. 2.18a shows the original oscillogram of the longitudinal flow velocity U ($= U_x$) recorded by S. Yokosi [66] in the Uji River.[18] The curves in Figs. 2.18b, c and d are the smoothed versions of this oscillogram, which correspond to $\Delta t = 6s$, $60s$ and $300s$ respectively. The longest period T_H, due to the passage of the largest eddies $\mathcal{E}_H$, is exhibited by the curve d. Using this

---

[18] Straight river region: $h = 2m$, $B = 100m$, $S = 0.00026$. The measurements were carried out at $y/h = 0.8$ of the flow center line. Time average longitudinal velocity at $y/h = 0.8$ was $u = 1.28m/s$.

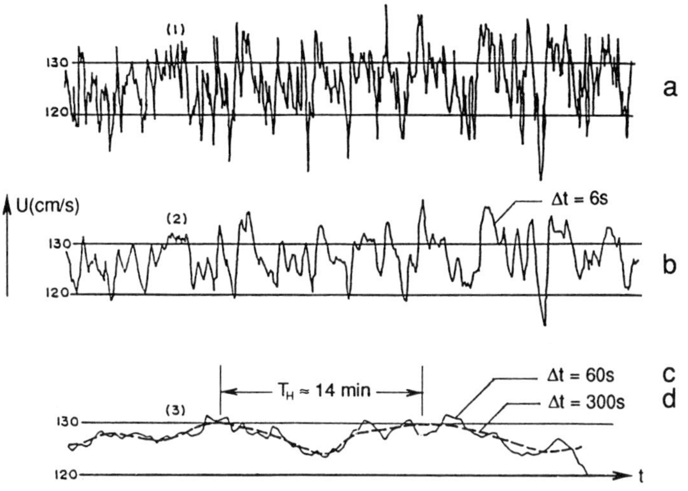

Fig. 2.18 (from Ref. [66])

diagram, or rather its extended version shown in Fig. 2.19, S. Yokosi determined $T_H \approx 14 min$, which together with $B = 100m$ and $v = 1.10m/s$,[19] yields

$$\frac{L_H}{B} = \frac{vT_H}{B} \approx \frac{(1.10)(14)(60)}{100} = 9.24 \approx 9.$$

This sample value of the ratio L_H/B is larger than the expected value $\approx$6, but it is certainly comparable with it.

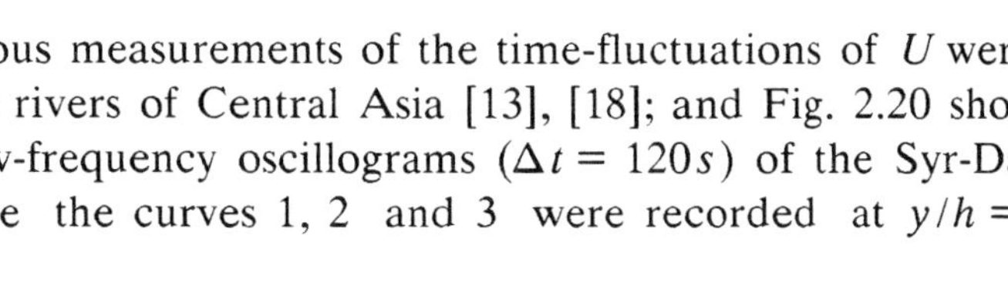

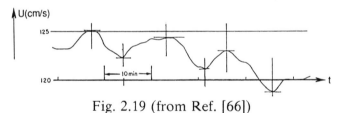

Fig. 2.19 (from Ref. [66])

Analogous measurements of the time-fluctuations of U were carried out also in some rivers of Central Asia [13], [18]; and Fig. 2.20 shows (the only available) low-frequency oscillograms ($\Delta t = 120s$) of the Syr-Darya River at Ak-Ajar. Here the curves 1, 2 and 3 were recorded at $y/h = 0.8$, 0.4 and

---

[19] Average flow velocity v is in coincidence with the time-average velocity u at $y/h = 0.368$ (see 1.4.1). Hence, $u = 1.28m/s$ recorded by S. Yokosi at $y/h = 0.8$ is, in fact, much nearer to u_{max} than to v. Identifying u with u_{max}, i.e. adopting $u_{max} \approx 1.28m/s$, and taking into account that $(u_{max} - v) = 2.5\, v_*$ (see (1.25)), where

$$v_* = \sqrt{gSh} = \sqrt{(9.81)(0.00026)(2)} \approx 0.07m/s\ ,$$

one determines $v = u_{max} - 2.5\, v_* \approx 1.10m/s$.

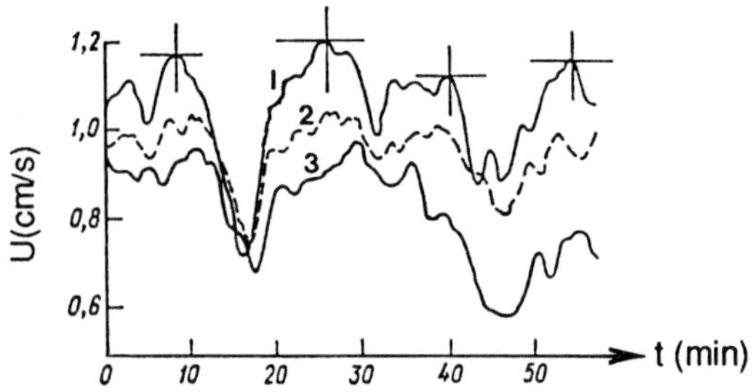

Fig. 2.20 (from Ref. [18])

0.2 respectively.[20] The average interval between the peaks is $\approx 13 min$. Identifying this value with T_H, and using $B = 100m$ and $v = 0.9m/s$, one determines

$$\frac{L_H}{B} \approx \frac{(0.9)(13)(60)}{100} = 7.02 \approx 7.$$

Being a property of turbulence, the dimensionless burst length L_H/B has a strong random component, and therefore a limited number of its samples, such as $\approx$9 and $\approx$7 above, cannot supply a reliable value for α. An adequate determination of α must rest on a large number of samples; yet only very few measurements, as those outlined above, have been conducted to date. A different approach to the determination of α is explained below.

ii- A very large number of measurements was carried out for the wavelength Λ_a of the alluvial bed forms known as alternate bars (Chapter 3). Using these data, it has been found that Λ_a and B are interrelated by the simple proportionality

$$\Lambda_a \approx 6B \tag{2.21}$$

(see Figs. 3.29a and b).

Now, alternate bars cannot be caused by vertical turbulence, for all of its large-scale lengths are proportional to h and they cannot possibly be "imprinted" on the mobile bed surface as the bed forms whose length is proportional to B (and is independent of h). Hence, alternate bars can be due *only* to horizontal turbulence, all of whose large-scale lengths in plan are proportional to B. The compatibility of the samples $\alpha \approx 9$ and $\approx$7 with $\Lambda_a/B \approx 6$ is an additional indication that Λ_a should be the "imprint" of the length L_H of horizontal bursts. Thus,

$$\alpha = \frac{L_H}{B} = \frac{\Lambda_a}{B} \approx 6. \tag{2.22}$$

---

[20] Straight river region: $h = 1.1m$, $B = 100m$ and $v = 0.9m/s$ (which is the time average of the curve 2 corresponding to $y/h = 0.4$ (≈ 0.368)).

In analogy to the relative length L/h of vertical bursts, the value $\alpha = L_H/B \approx 6$ may also be affected by k_s/h and/or Re_* (or by c) when they are sufficiently small. However, the scatter of the alternate bar data is too gross to reveal this possible influence and, for the present, one has no alternative but to consider (2.22) as applicable to all k_s/h and Re_*.

2.3.3 *Basic configuration of horizontal bursts*

Since in the case of an open-channel flow (having $B \gg h$) the largest value of u (at any z) is at the free surface, while the largest $\partial u/\partial z$ (at any y) is at the banks, the absolute maximum of $\partial u/\partial z$ must be at the intersection of banks with the free surface. Hence, the horizontal bursts must be expected to originate at some points (O_i) located in the neighbourhood of the upper corners of the flow cross-section. [In Ref. [63], M.A. Velikanov vividly describes how the several-meters-diameter free surface eddies are intermittently detaching from the banks of the Amu-Darya and are subsequently conveyed downstream].

Alternate bars are anti-symmetrical in plan with respect to the x-axis, and so must be the sequences $\beta_1, \beta_2, \ldots$, and $\beta_1', \beta_2', \ldots$, of horizontal bursts (issued from the right- and left-banks) which generate them. The anti-symmetrical arrangement of β_i and β_i', which are assumed to be "regular" in the sense of 2.2.6, is shown in Fig. 2.21.[21]

2.3.4 *Configuration criteria; "N-row" horizontal bursts*

i- Consider a steady and uniform rough turbulent flow in a rectangular open-channel: the bed is rigid and flat; its roughness is k_s. This flow, and thus the structure of its large-scale turbulence, is completely determined by the parameters ρ, Q (or v, v_*, etc.), h, B and k_s ([32], [35], [55]), and thus by the dimensionless variables

$$\frac{B}{h} \quad \text{and} \quad \frac{k_s}{h}\,.^{22} \tag{2.23}$$

Hence, for any Π_A of the structure of turbulence mentioned we have

---

[21] A detailed demonstration of the validity of anti-symmetrical arrangement of horizontal bursts can be found in Ref. [57]. This arrangement is consistent with the "chessboard-row" sequence of largest horizontal eddies ($\mathcal{E}_H$) in Ref. [18], where it is envisaged that the adjacent consecutive $\mathcal{E}_H$ rotate in opposite directions as they are conveyed downstream. In Ref. [18] it is also mentioned that certain attempts to associate alternate bars with the sequence of $\mathcal{E}_H$ have already been made in "some theoretical works" (not specified in [18]).

[22] The following is to be noted: 1) If $Re_* < \approx 70$, then Re_* is an additional variable. 2) The bed need not be literally rigid; if the mobile bed is covered by ripples, then k_s ($\sim D$) can be interpreted as "ripple roughness" K_s ($\sim \Delta_r$). 3) If B/h is large and the disk-like eddies e_H are rubbing the bed, then k_s affects the horizontal bursts directly; and not only via u, as it affects vertical bursts (2.2.2).

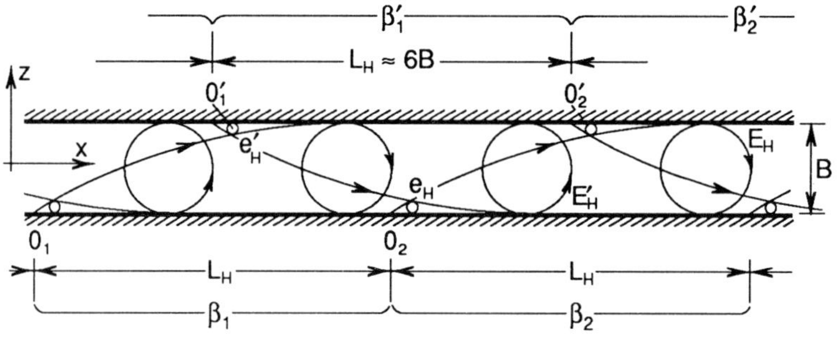

Fig. 2.21

$$\Pi_A = \phi_A\left(\frac{B}{h}, \frac{k_s}{h}\right). \tag{2.24}$$

Since a certain state or configuration of the large-scale turbulence can be specified by a certain constant value of one of its dimensionless properties Π_A, the occurrence of that state or configuration must be predictable by a criterion which is an interrelation between B/h and k_s/h. Indeed, if $\Pi_A = const_A$, then (2.24) becomes

$$\Pi_A = \phi_A\left(\frac{B}{h}, \frac{k_s}{h}\right) = const_A,$$

which yields

$$\frac{B}{h} = \psi_A\left(\frac{k_s}{h}\right). \tag{2.25}$$

It follows that a criterion related to the structure of the large-scale turbulence can be represented by a curve in the $(B/h; h/k_s)$-plane.

ii- Let ϵ_{max} be the thickness of the free (not contacting the bed) disk-like burst-forming eddies e_H at the end of a burst cycle, i.e. when $l_H =$ $= (l_H)_{max} = B$ (Figs. 2.22a and b). The criterion for whether or not the eddies are "rubbing" the bed surface can be reflected by the values of the ratio ϵ_{max}/h (which is one of the dimensionless properties (Π_A) of the structure of turbulence).

- If $(\epsilon_{max}/h) < 1$, then the eddies e_H, which are assumed to be "born" near the free surface, *are not* rubbing the bed at any stage of the burst cycle (Fig. 2.22b). Clearly, in this case the channel bed cannot be affected by the eddies e_H, and the large-scale horizontal turbulence does not have the potential to produce its bed forms.
- If $(\epsilon_{max}/h) > 1$, then the eddies e_H are rubbing the bed (Fig. 2.22c). In this case the channel bed can be affected by the eddies e_H, and the large-scale horizontal turbulence has the potential to produce its bed forms (which are the alternate or multiple bars (Chapter 3)).

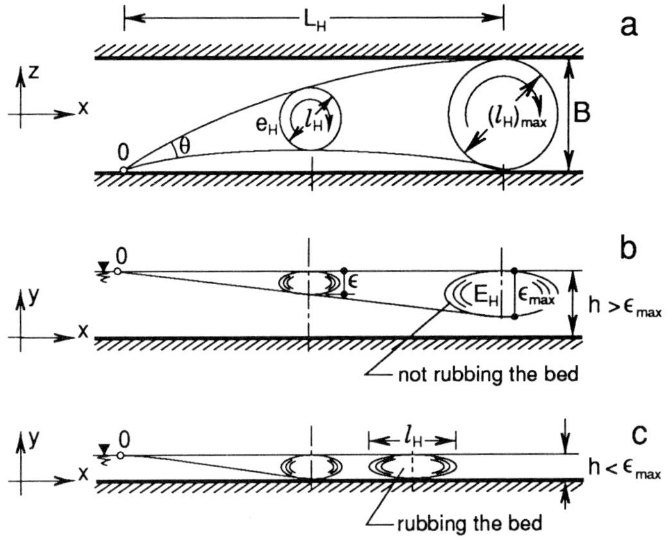

Fig. 2.22

Hence the line (curve) separating the regions, in the $(B/h;\, h/k_s)$-plane, where the horizontal turbulence can and cannot produce its bed forms is determined by a relation which can be expressed as

$$\frac{\epsilon_{\max}}{h} = \phi_\epsilon\left(\frac{B}{h}, \frac{k_s}{h}\right) = 1\,, \tag{2.26}$$

i.e.

$$\frac{B}{h} = \psi_\epsilon\left(\frac{h}{k_s}\right). \tag{2.27}$$

iii- Consider the arrangement of horizontal bursts in Fig. 2.21; it is present in all open-channel flows having sufficiently large relative depth h/B. If the burst-forming eddies e_H and e_H' are not rubbing the bed (i.e. if $(\epsilon_{\max}/h) < 1$), then the scheme in Fig. 2.21 is certainly valid. This, however, does not mean that it must necessarily be invalid if the eddies e_H and e_H' *are* rubbing the bed (i.e. if $(\epsilon_{\max}/h) > 1$). For the bed friction-effect (which increases with k_s/h and B/h) may not be as strong as to prevent these eddies from reaching the opposite banks (at which time they acquire their full size $(l_H)_{\max} = B$). If, on the other hand, the bed friction-effect *is* sufficiently strong, then it may prevent the eddies e_H and e_H' from reaching the opposite banks; and in this case they will meet each other in the midst of the flow, as shown schematically in Fig. 2.23a. Hence, with the increment of the bed friction-effect, the basic single-row configuration of horizontal bursts (in Fig. 2.21) degenerates into the double-row configuration (in Fig. 2.23a), where the largest size of horizontal eddies is $B/2$. But this means that the upper limit of validity of the arrangement in Fig. 2.21, which has the potential to generate alternate bars, can be reflected by $(l_H)_{\max} = B$ and thus by

$$\frac{(l_H)_{max}}{B} = \phi_1\left(\frac{B}{h}, \frac{k_s}{h}\right) = 1, \tag{2.28}$$

i.e.

$$\frac{B}{h} = \psi_1\left(\frac{h}{k_s}\right). \tag{2.29}$$

Clearly, the curve (2.29) is situated in the $(B/h;\, h/k_s)$-plane above the curve (2.27) $(\psi_1(h/k_s) > \psi_\epsilon(h/k_s))$.

iv- A further increment of the bed friction-effect (i.e. of B/h and/or k_s/h) must lead to a further reduction of the largest sizes of e_H and e_H'; and the double-row configuration in Fig. 2.23a must change into the triple-row, quadriple-row, ... , and N-row configurations, where the largest eddy sizes are $B/3$, $B/4$, ... and B/N. (Figs. 2.23b and c show the triple- and N-row configurations respectively). Since N is an integer, the transition from one configuration to the next, due to the continuous increment of the friction effect, must occur by "steps".

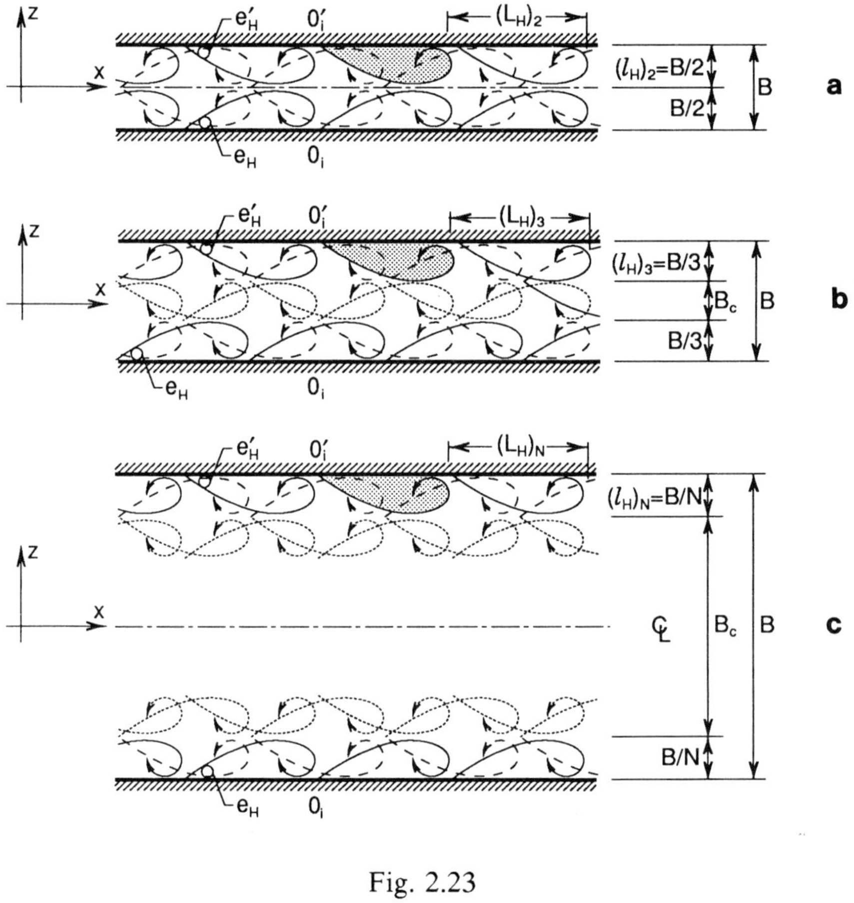

Fig. 2.23

All burst modules (the shaded areas in Figs. 2.23a, b and c) are roughly geometrically similar, and therefore if the length of horizontal bursts forming the N-row configuration is denoted by $(L_H)_N$, and the largest size of the eddies e_H by $(l_H)_N$ (which, at the same time, is the width of one row of the N-row configuration), then

$$\frac{(L_H)_1}{(L_H)_N} = \frac{(l_H)_1}{(l_H)_N} = N, \qquad (2.30)$$

where

$$(L_H)_1 = L_H \approx 6B \quad \text{and} \quad (l_H)_1 = (l_H)_{max} \approx B$$

are the characteristics of the single-row configuration in Fig. 2.21.

If $N > 2$, then there exists a region of the width $B - 2(l_H)_N = B - 2(B/N)$ which contains the large-scale horizontal eddies induced by the bursts issued from O_i and O_i': the linear dimensions of these "induced eddies" (which are shown in Figs. 2.23b and c by broken lines) are comparable with those of the "inductor bursts".[23] Since the sense of rotation of induced eddies alternates in the z-direction, their effect on the distribution of time average velocities along z cannot be significant. Hence, one can assume that the flow in the region $B - 2(B/N)$ is practically two-dimensional ($\partial u/\partial z \approx 0$ and $\tau_H \approx 0$ for any y), and thus that $B - 2(B/N)$ can be identified (as is done in Figs. 2.23b and c) with the width B_c of the central region (see Section 1.1):

$$B_c = B - 2(l_H)_N = B\left(1 - \frac{2}{N}\right). \qquad (2.31)$$

Being one of the dimensionless properties of turbulence, the number N of horizontal burst rows is also determined by the dimensionless variables B/h and k_s/h, i.e.

$$N = \phi_N\left(\frac{B}{h}, \frac{k_s}{h}\right), \qquad (2.32)$$

which yields

$$\frac{B}{h} = \psi_N\left(N, \frac{h}{k_s}\right). \qquad (2.33)$$

It follows that various configurations of horizontal bursts imply various regions, or zones, in the $(B/h;\, h/k_s)$-plane. The relation (2.33) can be taken as the equation of the line separating the zone of N-row configuration from

[23] The "induced" large-scale eddies in the rows within B_c are, of course, not quite the same as the bursts in the rows adjacent to the banks. Yet, for the sake of convenience, we will continue to use the term "N-row bursts".

the zone of $(N+1)$-row configuration. (The relation (2.29) is thus the special case of (2.33) which corresponds to $N=1$).[24]

2.3.5 *Formulation of "N-row" horizontal bursts*

i- Since the boundaries separating various configurations of horizontal bursts (in the $(B/h;\ h/k_s)$-plane) are, at the same time, the boundaries separating the corresponding configurations of bars (generated by these bursts), an attempt to penetrate the form of the function (2.33) should be worthwhile.

The ability of a burst-forming eddy e_H to advance along its trajectory s can be reflected by its kinetic energy E_k. If $\epsilon_{max} > h$ (Fig. 2.22c), then most of this enegy is spent to overcome the work W done by the friction force, which interacts betwen the lower face of the "disk" e_H and the bed. It would be reasonable to assume that the reduction of burst length by the factor $N = L_H/(L_H)_N$ depends on how the typical values of E_k and W (viz $\overline{E}_k$ and $\overline{W}$) compare with each other, i.e. it depends on the ratio $\overline{E}_k/\overline{W}$:

$$N = \phi_N(\overline{E}_k/\overline{W})\,. \tag{2.34}$$

But,

$$\overline{E}_k = \frac{1}{2}\rho\,(h\overline{A})u^2 \sim \rho h\overline{A}v^2 \quad \text{while} \quad \overline{W} \approx (\tau_0\overline{A})(L_H)_N \approx \rho\, v_*^2\,\overline{A}\,(L_H)_N$$

where $\overline{A}$ is the typical area of e_H in plan. Hence,

$$(\overline{E}_k/\overline{W}) \sim [h/(L_H)_N](v/v_*)^2 = [h/(L_H)_N]c^2. \tag{2.35}$$

In the case of rough turbulent flow under consideration, the friction factor c can be given by the power relation (1.28), viz $c = 7.66(h/k_s)^{1/6}$. Using this value of c and taking into account that $h/(L_H)_N = h/(L_H/N) \approx h/(6B/N)$, one can express (2.35) as

$$(\overline{E}_k/\overline{W}) \sim (h/B)(h/k_s)^{1/3}N. \tag{2.36}$$

Substituting (2.36) in (2.34), one determines

$$N = \phi_N[(B/h)/(h/k_s)^{1/3}]\,, \tag{2.37}$$

and consequently

$$\frac{B}{h} = \psi(N)\left(\frac{h}{k_s}\right)^{1/3}. \tag{2.38}$$

Hence if the flow is rough turbulent, then the boundaries separating the regions of N-row and $(N+1)$-row horizontal bursts (in a log-log $(B/h;\ h/k_s)$-plane) must be 1/3-inclined straight lines. The "level" of each

---

[24] As will be clarified in Chapter 3, the number of rows of multiple bars is the same as the number (N) of rows of horizontal bursts (which "imprint" them). Hence, the number of multiple bar rows cannot be specified at will (as it is frequently done in some papers on bed forms) – in every experiment, it is *determined* by the existing B/h and k_s/h.

line, as reflected by the increasing step function $\psi(N)$, can be revealed only by experiment.

It will be shown in Chapter 3 that the line separating the zone of alternate bars from that of multiple bars can be given by the relation

$$\frac{B}{h} \approx 31.5\left(\frac{h}{k_s}\right)^{1/3} \tag{2.39}$$

(which is Eq. (3.45) where D is replaced by $k_s/2$). The line representing (2.39) can thus be taken as the upper limit of existence of the single-row configuration of horizontal bursts ($N = 1$; $\psi(1) \approx 31.5$).

The relation (2.38) has no bearing on the form of the function (2.27) signifying the lower limit of existence of the single-row configuration of horizontal bursts which are rubbing the bed (lower limit of existence of alternate bars). For a given B, the rubbing must be expected to take place for those h which are smaller than the thickness ϵ_{max} of the free eddies E_H (Fig. 2.22c). Accordingly, the lower limit mentioned should correspond to that value of B/h which satisfies

$$\frac{B}{h} = \frac{(l_H)_{max}}{\epsilon_{max}} . \tag{2.40}$$

The ratio on the right is a dimensionless property of the horizontal macro-turbulence, and therefore it must possess a certain universal value. Consequently, the function $\psi_\epsilon(h/k_s)$ in (2.27) must tend to become a constant for sufficiently large values of h/k_s – only for small h/k_s should it exhibit a significant variation. Indeed, if k_s is large, then an interaction between the lower part of the eddy e_H and the disturbed (by k_s) flow at the bed must be expected to begin before e_H actually touches the bed surface; i.e. it should begin for the flow depths h that are larger than ϵ_{max}. It follows that the curve representing $B/h = \psi_\epsilon(h/k_s)$, and signifying the lower boundary of the existence region of bed-rubbing horizontal bursts, must be expected first to rise (for small h/k_s), and then tend to become parallel to the abcissa h/k_s.

ii- The relation (2.37) indicates that B/h and k_s/h determine N by combining themselves into a single variable

$$\omega = \left(\frac{B}{h}\right)\left(\frac{h}{k_s}\right)^{-1/3} \sim \left(\frac{B}{h}\right)\frac{1}{c^2} .^{[25]} \tag{2.41}$$

But this means that each of (2.37) and (2.32) can be symbolized by

$$N = \phi_N(\omega) , \tag{2.42}$$

[25] If the flow is not rough turbulent, then c is determined by both h/k_s and Re_*. Consequently, ω and N become functions of B/h, k_s/h and Re_* (which is in line with the statement in footnote 22).

while (2.38) implies

$$\omega = \psi(N). \tag{2.43}$$

Hence, (2.42) and (2.43) are inverse functions, and therefore it is sufficient to discuss only one of them: we take (2.42).

The relations above indicate that if $\omega < \approx 31.5$, then (2.42) reduces into $N = 1$, and (2.30) yields

$$(L_H)_1 \approx 6B \quad \text{and} \quad (l_H)_1 \approx B. \tag{2.44}$$

Consider the opposite extreme, viz $\omega \to \infty$, which can be interpreted as $B \to \infty$ while h and k_s remain constant (see (2.41)). Clearly, if B is "very large", then the horizontal bursts generated at one of the banks (by the finite τ_H acting there) cannot depend on B. But this means that when $B \sim \omega \to \infty$, then B must vanish from the expressions

$$(L_H)_N \approx 6B/N \quad \text{and} \quad (l_H)_N \approx B/N \tag{2.45}$$

which follow from (2.30). This, however, can be achieved only if the function $N = \phi_N(\omega)$ has the property

$$\lim_{\omega \to \infty} \phi_N(\omega) = (const)\,\omega\,. \tag{2.46}$$

Substituting $N = (const)\omega$ in (2.45), and reinterpreting ω according to (2.41), one realizes that when $B/h \sim \omega$ is "very large", then

$$(L_H)_N \approx 6h(\overline{const})\left(\frac{h}{k_s}\right)^{1/3} \quad \text{and} \quad (l_H)_N = h(\overline{const})\left(\frac{h}{k_s}\right)^{1/3} \tag{2.47}$$

(with $\overline{const} = (const)^{-1}$).

Hence the (so far unknown) step function $N = \phi_N(\omega)$ should be of such a nature as to be equal to unity when $\omega < \approx 31.5$, and as to become proportional to ω when $\omega \to \infty$.

iii- It follows that the linear dimensions of horizontal bursts (and of the induced eddies in B_c) do not scale always with B or always with h. When $\omega < \approx 31.5$, then they scale with B. With the increment of ω, the linear scale, which can be characterized by $B^n h^{1-n}$, shifts towards h (n decreases from unity to zero). And when $\omega \to \infty$, then the linear dimensions of horizontal structures scale with h.[26]

---

[26] It should be remembered that the present analysis corresponds to a rectilinear open-channel with straight parallel banks. A natural river is, as a rule, irregular; and any irregularity in plan (sudden change in direction, salient points, confluence, bifurcation, etc.) may generate additional horizontal eddies whose linear dimensions scale with B. These eddies, which can occur for any $\omega \sim B/h$, may in some cases "eclipse" the turbulent structures considered here.

From the considerations above, it also follows that when $(B/h) \sim \omega$ is comparatively small, then the open-channel flow manifests itself as a "correlated totality" – the flow at the left bank "knows" what the flow at the right bank "is doing". e.g. if $\omega < \approx 31.5$, then the burst sources O_i and O_i' at each bank "organize themselves" so that O_i' is exactly in the middle of $\overline{O_iO_{i+1}}$ (Fig. 2.21); a possible (accidental) increment of flow velocities at the right bank is compensated by the corresponding velocity decrement at the left bank, etc. With the increment of ω this lateral correlation deteriorates, and when ω is sufficiently large, it vanishes completely [24], [63], [10].

The induced eddies tend to perpetuate themselves: they "diffuse" in the lateral direction z (away from the banks). However, this lateral diffusion cannot go on along z indefinitely: the bed friction prevents it. But this means that if B/h is sufficiently large, then there must exist such a part b_c of the central region B_c where the large-scale horizontal turbulence is no longer present; i.e. where we have vertical turbulence only (Fig. 2.24).[27]

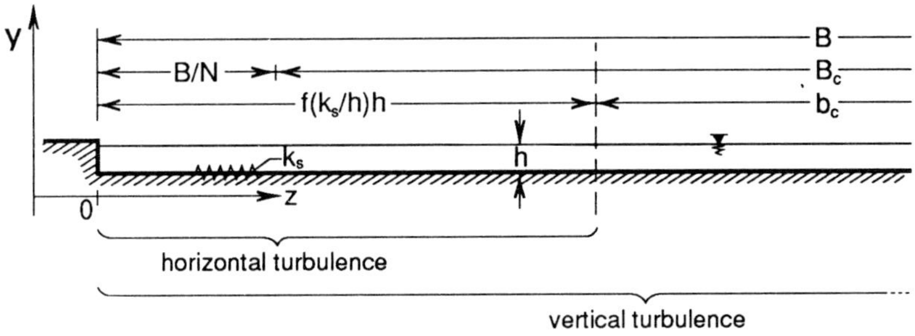

Fig. 2.24

If B/h is sufficiently large and b_c exists, then the distances $(B - b_c)/2$ must be proportional to h, the proportionality factor being a function of k_s/h. i.e.

$$b_c = B - 2f(k_s/h)h, \tag{2.48}$$

where the (unknown) function $f(k_s/h)$ satisfies

$$(B/h)/N < f(k_s/h) < (B/h)/2 \tag{2.49}$$

(the left-hand side of (2.49) is due to $b_c < B_c$; see (2.31)).

Consider now the expression of B_c, for which we have the relations (1.1) and (2.31). Equating these relations, one obtains

$$m = \frac{(l_H)_N}{h} = \frac{B}{h}\frac{1}{N},$$

where N is determined by B/h and c ((2.41) and (2.42)). Hence, in general, m is determined also by B/h and c, and it reduces into a function of only c when $(B/h) \to \infty$ (see (2.47)). In

---

[27] The central region B_c is two-dimensional with regard to the *time average* velocities u, which do not depend on the the velocities u_i' fluctuating around them. Hence, the fact that the large-scale horizontal eddies and thus the low-frequency fluctuations (u_i') caused by them progressively decrease along z, has no bearing on the two-dimensionality of u in B_c.

the present chapter, turbulence was studied for the simplest case of an open-channel: vertical banks ($\phi = \pi/2$) and bank roughness (k_s') equal to the bed roughness k_s. In reality, ϕ and $(k_s')/k_s$ are the additional variables and m must be treated as

$$m = \phi_m(B/h, c, (k_s')/k_s, \phi). \tag{2.50}$$

The analogous is valid for the function $f(k_s/h)$ (in (2.48)).

In the considerations above, which correspond to large B/h, the distance mh was identified with the width $(l_H)_N$ of horizontal burst rows corresponding to $N \geq 3$. If, however, B/h is not large, and vertical turbulence dominates, then mh is simply the distance affected by side walls. Clearly, in this case m cannot depend on B/h, and it varies, in the neighbourhood of ≈ 2.5 (see [65]), depending on c, $(k_s')/k_s$ and ϕ only.

References

1. Aubry, N. et al: *The dynamics of coherent structures in the wall region of a turbulent boundary layer.* J. Fluid Mech., Vol. 192, 1988.
2. Bandyopadhyay, P.R., Watson, R.D.: *Structure of rough-wall turbulent boundary layers.* Phys. Fluids, Vol. 31, July 1988.
3. Blackwelder, R.F.: *Analogies between transitional and turbulent boundary layers.* Phys. Fluids, Vol. 26, Oct. 1983.
4. Blackwelder, R.F.: *The bursting phenomenon in bounded shear flow.* Von Karman Inst. for Fluid Dyn., Lecture Series 1983-03, 1983.
5. Blackwelder, R.F., Eckelmann, H.: *Streamwise vortices associated with the bursting phenomenon.* J. Fluid Mech., Vol. 94, 1979.
6. Blackwelder, R.F.: *The bursting process in turbulent boundary layers.* Lehigh Workshop on Coherent Structures in Turbulent Boundary Layers, 1978.
7. Blackwelder, R.F., Kovasznay, L.S.G.: *Time scales and correlations in a turbulent boundary layer.* Phys. Fluids, Vol. 15, 1972.
8. Brown, G.L., Thomas, A.S.W.: *Large structure in a turbulent boundary layer.* Phys. Fluids, Vol. 20, Oct. 1977.
9. Cantwell, B.J.: *Organised motion in turbulent flow.* Ann. Rev. Fluid Mech., Vol. 13, 1981.
10. Chitale, S.V.: *Theories and relationships of river channel patterns.* J. Hydrology, 19, 1973.
11. Coleman, J.M.: *Brahmaputra river; channel processes and sedimentation.* Sediment. Geol., Vol. 3, 1969.
12. Coles, D.: *Coherent structures in turbulent boundary layers.* Perspectives in turbulent studies, Int. Symposium DFVLR Res. Center, Gottingen, May 1987.
13. Dementiev. M.A.: *Investigation of flow velocity fluctuations and their influences on the flow rate of mountainous rivers.* (In Russian) Tech. Report of the State Hydro-Geological Inst. (GGI), Vol. 98, 1962.
14. Falco, R.E.: *Coherent motions in the outer region of turbulent boundary layers.* Phys. Fluids, Vol. 20, Oct. 1977.
15. Fiedler, H.E.: *Coherent structures in turbulent flows.* Prog. Aerospace Sci., Vol. 25, 1988.
16. Gad-el-Hak, M., Hussain, A.K.M.F.: *Coherent structures in a turbulent boundary layer.* Phys. Fluids, Vol. 29, July 1986.
17. Grass, A.J.: *Structural features of turbulent flow over smooth and rough boundaries.* J. Fluid Mech., Vol. 50, 1971.
18. Grishanin, K.V.: *Dynamics of alluvial streams.* Gidrometeoizdat, Leningrad, 1979.
19. Guoren, D.: *General law of laminar and turbulent flows in pipes and open channels.* Scientia Sinica (Series A), Vol. XXV, Oct. 1982.
20. Guoren, D.: *The structure of turbulent flow in channels and pipes.* Scientia Sinica, Vol. XXIV, May 1981.
21. Hinze, J.O.: *Turbulence.* (2nd ed.), McGraw-Hill, 1975.

22. Hunt, J.C.R.: *Gas transfer at Water Surfaces.* Edited by W. Brutsaert and G.H. Jirka (Reidel, Dordrecht), 1984.
23. Hussain, A.K.M.F.: *Coherent structures – reality and myth.* Phys. Fluids, Vol. 26, Oct. 1983.
24. Ibad-Zade, Y.A.: *Sediment transport in open channels.* (In Russian) Stroyizdat, 1974.
25. Jackson, R.G.: *Sedimentological and fluid-dynamics implications of the turbulent bursting phenomenon in geophysical flows.* J. Fluid Mech., Vol. 77, 1976.
26. Johansson, A.V., Alfredsson, P.H.: *Velocity and pressure fields associated with near-wall turbulence structures.* Int. Seminar on Near-Wall Turbulence, Dubrovnik, May 1988.
27. Kim, H.T., Kline, S.J., Reynolds, W.C.: *The production of turbulence near a smooth wall in a turbulent boundary layer.* J. Fluid Mech., Vol. 50, 1971.
28. Kinoshita, R.: *An analysis of the movement of flood waters by aerial photography, concerning characteristics of turbulence and surface flow.* Photographic Surveying, Vol. 6, 1967 (in Japanese).
29. Klaven, A.B., Kopaliani, Z.D.: *Laboratory investigations of the kinematic structure of turbulent flow past a very rough bed.* Tech. Report of the State Hydro-Geological Inst. (GGI), Vol. 209, 1973.
30. Klaven, A.B.: *Investigation of structure of turbulent streams.* Tech. Report of the State Hydro-Geological Inst. (GGI), Vol. 136, 1966.
31. Kondratiev, N., Popov, I., Snishchenko, B.: *Foundations of hydromorphological theory of fluvial processes.* (In Russian) Gidrometeoizdat, Leningrad, 1982.
32. Landau, L.D., Lifshitz, E.M.: *Fluid Mechanics.* Pergamon Press, Oxford, 1986.
33. Lu, L.J., Smith, C.R.: *Use of flow visualization data to examine spatial-temporal velocity and burst-type characteristics in a turbulent boundary layer.* J. Fluid Mech., Vol. 232, 1991.
34. Lu, S.S., Willmarth, W.W: *Measurements of the structure of the Reynolds stress in a turbulent boundary layer.* J. Fluid Mech., Vol. 60, 1973.
35. Monin, A.S., Yaglom, A.M.: *Statistical Fluid Mechanics: Mechanics of Turbulence.* Vol. 1 and 2, The MIT Press, 1975
36. Muller, A.: *Secondary flows in an open channel.* Proc. XVII Congress IAHR, No. B.A.3, 1979.
37. Nakagawa, H.: *Study on interaction between flowing water and sediment transport in alluvial streams. Part III. Stochastic model for bed material load.* Report of Grant-in-Aid for Scientific Research (B), Japan, 1986.
38. Nakagawa, H., Nezu, I.: *Structure of space-time correlations of bursting phenomena in an open-channel flow.* J. Fluid Mech., Vol. 104, 1981.
39. Nakagawa, H., Nezu, I., Tominaga, A.: *Spanwise streaky structure and macroturbulence in open-channel flows.* Memoirs, Fac. of Engrg., Kyoto Univ., Vol. XLIII, Part 1, Jan. 1981.
40. Nakagawa, H., Nezu, I.: *Bursting phenomenon near the wall in open-channel flows and its simple mathematical model.* Memoirs, Fac. of Engrg., Kyoto Univ., Vol. XL, Part 4, Oct. 1978.
41. Nakagawa, H., Nezu, I.: *Prediction of the contributions to the Reynolds stress from bursting events in open-channel flows.* J. Fluid Mech., Vol. 80, 1977.
42. Nakagawa, H., Nezu, I.: *Turbulence of open-channel flow over smooth and rough beds.* Proc. of JSCE, Vol. 241, Sept. 1975.
43. Nakagawa, H., Nezu, I.: *On a new eddy model in turbulent shear flow.* Proc. of JSCE, Vol. 231, Nov. 1974.
44. Nezu, I., Nakagawa, H.: *Numerical calculation of turbulent open-channel flows in consideration of free-surface effect.* Memoirs, Fac. of Engrg., Kyoto Univ., Vol. XLIX, April 1987.
45. Nezu, I., Rodi, W.: *Open-channel flow measurements with a laser doppler anemometer.* J. Hydr. Engrg., ASCE, Vol. 112, No. 5, May 1986.
46. Nezu, I., Rodi, W.: *Experimental study of secondary currents in open channel flow.* Proc. XXI Congress IAHR, Melbourne, Aug. 1985.
47. Nezu, I., Nakagawa, H.: *Cellular secondary currents in straight conduits.* J. Hydr. Engrg., ASCE, Vol. 110, No. 2, Feb. 1984.

48. Nezu, I., Nakagawa, H., Tominaga, A.: *Secondary currents in a straight channel flow and the relation to its aspect ratio.* 4th Int. Symp. on Turbulent Shear Flow, Karlsruhe, 1983.
49. Nychas, S.G., Hershey, H.C., Brodkey, R.S.: *A visual study of turbulent shear flow.* J. Fluid Mech., Vol. 61, 1973.
50. Offen, G.R., Kline, S.J.: *A proposed model of the bursting process in turbulent boundary layers.* J. Fluid Mech., Vol. 7, 1975.
51. Offen, G.R., Kline, S.J.: *Combined dye-streak and hydrogen-bubble visual observations of a turbulent boundary layer.* J. Fluid Mech., Vol. 62, 1974.
52. Rao, K.N., Narasimha, R., Narayanan, M.A.B.: *The bursting phenomenon in a turbulent boundary layer.* J. Fluid Mech., Vol. 4, 1971.
53. Rashidi, M., Banerjee, S.: *Streak characteristics and behavior near wall and interface in open channel flows.* J. Fluids Engrg., Vol. 112, June 1990.
54. Rashidi, M., Banerjee, S.: *Turbulence structure in free-surface channel flows.* Phys. Fluids, Vol. 31, Sept. 1988.
55. Schlichting, H.: *Boundary layer theory.* McGraw Hill Book Co. Inc., Verlag G. Braun (6th edition), 1968.
56. Sherenkov, I.A.: *Applied problems of plan hydraulics of tranquil flows.* (In Russian) Energy Publishing House, Moscow, 1978.
57. Silva, A.M.F.: *Alternate bars and related alluvial processes.* M.Sc. Thesis, Dept. of Civil Engrg., Queen's Univ., Kingston, Canada, 1991.
58. Smith, C.R., Walker, J.D.A, Haidari, A.H., Sobrun, U.: *On the dynamics of near-wall turbulence.* Phil. Trans. R. Soc. Lon. A, Vol. 336, 1991.
59. Tsujimoto, T.: *Longitudinal stripes of sorting due to cellular secondary currents.* J. Hydroscience and Hydraulic Engrg., Vol. 7, No. 1, Nov. 1989.
60. Utami, T., Ueno, T.: *Experimental study on the compound meandering channel flow using flow visualization and picture processing.* J. Hydroscience and Hydraulic Engrg., Vol. 9, No. 1, May 1991.
61. Utami, T. et al: *On the mechanism of secondary flow in prismatic open channel flow.* Flow Visualisation II. Proc. Second Int. Symp. on Flow Visualisation, Bochum, Sept. 9-12, 1980.
62. Utami, T., Ueno, T.: *Lagrangian and Eulerian measurement of large scale turbulence.* Flow Visualisation I. Proc Int. Symp. on Flow Visualisation, Tokyo, Oct. 1977.
63. Velikanov, M.A.: *Dynamics of alluvial streams.* (In Russian) Gostechizdat, Moscow, 1955.
64. Willmarth, W.W., Lu, S.S.: *'Reynolds stress structure in turbulent boundary layer.* Proc. Symp. on Turbulent Diffusion in Environmental Pollution, Charlottesville, Virginia, April 1974.
65. Yalin, M.S.: *Mechanics of sediment transport.* Pergamon Press, Oxford, 1977.
66. Yokosi, S.: *The structure of river turbulence.* Bull. Disaster Prevention Res. Inst., Kyoto Univ., Vol. 17, Part 2, No. 121, Oct. 1967.

CHAPTER 3

BED FORMS AND FRICTION FACTOR

3.1 Classification of Bed Forms

3.1.1 *Quasi-uniformity of flow and bed forms*

Consider a sediment transport experiment (run) which commences, at the instant $t = 0$, on the flat surface of a mobile bed: $\eta = (v_*/v_{*cr})^2 > 1$; $q_s = q_{sb} > 0$. In order to reveal what might subsequently happen to this flat bed surface, we invoke the transport continuity equation (1.76). Substituting in this equation $p_s = 0$ (no material is deposited from an external source) and considering that in the case of a flat bed $\partial y_b/\partial x = 0$, one determines the following special form of (1.76):

$$\frac{\partial q_s}{\partial x} + \frac{dy_b}{dt} = 0 \tag{3.1}$$

$$(q_s = f_s(x, t);\ y_b = f_b(x, t)).$$

Here dy_b/dt is the speed V_b of the vertical displacement of a point of the bed surface. The relation (3.1) indicates that if the initial flow is completely uniform, i.e. if its properties u and τ, and consequently q_s, do not vary along x, then the flat bed surface will remain as it is ($\partial q_s/\partial x = dy_b/dt = 0$). Similarly, if $\partial q_s/\partial x = const$ (or $\partial q_s/\partial x = f_s(t)$), then the bed surface will move up or down, but it will still remain undeformed (flat). If, however, the initial flow is quasi-uniform, i.e. if it consists of a sequence of identical non-uniform regions R_i, each of the same length L, then the bed-load rate q_s at $t = 0$ is a periodic (along x) function

$$q_s = f_s(x, 0) = f_s(x + L, 0) \quad (\text{with } (q_s)_{av} = const), \tag{3.2}$$

which renders the (initial) vertical displacement velocity V_{b0} also a periodic function:

$$V_{b0} = -(\partial q_s/\partial x)_{t=0} = f_{V_b}(x, 0) = f_{V_b}(x + L, 0) \tag{3.3}$$

$$(\text{with } (V_{b0})_{av} = 0).$$

Clearly, in this case the flat initial surface of the mobile bed, viz

$$y_b = f_b(x, 0) = 0\,, \tag{3.4}$$

cannot remain undeformed. Indeed, after the passage of a small interval of time δt, the plane (3.4) will turn into the surface

$$\delta y_b = V_{b0}\,\delta t\,, \tag{3.5}$$

which is certainly periodic, for V_{b0} is periodic:

$$\delta y_b = f_{V_b}(x, 0)\delta t = f_b(x, \delta t) = f_b(x + L, \delta t) \quad (\text{with } (\delta y_b)_{av} = 0). \tag{3.6}$$

The average plane of the periodically deformed ("wavy") surface at $t = \delta t$ is the same as the plane of the initial bed at $t = 0$: $(\delta y_b)_{av} = 0$ ($= f_b(x, 0)$).

Hence, if the flow at $t = 0$ is quasi-uniform, then the initially flat surface of a mobile bed will eventually deform into an undulated one: it will be covered by *bed forms* (or *sand waves*). Each bed form i is one "erosion-deposition" sequence (as shown schematically below), its length Λ_i being identical to the period L_i of the quasi-uniform flow:

$$\Lambda_i \equiv L_i\,. \tag{3.7}$$

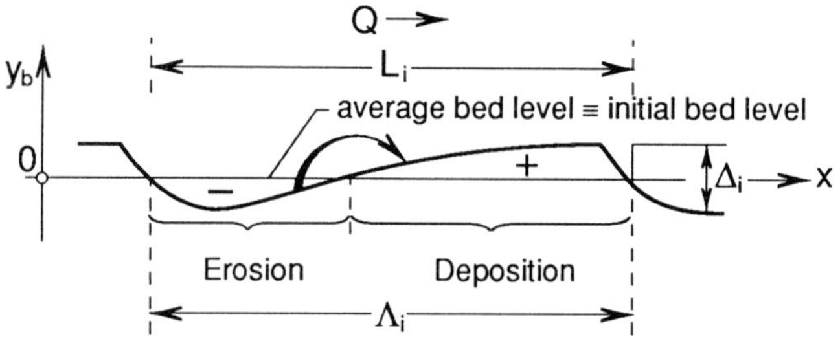

A basically uniform open-channel flow becomes quasi-uniform either because of the structures of turbulence (bursts, viscous flow undulations at the bed, etc.) contained in it, or because of the waves which may occur on its free surface.[1]

Since the granular skin roughness k_s is proportional to the typical grain size D (Section 1.6), any function of k_s introduced in the preceding parts of the book will be interpreted in this chapter in terms of D.

---

1 The flow can also possess a certain periodicity in its lateral direction (z), caused by the secondary currents. Since the lateral period of secondary currents is $\approx 2h$ (see 2.2.4), the lateral distance between the *longitudinal ridges* generated by them must also be $\approx 2h$ (and experiment shows that this is indeed so). However, the role of longitudinal ridges in fluvial processes is rather limited, and therefore they will not be elaborated here. Interested readers are referred to the outstanding recent works [33], [34], [42], [43], [44].

3.1.2 *Definition of bed forms and their general properties*

i- *Antidunes*

If $Fr > \approx 0.6$ [4], [24], [54], then standing waves occur on the free surface of an open-channel flow. The length L_g of these (gravity) waves is determined by the form

$$L_g = \Phi_g(Fr)\, h\,, \tag{3.8}$$

where $\Phi_g(Fr)$ can be approximated by $6Fr$. (More refined expressions of $\Phi_g(Fr)$ can be found in Refs. [24], [3], [54]).

The relative height of standing waves increases with Fr, and when $Fr > 1$ (i.e. when the flow is supercritical) they are usually able to deform the initially flat surface of a mobile bed as to produce "their" bed forms – *antidunes*. Hence, using (3.7), one can write for the length Λ_g of antidunes

$$\Lambda_g \equiv L_g = \Phi_g(Fr)\, h. \tag{3.9}$$

ii- *Dunes and bars*

Consider now the bed forms which are due to the quasi-uniformities caused by the structures of turbulence. From the content of Chapter 2, it should be clear that the length (along x) of a turbulent structure of an open-channel flow is proportional to one of the characteristic lengths h, B or ν/v_*. Consequently, the same must be valid for the length of bed forms "imprinted" by one of those structures. In this paragraph we will consider only the bed forms produced by the largest turbulent structures, viz bursts. In Chapter 2 it has also been shown that all linear dimensions of vertical bursts are proportional to h; those of horizontal bursts, to B/N (where N is the number of burst-rows). In this text, *dunes* and *bars* will be identified with (or defined as) those bed forms which are produced by vertical bursts and horizontal bursts respectively. If $N = 1$, i.e. if we have the basic single-row configuration of horizontal bursts (Fig. 2.21), then the resulting bars are *single-row bars* or *alternate bars*. If $N > 1$ (Fig. 2.23), then the imprinted bed forms are *N-row bars* or *multiple bars*.

1- If the flow past the flat initial bed is rough turbulent, then no viscous influence is present and the length of vertical bursts is $L \approx 6h$ (2.2.1 (i)). Hence, according to the definitions above (and Eq. (3.7)), the length Λ_d of dunes generated by a rough turbulent flow should be given by

$$\Lambda_d \equiv L \approx 6h. \tag{3.10}$$

As is known from research on sediment transport, the length of dunes produced by rough turbulent flows can be expressed indeed as $\approx 6h$ (e.g. $\Lambda_d \approx 2\pi h$ [54]; $\Lambda_d = 7h$ [17]; $\Lambda_d = 5h$ [57]; ... etc.).

If the flow past the flat initial bed is not rough turbulent, then the ratio Λ_d/h may be affected by $X = v_* k_s/\nu$ ($\sim Re_*$) and $Z = h/D$ ($\sim h/k_s$), and (3.10) must be generalized into

$$\Lambda_d = \Phi_d(X, Z)\, 6h. \tag{3.11}$$

The form (3.11) is consistent with experimental measurements: the family of experimental curves in Fig. 3.18 indicates that, in general, the ratio Λ_d/h ($= (\Lambda_d/D)/Z$) varies as a function of X and Z, and that it becomes ≈ 6 only when $X > \approx 35$ (for any Z).[2] The deviation of (Λ_d/h)-values from ≈ 6 with the decrement of $X \sim Re_*$ is (as can be inferred from Fig. 3.18) much larger than what one would expect on the basis of explanations given in Section 2.2. The reason for this will be clarified in 3.1.3 (iv).

2- In 2.3.2 it has been explained that the length $L_H = (L_H)_1$ of horizontal bursts corresponding to the basic configuration $N = 1$ can be identified with the length Λ_a of alternate bars, viz $\approx 6B$. Hence we have

$$\Lambda_a \equiv (L_H)_1 \approx 6B, \tag{3.12}$$

which is in line with the above definition of alternate bars (and with (3.7)). It should be pointed out here that (in analogy to Λ_d/h) the relative bar length Λ_a/B may also be affected by X and/or Z if the initial flow is not rough turbulent. However, the existing plots of Λ_a (Figs. 3.29a and b) are too scattered to supply any information on this score, and we will use (3.12) for all regimes of the initial flow.

3- If $N > 1$, then $(L_H)_N = (L_H)_1/N$ (see (2.30)) and (3.12) must be replaced by

$$(\Lambda_b)_N \equiv (L_H)_N \approx 6B/N. \tag{3.13}$$

Experiment shows that the length of multiple bars decreases indeed with the increment of N (see [9], [18], [20], etc.). Being an outcome of flow past the flat initial bed (at $t = 0$), the number of burst rows, N, is determined by B/h, $h/k_s \sim Z$ and $Re_* \sim X$ (see 2.3.5). More specifically, it is determined by the dimensionless combination $\omega \sim (B/h)/\bar{c}^2$ (Eqs. (2.41) and (2.42)), where $\bar{c} = \bar{\Phi}_c(Re_*, h/k_s) = \bar{\Psi}_c(X, Z)$.[3]

It has also been demonstrated in 2.3.5 that if $(B/h) \sim \omega \to \infty$, then $N = (const)\,\omega \sim B/h$. In this case (3.13) yields $(\Lambda_b)_\infty \sim h$, whereas for $N =$

---

[2] Usually $k_s \approx 2D$, and therefore $X \approx 35$ means $Re_* \approx 70$, which is the lower limit of rough turbulent flows (see e.g. [37]).

[3] In 2.3.5 the analysis was confined to rough turbulent flows only, and therefore $\bar{c}$ was expressed (independently of Re_*) as $\bar{c} \sim (h/k_s)^{1/6} = \bar{\psi}_c(Z)$.

$= 1$ it gives[4] $(\Lambda_b)_1 = \Lambda_a \sim B$ (the linear scale of bars shifts from B to h with the increment of B/h, and thus of N).

iii- *Ripples*

Ripples will be defined, in this text, as the bed forms "imprinted" by the viscous flow structures (undulations) at the bed of a turbulent flow (see 2.2.5). Since the average wave length λ_x of the quasi-periodic structures is given by (2.19), we have for the length λ_0 of ripples emerging at $t = 0$

$$\lambda_0 \equiv \lambda_x \approx 2000 \, \frac{\nu}{\upsilon_*} \, . \tag{3.14}$$

It is important to realize that (3.14) corresponds to the case of a flat bed and a *clear* fluid (zero concentration of sediment). Yet, when ripples are forming, the bed roughness changes, and the flow at the bed becomes saturated with transported grains. These factors affect the flow at the bed and consequently the value (≈ 2000) of its dimensionless wave length $\lambda_x \upsilon_*/\nu$.[5] Considering that the concentration C_ϵ of transported solids is one of the properties of the two-phase motion at the bed, and thus that it is itself a certain function of ξ and η (see (1.67) and (1.68)), one determines on the basis of (3.14) the following relation for the (developed) ripple length Λ_r at T_r:

$$\frac{\Lambda_r \upsilon_*}{\nu} \equiv \frac{\lambda_x' \upsilon_*}{\nu} = \phi_1(C_\epsilon) = \phi_2(\xi, \eta) \, . \tag{3.15}$$

Here λ_x' is the wave length of viscous structures at $t = T_r$: the functions on the right satisfy

$$\phi_1(0) = \phi_2(\xi, 1) \approx 2000 \, . \tag{3.16}$$

Fig. 3.1 shows that the experimental values of $\Lambda_r \upsilon_*/\nu$ form indeed a family of curves, as implied by (3.15). Note also that when $\eta \to 1$ (clear fluid), then $\Lambda_r \upsilon_*/\nu$ approaches ≈ 2400, which is comparable with ≈ 2000.
Since

$$\frac{\Lambda_r \upsilon_*}{\nu} = \frac{\Lambda_r}{D} \, X \, , \quad Y = \frac{X^2}{\xi^3} \quad \text{and} \quad Y_{cr} = \Psi(\xi) \tag{3.17}$$

(see 1.3.2), the relation (3.15) can be expressed as

$$\frac{\Lambda_r}{D} = \frac{1}{X} \, \phi_2\!\left(\xi, \frac{X^2}{\xi^3 \Psi(\xi)}\right) = \Phi_r(X, \xi) \, , \tag{3.18}$$

---

[4] In the following, depending on the purpose, both Λ_a and $(\Lambda_b)_1$ will be used to denote the length of alternate bars ($\Lambda_a = (\Lambda_b)_1$).

[5] This is not so in the case of bed forms defined previously, for the flow structures producing them are not confined to the neighbourhood of the bed (where the concentration C is much larger than elsewhere).

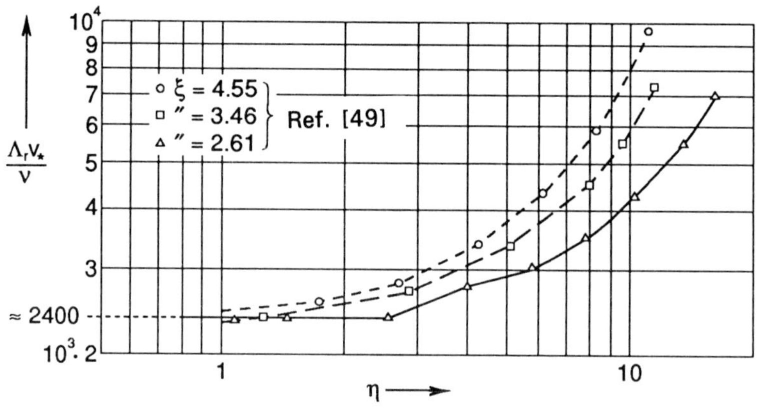

Fig. 3.1 (from Ref. [49])

i.e. as

$$\Lambda_r = \Phi_r(X, \xi)\, D. \tag{3.19}$$

As will be seen later (Fig. 3.24), the function $\Phi_r(X, \xi)$ varies around ≈ 1000; hence the reason for the approximate relation

$$\Lambda_r \approx 1000\, D \tag{3.20}$$

proposed in the earlier works of the author [57], [54].

The relation (3.19) indicates that the flow affects the ripple length Λ_r only by means of v_* (which appears in X): no $h \sim Z$ is present in this relation. Thus, the fact that the length of ripples does not depend on the flow depth $h \sim Z$ *follows* from their present definition (it is not explicitely stated in it).[6]

3.1.3 *Additional remarks*

i- No aspects peculiar only to open-channel flows were used in the present definition of dunes and ripples. Hence, these bed forms must be expected to occur in any turbulent shear flow having vertical bursts and viscous flow structures at the bed. And, as is well known, dunes and ripples are encountered indeed in closed conduits as well as in deserts. In the latter case, the boundary-layer flow of air (wind) does not have any distinct upper limit and the determination of the burst length L may be very difficult or even impossible. This, however, has no bearing on the fact that Λ_d is nonetheless due to L.

---

6 In the literature, bed forms are usually defined without taking into account their origins and without indicating how their size Λ_i is related to the characteristic lengths of the flow. They are referred to as "large" or "small"; yet it is not mentioned "in comparison to what" are they large, or how small is "small". In this text we will use only the definitions introduced in the preceding three paragraphs.

ii- No side walls or banks are needed for the occurrence of vertical bursts. Yet they are needed for the occurrence of alternate and multiple bars (which are generated by horizontal bursts issued from the side walls or banks). Hence there are no "desert bars" in analogy to desert dunes and ripples.

iii- Let Λ_i and Λ_j be the lengths of two different types of bed forms. If these lengths are comparable ($\Lambda_i \approx \Lambda_j$), then such bed forms are mutually exclusive: only the bed forms of the "stronger agent" materialize. For example, the lengths of dunes and antidunes are comparable, and therefore they cannot coexist: we have *either* dunes *or* antidunes. If $Fr < \approx 1$, then the impact of vertical bursts is stronger than that of the standing waves, and we have (only) dunes; if $Fr > \approx 1$, then it is the other way around, and we have (only) antidunes.
If Λ_i and Λ_j are substantially different ($\Lambda_i \ll \Lambda_j$), then such two bed forms can coexist (as i superimposed on j). Consider e.g. $\Lambda_r \approx 1000\,D$ and $\Lambda_d \approx 6h$; they yield $\Lambda_d / \Lambda_r \approx Z/166$. If Z is by multiple times larger than 166, then both dunes *and* ripples (superimposed on the dunes) can be present.[7] Similarly, if $\Lambda_a / \Lambda_d \approx B/h$ is sufficiently large, then alternate bars and dunes (superimposed on bars) may occur simultaneously. And if $\Lambda_r \ll \Lambda_d \ll \Lambda_a$, then we can even have ripples *and* dunes *and* bars superimposed on each other (the Tagus at Santarém).

iv- As is well known [55], [56], and as can be readily inferred from (3.1), the duration of development T_i of a bed form i can be characterized by the proportionality

$$T_i \sim \frac{\Delta_i \Lambda_i}{q_s}, \tag{3.21}$$

where q_s is an averaged value of the transport rate $q_s = f_s(x, t)$ and $(\Delta_i \Lambda_i)$ is ($\approx$twice) the area of the developed bed form profile. Consider the case of dunes and ripples superimposed on them: ($\Lambda_r \ll \Lambda_d$). These bed forms are generated by the same flow and thus by the same q_s. Assuming that their steepness is comparable, one obtains from (3.21)

$$\frac{T_r}{T_d} \sim \left(\frac{\Lambda_r}{\Lambda_d}\right)^2, \tag{3.22}$$

which indicates that if dunes are, for example, 5 or 10 times longer than ripples (superimposed on them), then ripples will develop 25 or 100 times

---

7 Vertical bursts are always present in a turbulent open-channel flow; yet the viscous flow undulations at the bed, which generate ripples, are present only if X is sufficiently small. Thus for a given appropriate η, a "large enough Z" is merely a necessary condition for superimposition of ripples on dunes (as implied by "*can* be present"). The necessary and sufficient condition for the superimposition mentioned is "large enough Z" *and* "small enough" X (smaller than $\approx$30, say).

faster than dunes. But this means that in most cases ripples are already nearly developed when the development of dunes has only just started. And this, in turn, means that (in the case of dunes+ripples) dunes do not really form on a flat bed having granular roughness $k_s \sim D$: they form on a "flat" bed having (the earlier developed) ripple roughness $K_s \sim \Delta_r$. (Clearly, exactly the same can be said for the case of dunes+bars (if $\Lambda_d \ll (\Lambda_b)_N$)). The change of the initial bed roughness from $k_s \sim D$ to $K_s \sim \Delta_r$ ($\gg D$) must, to some extent, affect the length of vertical bursts and consequently the length of dunes imprinted by them. On the other hand, the steepness of ripples increases with the prominence of viscosity at the bed, and therefore $K_s \sim \Delta_r$ is a direct (non-decreasing) function of the grain size Reynolds number $X \sim \nu^{-1}$. Hence, the influence of X on Λ_d (mentioned in 3.1.2 (ii)) is not so much because the burst length L ($= \Lambda_d$) is directly affected by X, but because it is directly affected by the (earlier developed) ripple roughness $K_s \sim \Delta_r$ – which is directly affected by X.

3.2 Formation of Bed Forms

3.2.1 *Bed forms caused by bursts (dunes and bars)*

i- *Discontinuity and burst sequences*

The fact that "bursts are randomly distributed in space and time" (2.1.1) implies that, under completely uniform conditions of flow and bed surface, the probability (or frequency) of occurrence of bursts in a region Δx, during a time interval Δt, does not depend on where this region is located on the x-axis. Clearly such a homogeneous, or uniform, distribution of bursts along x can never yield a wave-like deformation of the bed surface: for this deformation to occur, the conversion of uniform conditions to the quasi-uniform ones is required (3.1.1). In practice, such a conversion is realized by means of a *local discontinuity*, d say, (at the section $x = 0$) on the bed surface or banks. Under laboratory conditions, d can be the beginning of mobile bed or banks, an accidental ridge across the sand bed surface, etc. The presence of d causes the behaviour of flow and its turbulent structure at $x = 0$ to be different from its "usual" behaviour at any other x. For example, if d is a ridge, then it will certainly "promote" the eddy shedding, and the burst-forming eddies e will be generated at $x = 0$ more frequently than anywhere else. One can say that the discontinuity section $x = 0$ manifests itself as the "location of preference" as far as the generation of bursts is concerned. In the case of vertical bursts, the increment of the frequency of bursts generated at $x = 0$ inevitably leads to the increment of that frequency also at the sections $x = L, 2L, \ldots$, etc., for the break-up of one burst (of the length L) triggers the "birth" of the next (2.1.1). The same can be said with regard to the horizontal bursts (of the length L_H) issued at the banks. It follows that it is not just the single bursts, but the *sequences* of bursts which are generated at $x = 0$ more frequently than elsewhere. The dunes and bars, which invariably originate at the discontinuity-sections, are the imprints of the burst sequences.

One may wonder how an attenuating (along x) sequence of bursts can produce bed forms of the same height throughout the length of a (long) channel; why is the influence of this sequence not confined to a limited x-region downstream of the discontinuity only? The answer to this question lies in the fact that the first emerging infinitesimal "steps" $(\Delta_t)_I$, $(\Delta_t)_{II}$, $(\Delta_t)_{III}$, ... (in Fig. 3.2) act themselves as new discontinuities, and create (just like the original d) their own new steps on their downstream side: the activity started by d perpetuates and propagates along x.[8]

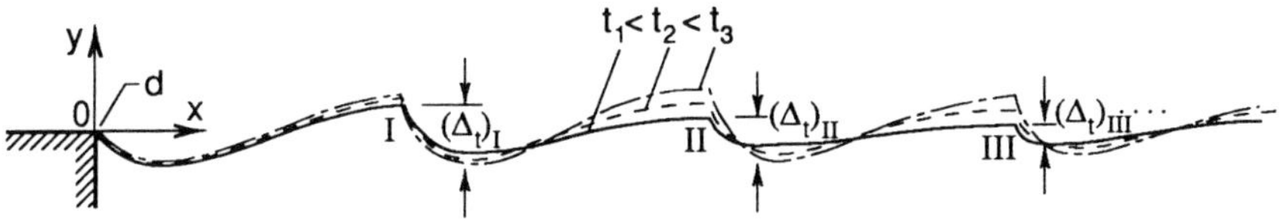

Fig. 3.2

ii- *Dunes*

1- The equality $\Lambda_d = L \approx 6h$ suggests that the occurrence of dunes merely causes the vertical burst modules to be "tilted" by an angle ϕ – without altering their length and configuration (Fig. 3.3).[9] The same applies to the horizontal burst modules and bars, whose steepness is, in general, even smaller than that of dunes. And also to the standing waves and antidunes: the growth of antidunes in the y-direction cannot cause the "nesting" system of surface waves to elongate or contract in the x-direction. Only ripples are the exceptions – their growth alters the viscous flow (at the bed) producing them. On the basis of the aforementioned, one can assert that the flow and the large-scale bed forms created by it ($\Lambda_i \sim h$ or B) are compatible with each other. This compatibility cannot be expected to be present if the flow takes

---

8 It follows that although the duration of development T_d of a certain number of dunes is meaningful, the duration of development $(T_d)_{ch}$ of dunes in a channel is meaningless, if the channel length (L_{ch}) is not specified. For, as one can easily infer, $(T_d)_{ch}$ increases in proportion to L_{ch}.

9 The "tilt" (ϕ) cannot be substantial (for the dune steepness Δ/Λ is always less than ≈ 0.06 (Fig. 3.19)). Nonetheless, it causes the macroeddy E to be pressed against the free surface, and consequently it enhances the occurrence of "boils"; the larger the dune steepness, the more prominent are boils. (Boils are never observed if dunes (and thus "tilts") are not present: more on the topic in the next paragraph).

place past irregularities of an independent origin.[10]

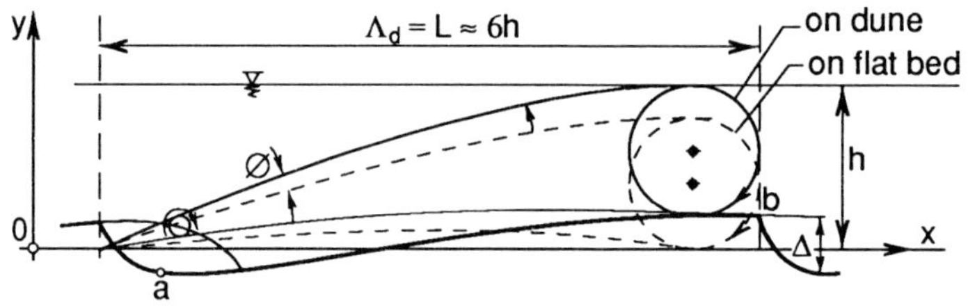

Fig. 3.3

2- If the dune length Λ_d is the imprint of the burst length $L \approx 6h$, why then do the dunes not originate as to possess the length $\Lambda_d = L$ in the first place? Why does their length, Λ_t say, "grow" during the time interval $0 < t < T_d$, starting from a much smaller value, $\Lambda_0 \approx h$ say (just after $t = 0$), and acquiring the value $\Lambda_d = L$ only at $t = T_d$? The answer to this question lies in the fact that the bursts of the length $L \approx 6h$ (in short the bursts L) are not the only bursts of the flow: they are merely the largest bursts (see 2.2.3). Let

$$L_0 \approx h < L_1 < L_2 < \cdots < L_i < \cdots < L \approx 6h \tag{3.23}$$

be the symbolic "hierarchy" of vertical burst sequences (Fig. 3.4). The smaller

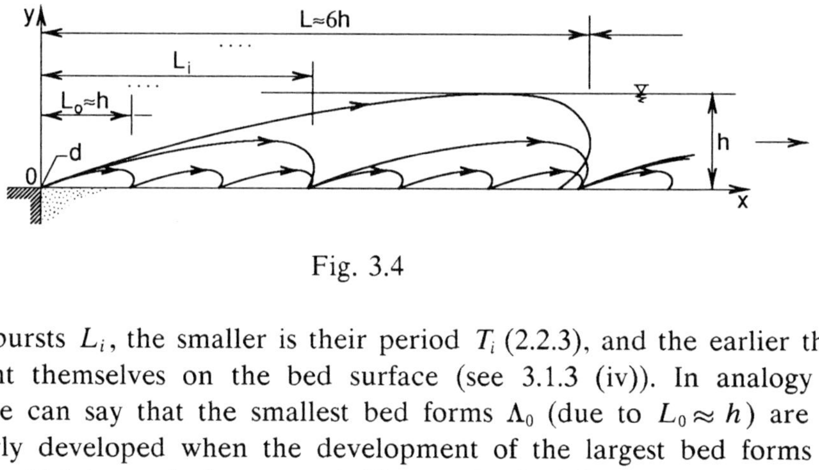

Fig. 3.4

the lesser bursts L_i, the smaller is their period T_i (2.2.3), and the earlier they can imprint themselves on the bed surface (see 3.1.3 (iv)). In analogy to ripples, one can say that the smallest bed forms Λ_0 (due to $L_0 \approx h$) are already nearly developed when the development of the largest bed forms Λ_d (due to $L \approx 6h$) has only just started. Hence, the bed first becomes covered by the shortest bed forms Λ_0. With the passage of time, the sequences of larger bursts L_i begin to "catch up" with their bed form production. The

---

[10] Very often, rigid boundaries are used (rigid dunes, meander loops, etc.) to reveal the characteristics of flow past these features. From the aforementioned, it should be clear that the information supplied by such studies can be misleading, if the shape and size of the rigid boundaries used are not identical to those which the flow under study would have created itself (in an alluvium of comparable skin roughness).

emergence of bed forms Λ_i is associated with the elimination of the previous bed forms Λ_{i-1}. The transition from Λ_{i-1} to Λ_i occurs, as is well known, by *coalescence*: a series of n (smaller) bed forms Λ_{i-1} coalesce as to form $n-1$ (larger) bed forms Λ_i. Thus the length of bed forms "grows" from Λ_{i-1} to Λ_i. This pseudo-growth terminates (at $t = T_d$) when the largest bed forms, viz the dunes Λ_d, are produced by the largest, or usual, bursts $L \approx 6h$. The same is valid (*mutatis mutandi*) for bars. Fig. 3.5 shows the "growth" of alternate bars, where the smaller (earlier) bed forms are due to the lesser horizontal bursts $(L_H)_i$.

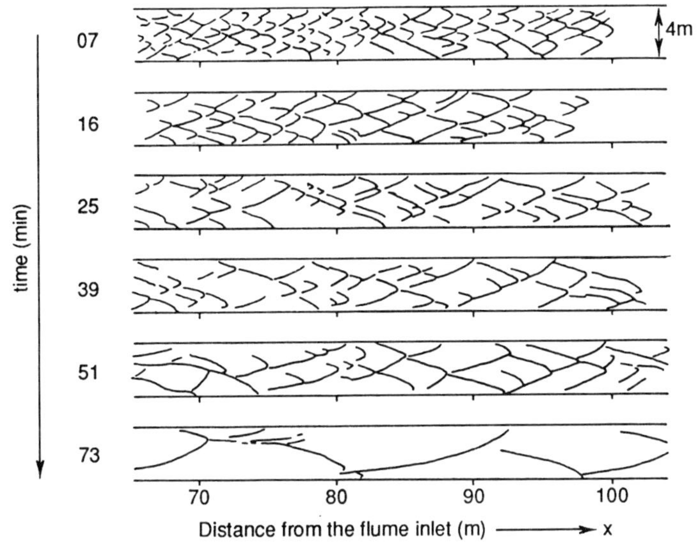

Fig. 3.5 (from Ref. [18])

iii- *Additional information on dunes*

1- In item (2) of the preceding paragraph, the length L_0 of the smallest lesser bursts was approximated by the flow depth h. This was done on the basis of laboratory experiments reported in Ref. [5]. These experiments were carried out for $D = 0.5mm$ and $1mm$ sand, and $D = 1mm$ Bakelite. The ratio Λ_0/Λ_d, just after $t = 0$, was $\approx 1/6 \approx 0.17$ for all three materials used; hence, $L_0 \approx h (= \Lambda_0)$ was adopted in (3.23). Fig. 3.6 shows, as example, the time growth of relative dune length for $D = 0.5mm$ sand. (No explanation can be offered by the author as to why L_0 should be equal (or comparable with) h).

2- From special flow visualization experiments reported in Refs. [30], [31], it appears that the burst-forming eddies e of the flow past dunes originate at the interface b_0S between the main flow and the "roller" R at the downstream dune face (Fig. 3.7). Thus, the occurrence of dunes renders the large-scale vertical turbulence more regular. Indeed, the bursts L are no

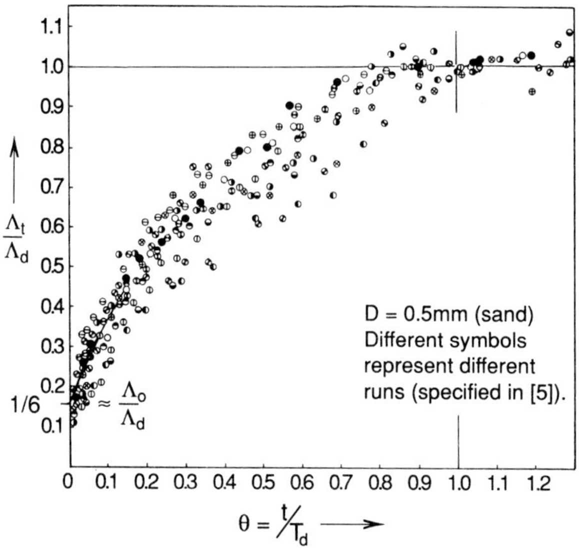

Fig. 3.6 (from Ref. [5])

longer distributed randomly in space and time: they initiate predominantly at the well defined locations (b_0S) and they terminate, by the break-up's of their macroturbulent eddies E, at the dune crests b. This "regularization" of turbulent flow by dunes was apparently first pointed out in Ref. [29].

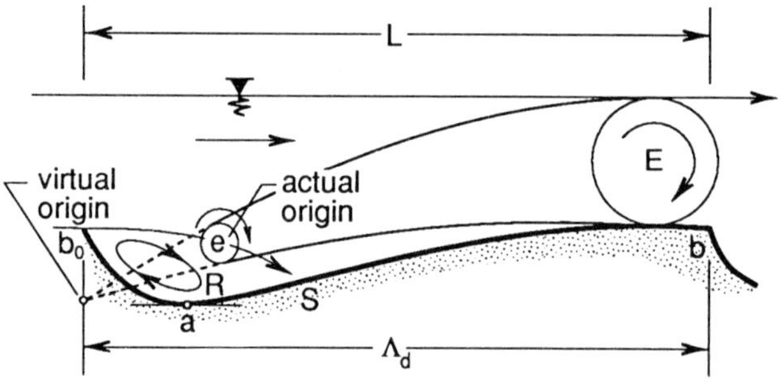

Fig. 3.7

3- In Chapter 2 it has already been mentioned that boils are caused by the impingement of macroturbulent eddies E on the free surface. The burst module extends along the upstream faces of dunes ab, and the fact that the dune steepness causes this module to be tilted (see Fig. 3.3) means that the macroturbulent eddy E, at b, is "pressed" against the free surface. This may cause the eddy E to protrude through the free surface as to form the boil: the more so, the steeper are dunes. From observations and measurements carried out in natural rivers, it is known [21], [26], [6] that boils occur indeed (intermittently) above the dune crests and that their prominence increases with dune steepness. The average frequency of the occurrence of boils is in coincidence with the average frequency of bursts.[11]

4- It has been explained in 3.2.1 (i) that the local discontinuity (d) on the flow bed at $x = 0$ causes the increment of vertical burst generations there; which, in turn, leads to the generation of dunes.

Let us suppose that the discontinuity d extends along z throughout the flow width B (Fig. 3.8). The maximum width of a vertical burst in plan, and

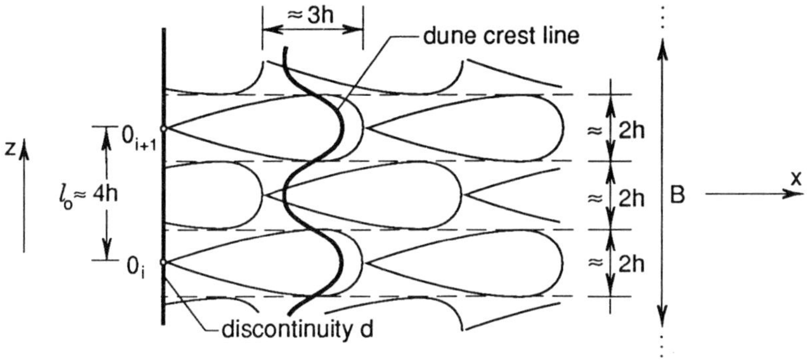

Fig. 3.8

thus the width of a "burst-channel" accommodating a sequence of bursts (see Fig. 2.11), is $\approx 2h$. If the burst sequences in all the adjacent channels were in phase, then the distance l_0 between the "point sources" O_i and O_{i+1} on the discontinuity-line d would also be $\approx 2h$. However, the high- and low-speed regions of a wide turbulent flow have, ideally, a chessboard-like arrangement in plan (Fig. 2.1), and therefore the burst-sequences in the adjacent burst channels are out of phase along x by $L/2$ (see Fig. 2.11). One would expect that with the increment of B/h and h/D the flow-dune system should tend to organize itself so as to conform to this fundamental chess-board-like arrangement. Consequently, l_0 should tend to be $\approx 4h$ as shown in Fig. 3.8. This (highly idealized) sketch helps to understand why, in the case of a wide flow, the dune crests are curvilinear in plan. (This sketch also suggests that dune crest-lines should be symmetrical or anti-symmetrical with

[11] Recall from Chapter 2 that the cross-section of a burst module consists of two opposite rotating eddies which grow with the passage of time in the $(y; z)$-plane (Fig. 2.10). Suppose now that dye is injected into a point inside one of these eddies – inside the eddy in the region $z > 0$, say. In this case the burst will be visualized as a clockwise rotating spiral which grows in its diameter as it advances along its upward-downstream inclined axis. This may explain why in some earlier works boils were attributed to the "upward tilted streamwise vortices". An informative account of boils is given in I. Nezu and H. Nakagawa: *Turbulent structures over dunes and their role on suspended sediments in steady and unsteady open channel flows.* Proc. Int. Symp. on the Transport of Suspended Sediment and its Mathematical Modelling, Florence, Sept. 1991.

respect to the x-axis, depending on whether B is an odd or even multiple of $2h$).

Consider a stationary and uniform two-dimensional turbulent flow past a flat initial bed. Let $U = f_U(x, t)$ be the fluctuating (along x and t) longitudinal flow velocity at a level y near the bed, and

$$R_{UU}(x) = e^{-ax} \cos\left(2\pi \frac{x}{L}\right) \tag{3.24}$$

its autocorrelation function along x. The emergence of bed forms due to vertical turbulence can also be explained (and was explained in the past [45], [46], [54]) with the aid of this function. It is assumed that bed forms are caused by those longest (along x) U-fluctuations, whose frequencies are confined to a narrow band including $2\pi/L$. The function $R_{UU}(x)$ and its graph in Fig. 3.9 indicate that if the (fluctuating) values of U at a section $x = 0$, say, ex-

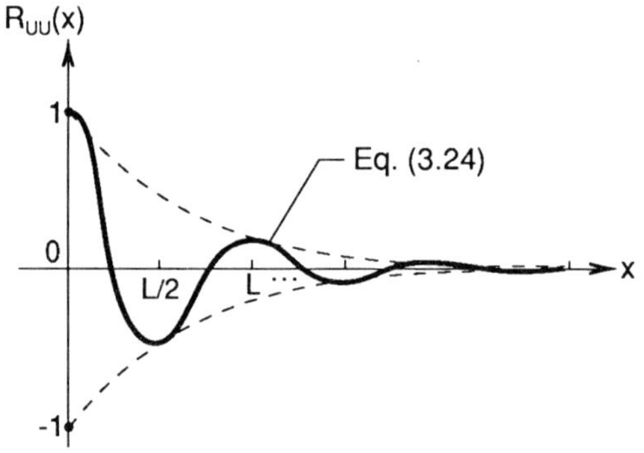

Fig. 3.9

hibit a systematic deviation (e.g. an increment) from their "regular" values (which can be caused e.g. by a discontinuity d at $x = 0$), then a tendency for an analogous deviation (increment) must be present also at the sections $x = L, 2L, \ldots$, whereas at the sections $x = L/2, 3L/2, \ldots$ a tendency for the decrement of U must be present. All these trends must, of course, progressively die out with the increment of x (due to the damping factor e^{-ax}). This repetitive (along x) behaviour of U leads to the analogous behaviour of τ and $q_s = q_{sb}$; and consequently to the generation of bed waves of the length $\Lambda \equiv L$. However, this mathematical method is inherently incapable of supplying the value of Λ ($\equiv L$) itself: it operates with the *unknown* L. And yet, it is exactly the magnitude of Λ which is the focus of interest in any study of alluvial forms. The same drawback is present also in the stability analysis, which is frequently used today.

Furthermore, the stability methods rely on a number of notions which are "taken for granted"; and yet some of them are not obvious at all, at least as far as fluvial hydraulics is concerned. For example, why should the initial perturbation be necessarily a periodic function of x? How can such a perturbation (which in reality occurs (*if* it occurs) for a real fluid and a rough bed) extend throughout the (infinitely long) x-axis − with no decrement of its amplitude, nor any change of its period? What *is* the cause of such an extraordinary perturbation? Why should one express *every* perturbation, i.e. irrespective of its physical origin, by the same sine- or cosine-functions (although there is an infinite manifold of continuous periodic functions)? ... , etc. The consideration of these questions would certainly be worthwhile.

iv- *Bars*

1- From the content of 2.3.4, it should be clear that although horizontal bursts are always present in an open-channel flow, they cannot always produce "their" bed forms, viz bars. Only if B/h is sufficiently large ($B/h > \psi_\varepsilon(h/k_s)$) and the burst-forming eddies e_H are rubbing the bed (Fig. 2.22c), can the bars be produced.

2- Fig. 3.10a shows the basic arrangement of horizontal bursts in a simplified manner: the discontinuity d is the origin (O_1) of the burst sequence.

If the burst-forming horizontal eddies e_H and e_H' were moving without interfering with each other, then the boundaries of their modules would be the lines $\overline{O_i P_i}$ and $\overline{O_i' P_i'}$ (Figs. 3.10a and b). In reality, however, their motions interfere, and therefore $\overline{O_i P_i}$ and $\overline{O_i' P_i'}$ must deform, eventually, into a sequence of non-intersecting lines l_i and l_i' whose end points are O_i' and O_{i+1} (Fig. 3.10c). To put it differently, each line $\overline{O_i P_i}$ must deform so that its end point P_i is displaced upstream to the location O_i' (the analogous applies to each $\overline{O_i' P_i'}$).

Apparently the eddies e_H and e_H', which originate at O_i and O_i' near the free surface, erode the bed by the "tornado action", and therefore it is only natural that the erosion occurs at the beginning of burst modules; the deposition, at their ends. If the bursts were not interfering, then the material eroded from the shaded regions would have been deposited in the areas $\overline{D}_i$

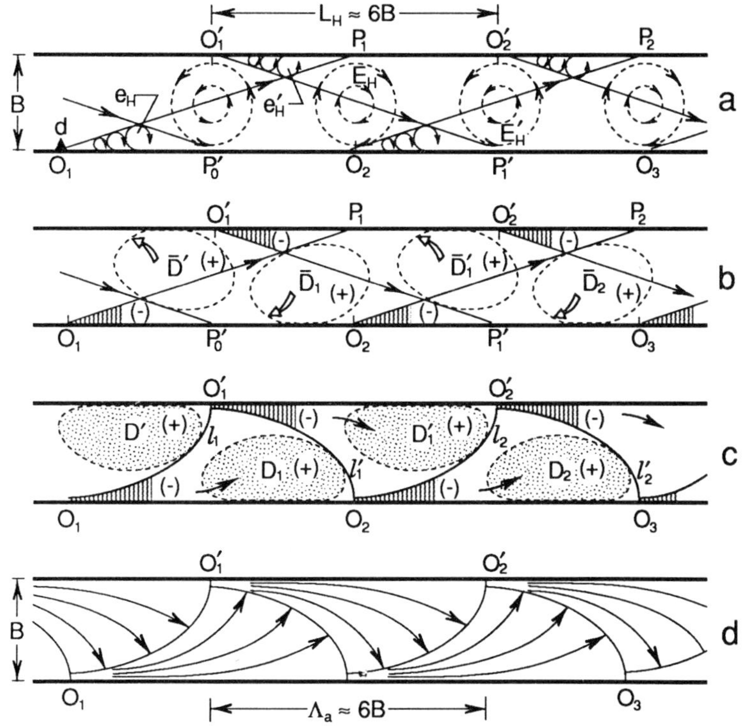

Fig. 3.10

and $\overline{D}_i'$ near the end of the burst modules (Fig. 3.10b). However, in actual case, the depositions cannot occur as far downstream, for the excessive advance of $\overline{D}_i$, say, will be "checked" by the stream of eddies e_H' issued from O_i': this stream will simply wash away any excessive material. Hence, the actual depositions will occur in the locations D_i and D_i' (Fig. 3.10c), which are not as far downstream as the "intended" depositions $\overline{D}_i$ and $\overline{D}_i'$. As a result of this, the downstream boundary of each deposition D_i and the curved boundary of the (shaded) "channel" eroded by e_H' from O_i' will form a single line l_i' connecting O_i' and O_{i+1}. (The same applies to l_i, connecting O_i and O_i'). The lines l_i and l_i' emerging in this way are the crest-lines of alternate bars, each bar being an erosion-deposition-surface confined between l_i and l_i'. Fig. 3.10d shows schematically the time average streamlines of the flow past the bar surfaces.

In analogy to dunes, the occurrence of alternate bars leads to the "regularization" of horizontal turbulence: the presence of "holes" at O_i and O_i' enhances the roll-up of e_H and e_H' at the free surface above them.

One would expect that multiple bars (caused by the horizontal bursts adjacent to the banks and the induced eddies) are formed in a similar manner.

3.2.2 *Antidunes and ripples*

Consider the time average streamlines of a flow having standing waves on its surface. The streamlines under the troughs are closer to each other, those under the crests are further apart from each other than the parallel streamlines of a strictly uniform flow. Hence, the values of u and consequently of τ_0 and q_s vary periodically along x. This causes the initially flat bed to deform into an undulated one − to be covered by antidunes. Similarly, the high- and low-velocity zones under the troughs and crests of the undulated viscous flow at the bed (Figs. 2.13 and 2.14) cause the periodicity of q_s, leading to the formation of ripples. In other words, ripples are "imprinted" by the viscous flow at the bed in much the same way (strange to say) as antidunes are imprinted by standing waves. Since the *standing* waves are stationary, so are the periodic (along x) congestions and recessions of streamlines. Hence, antidunes can originate anywhere on the flow bed: no need for a local discontinuity to initiate their formation. The viscous flow structures at the bed are not permanent. Yet, their deformation and displacement is comparatively slow, and they may persist at an area for a duration long enough to permit the initiation of ripples there. This is why ripples can also initiate spontaneously anywhere on the flow bed (in addition to their systematic initiation at a discontinuity). The emergence of ripples renders the bed rougher. Consequently, the viscous influence at the bed decreases and any Reynolds number related to the flow near the bed increases. As a result of this, the plan dimensions of viscous structures (including λ_x) become larger. The larger λ_x prompt the emerged ripples to increase in their size as to match them. The increment of ripple size (by coalescence) leads, in turn, to the further increment of λ_x, and so on − until the equilibrium state is reached, i.e. until the ripple steepness corresponding to the existing η is achieved. The length of ripples does not grow during their development as much as the length of dunes. The factor of the pseudo-growth of ripples is usually less than ≈ 2 (see photographs in Figs. 5.7 and 5.12 of Ref. [38]: see also Refs. [54] and [1]). The research on ripples [2], [11], [12], [13], [47] confirms that they are due to the viscous structures at the bed: a detailed description of the formation of ripples can be found in Ref. [38].

3.3 Steepness of Bed Forms

3.3.1 *Transport continuity equation and bed form steepness*

i- In this section we will deal with the bed forms produced by turbulent structures only (antidunes excluded). Experiment shows that if η (> 1) is "sufficiently small" (smaller than a certain η_2, introduced in 3.3.3), then all grains tranported over the bed form crest slide down its abrupt downstream face, which is inclined by the angle of repose ϕ. In this case, no grains are "flying" forward so as to "land" on the upstream face of the next bed form. Hence if $1 < \eta < \eta_2$, then $p_s = 0$ and the transport continuity equation (1.76) is of the form

$$\frac{\partial q_s}{\partial x} = -\frac{dy_b}{dt} + U_b \frac{\partial y_b}{\partial x}, \tag{3.25}$$

where q_s is the bed-load ($q_s = q_{sb}$). It is intended to reveal how this relation is affected by the time growth of the bed form height Δ_t: the initial bed is flat.

At $t = 0$ we have $\Delta_t = 0$ and thus $\partial y_b / \partial x = 0$. Hence, at $t = 0$ and at the instants "just after" $t = 0$, Eq. (3.25) reduces into its (first) special form

$$\frac{\partial q_s}{\partial x} = -\frac{dy_b}{dt} + 0 \tag{3.26}$$

(which is Eq. (3.1) used to formulate the emergence of bed forms).

With the passage of time, Δ_t increases. Consequently, $\partial y_b / \partial x$ also increases (from zero onwards), while dy_b / dt decreases (for the larger is Δ_t, the smaller is its growth rate, and thus the smaller is the vertical displacement velocity $V_b = dy_b / dt$ of a point on the bed form surface). This means that, with the increment of t, the special form (3.26) must gradually turn into the general form (3.25) (where the first and second terms are comparable). The further increment of t, i.e. the further growth of Δ_t, causes the first term to become smaller than the second, and when the growth of Δ_t stops, the first term reduces to zero. Hence when the bed form is developed, i.e. when $t = T$ and $\Delta_t = \Delta$, the general form (3.25) acquires its (second) special form

$$\frac{\partial q_s}{\partial x} = 0 + U_b \frac{\partial y_b}{\partial x}. \tag{3.27}$$

This form remains valid for all $t > T$; and it is thus applicable to the developed bed forms only.

ii- Since the bed forms originate from the flat bed ($y_b = 0$) at $t = 0$, the initial infinitesimal increments δt and δy_b are equal to the (small) values of t and y_b themselves ($t = 0 + \delta t$; $y_b = 0 + \delta y_b$). But this means that the initial vertical displacement velocity V_{b0} of a point on the bed plane $y_b = 0$ at $t = 0$ can be expressed as

$$V_{b0} = \lim_{\delta t \to 0} \frac{\delta y_b}{\delta t} = \lim_{t \to 0} \frac{y_b}{t}. \qquad (3.28)$$

Here, the function $y_b = f_b(x, t)$ signifies the contour of the infinitesimal bed form profile, or the "y_b-diagram", at the instant $t = \delta t$ (Fig. 3.11).

Consider the first special form (3.26) which corresponds to $t = 0$. The vertical displacement velocity $V_b = dy_b/dt$ on its right-hand side is, in fact, V_{b0}, and therefore

$$-\frac{\partial q_s}{\partial x} = \lim_{t \to 0} \frac{y_b}{t}, \qquad (3.29)$$

which indicates that the infinitesimal bed form-profile emerging just after $t = 0$ is in "inverse phase" with the x-derivative of the q_s-diagram (Fig. 3.11).[12]

The second special form (3.27) indicates that the x-derivatives of y_b and q_s are proportional to each other, and thus that the developed bed form-profile is "in phase" with the q_s-diagram itself (Fig. 3.11).

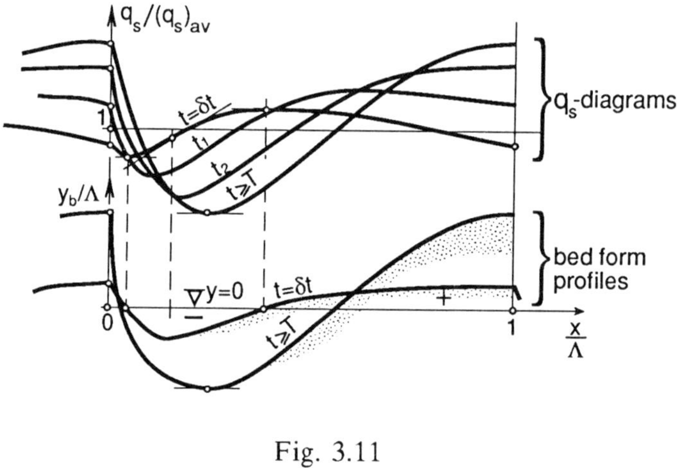

Fig. 3.11

[12] It is relevant (with regard to (3.29)) that the origin of y_b is at the level of the flat initial bed at $t = 0$ or, which is the same, at the average level of the (undulated) bed at $t \gg 0$ (see 3.1.1). Otherwise, the bed point which is not moving in vertical direction will not coincide with $\partial q_s/\partial x = 0$.

Hence, the bed-load diagram deforms substantially during the growth of the bed form height: just after $t = 0$, it is the (minus) x-derivative of the q_s-diagram which "nests" with the bed profile; at $t \geq T$, it is the q_s-diagram itself. This time variation of the q_s-diagram is illustrated schematically in Fig. 3.11.

3.3.2 *Some aspects of the bed form steepness*

i- Consider the turbulent flow past the "barrier" $b'd'$ in Fig. 3.12a. The boundary ab of the main flow past the bed form in Fig. 3.12b is remarkably similar to its counterpart $a'b'$ in Fig. 3.12a (as if the bed form comes into being because the dead area $a'b'd'$ is filled with the deposited sediment). This similarity suggests that the shape of bed forms and consequently their steepness Δ/Λ are of the flow-induced origin. One can say that the flow (specified by some given values of $Re_* \sim X$ and $h/k_s \sim Z$) "tries" to shape its lower boundary so that it possesses a certain *ideal steepness* $(\Delta/\Lambda)_* = \phi_*(X, Z)$. Whether or not the flow will actually "succeed" in this endeavour, depends on the magnitude of the existing transport rate. It will be demonstrated later on that the ideal steepness $(\Delta/\Lambda)_*$ can be achieved only for a limited ("cooperative") range of $q_s = q_{sb}$, or η.

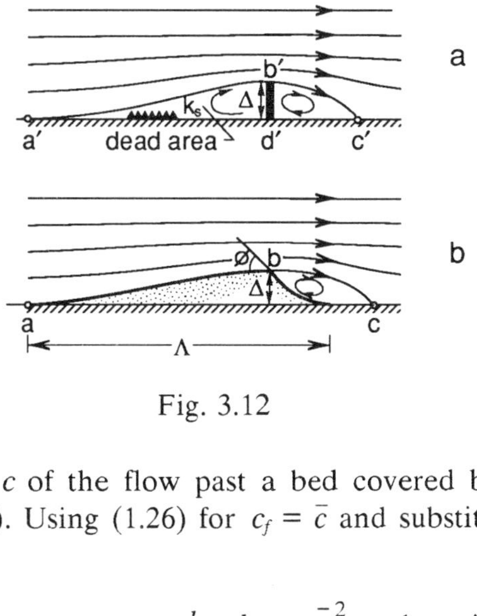

Fig. 3.12

ii- The friction factor c of the flow past a bed covered by ripples or dunes is determined by (1.35). Using (1.26) for $c_f = \bar{c}$ and substituting $k_s \approx 2D$, one can express (1.35) as

$$\frac{1}{c^2} = A_f + A = \left[2.5 \ln\left(\frac{b_s}{2}\frac{h}{D}\right)\right]^{-2} + \frac{1}{2}\left(\frac{\Delta}{\Lambda}\right)^2 \frac{\Lambda}{h} . \qquad (3.30)$$

It is intended to reveal how the variations of c, h/D and Δ/Λ are interrelated, for some given (constant) values of Λ/h and b_s. We assume, in accordance with reality, that A_f and A can acquire comparable values (albeit at different stages). Experiment shows that the variation of c is rather limited: its variation is at the most by a factor of ≈ 4, say (note e.g. from Figs. 3.34 to 3.37 that

the c-range is roughly $\approx 7 < c < \approx 25$). In contrast to this, Δ/Λ and D/h (which can be zero and nearly zero respectively) may vary over several log-cycles. Hence, a substantial increment of D/h must, according to (3.30), be compensated mainly by a decrement of Δ/Λ. But this is a different way of saying that, for the same remaining conditions, the steepness of dunes and ripples must, in general, decrease with the increment of the grain size D ($\sim k_s$). This conclusion, which will be frequently referred to, is in agreement with experiment (see e.g. Fig. 3.19, and also recall that there are no prominent dunes or ripples in gravel bed rivers).
In the case of alternate bars, h/D and Δ/Λ are "small".[13] Consequently, A is always negligible in comparison to A_f ($c \approx \bar{c}$); and the above discussion is not applicable. In fact, as will become apparent in 3.5.3 (ii), the steepness of alternate bars increases with the increment of D/h.

iii- Consider a "run" conducted on a given sand bed: q, S and D are specified. The growth of bed forms means the growth of the total bed roughness K_s and thus the reduction of the friction factor c. But, as can be inferred from (1.31), the product of c^2 with S is the Froude number ($Fr = v^2/gh = c^2 S$). Consequently, the growth of bed forms is associated with the decrement of Fr. In the next chapter, we will see that the self-formation of an alluvial channel is motivated by the trend $Fr \to \min$. Hence, the development of bed forms is consistent with (though not motivated by)[14] this trend.

3.3.3 *Bed form steepness and flow intensity η*

It has been stated in 3.3.2 (i) that the ideal bed form steepness $(\Delta/\Lambda)_*$ can be achieved only for a limited range of q_{sb} and thus of η. The present paragraph deals with this topic.

As is well known (and as can be inferred from (3.27)), the average (along x) bed-load rate q_{sb} can be expressed in terms of the developed two-dimensional bed form properties as

$$q_{sb} = \alpha_s U_b \Delta \qquad (\text{with } \alpha_s \approx 1/2). \tag{3.31}$$

This relation indicates that if η is just larger than unity, and thus if q_{sb} is just larger than zero, then Δ, and consequently Δ/Λ, must also be just larger than zero (for $U_b \sim v_*$ is always finite ($v_* > v_{*cr} \gg 0$)). Hence the ideal bed form steepness $(\Delta/\Lambda)_*$, which has a certain finite value, cannot be achieved for the values of η which are just larger than unity.

If Δ/Λ is small and consequently $c/\bar{c} \approx 1$, then $q_s = q_{sb}$ can be given by the flat-bed version of Bagnold's bed-load formula (1.57), viz

---

[13] We have $(\Delta/\Lambda)_a = \delta_a \leq 0.02$ (Fig. 3.30).

[14] The development of bed forms cannot be attributed to $Fr \to \min$, at least because they occur also in flows which do not have a free surface.

$$q_{sb} = \beta \frac{u_b}{\gamma_s} [\tau_0 - (\tau_0)_{cr}] . \quad (3.32)$$

Identifying (3.32) with (3.31), treating Λ as independent of η and considering that $U_b \sim v_* \sim u_b$, one determines

$$\frac{\Delta}{\Lambda} \sim (\eta - 1) , \quad (3.33)$$

which indicates that when η is just larger than unity the steepness Δ/Λ of the developed bed forms increases as a linear function of η (straight line σ in Fig. 3.13). With the increment of Δ/Λ, this linear relation must progressively deteriorate, for if the bed is no longer flat, then q_{sb} varies also with $c/\bar{c}$, which, in turn, decreases with the increment of Δ/Λ. Consequently, the straight line σ must degenerate into a convex curve, l_{01} say (Fig. 3.13). Eventually, at a certain $\eta = \eta_1$, the curve l_{01} reaches the level of the ideal steepness, and for $\eta > \eta_1$ we have $(\Delta/\Lambda) = (\Delta/\Lambda)_*$ (horizontal line segment l_{12} in Fig. 3.13). However, the ideal steepness too, cannot be maintained indefinitely, and the reason for this can be explained as follows.

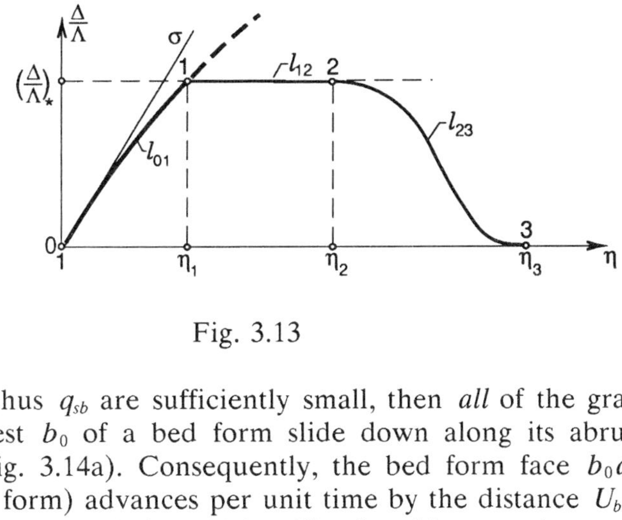

Fig. 3.13

When η and thus q_{sb} are sufficiently small, then *all* of the grains transported over the crest b_0 of a bed form slide down along its abrupt downstream face b_0a (Fig. 3.14a). Consequently, the bed form face b_0a (and, in fact, the whole bed form) advances per unit time by the distance U_b, the area Ω "swept" in the process being $U_b\Delta$. Clearly, this area is equal to the transport rate $(q_{sb})_{max}$ at the dune crest, and one can write

$$(q_{sb})_{max} = U_b \Delta . \quad (3.34)$$

If, on the other hand, η and thus $q_{sb} = \alpha_s(q_{sb})_{max}$ are not sufficiently small, then *not all* of the grains transported over the bed form crest slide down its downstream face b_0a – some of them "fly" forward, and land on the upstream face of the next bed form, or even further (Fig. 3.14b). As a result of this, instead of the area Ω (per unit time) we have a smaller area Ω', and consequently instead of U_b, a smaller migration velocity U_b'. The area-difference $\Omega - \Omega'$ is transferred into the deposition area $\omega = \omega_1 + \omega_2$, where the (larger) part ω_2 extends along the upstream face of the next bed form, the

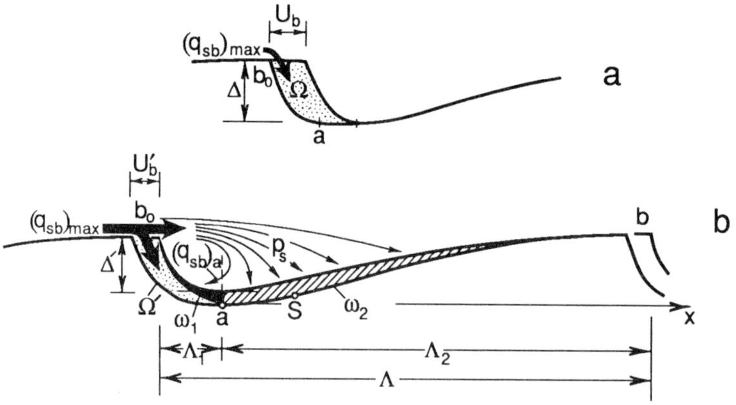

Fig. 3.14

smaller (black) part ω_1 being on the downstream face b_0a. Since the landed material is concentrated mainly around the stagnation point S, the largest thickness of the area ω is in the neighbourhood of S. Clearly, the emergence of ω is accompanied by the reduction of the bed form height (from Δ to Δ'), and thus of the bed form steepness. But this means that the ideal steepness $(\Delta/\Lambda)_*$ cannot be maintained for η larger than a certain η_2, for which some of the transported grains begin to fly to the next bed form: the value of Δ/Λ must progressively decrease with the increasing values of η ($> \eta_2$), as shown schematically in Fig. 3.13 by the curve l_{23}. Eventually, when $\eta = \eta_3$, say, all of the grains fly forward. In this case the bed forms vanish ($\Delta \to 0$) and we have the "flat bed at advanced stages".[15]

The conditions described above can be formulated in the following manner. Consider again the transport continuity equation (1.76) and apply it to the developed dunes ($V_b = dy_b/dt = 0$) corresponding to $\eta_2 < \eta < \eta_3$. In this case, p_s is no longer zero: it varies as a function of x ($p_s = p_s(x)$). Hence, Eq. (1.76) is of the form

$$\frac{\partial q_{sb}}{\partial x} = U_b \frac{\partial y_b}{\partial x} + p_s(x) . \tag{3.35}$$

Taking the origin of x at the trough a (Fig. 3.14b) and integrating (3.35) along $0 < x < \Lambda_2$, one obtains

$$(q_{sb})_{\max} - (q_{sb})_a = U_b\Delta + \int_0^{\Lambda_2} p_s(x)\,dx . \tag{3.36}$$

Here $(q_{sb})_{\max}$ and $(q_{sb})_a$ are the bed-load rates passing through the vertical sections at the crest and trough respectively, the integral on the right being

---

[15] In reality, the transition from each line to the next in Fig. 3.13 occurs "smoothly"; i.e. the line complex $[l_{01}, l_{12},$ and $l_{23}]$ is a single continuous curve (see bed form steepness plots in Section 3.5).

that part ω_2 of the total deposition area ω which is to the right of a. Consider now the area ω_1 which is to the left of a. It is due to the particles forming the (leftward-directed) transport rate $(q_{sb})_a$:[16]

$$(q_{sb})_a = \int_{-\Lambda_1}^{0} p_s(x)dx = \omega_1 . \tag{3.37}$$

Consequently, (3.36) implies

$$(q_{sb})_{max} = U_b\Delta + \omega , \tag{3.38}$$

where

$$\omega = \omega_1 + \omega_2 = \int_{-\Lambda_1}^{0} p_s(x)dx + \int_{0}^{\Lambda_2} p_s(x)dx = \int_{\Lambda} p_s(x)dx . \tag{3.39}$$

Since ω is due to $p_s(x)$, while $p_s(x)$ is due to $(q_{sb})_{max}$, the value of ω is an increasing function of $(q_{sb})_{max}$. Owing to dimensional reasons, the deposition area per unit time ω must, in fact, be proportional to $(q_{sb})_{max}$, the dimensionless proportionality factor being determined by the transport intensity η. i.e.

$$\omega = \phi(\eta)(q_{sb})_{max} \tag{3.40}$$

where the function $\phi(\eta)$, which is unknown at present, must have the property

$$\phi(\eta) = 0 \text{ if } \eta < \eta_2 \quad \text{and} \quad \phi(\eta) = 1 \text{ if } \eta > \eta_3 . \tag{3.41}$$

Using (3.40), one can express (3.38) as

$$[1 - \phi(\eta)]\,(q_{sb})_{max} = U_b\Delta . \tag{3.42}$$

When $\eta < \eta_2$, then $\phi(\eta) = 0$ and (3.42) reduces into its special form (3.34). When $\eta > \eta_3$, then $\phi(\eta) = 1$ and (3.42) reduces into $U_b\Delta = 0$, implying the flat bed at advanced stages.

The relation (3.42) may help to realize that the migration velocity U_b is not an independent characteristic of sediment transport as the bed-load rate q_{sb} or the bed form height Δ, say. Indeed, if $\eta < \eta_2$ (and $\phi(\eta) = 0$), then U_b is completely determined by $q_{sb} \approx$ $\approx (q_{sb})_{max}/2$ and Δ; only if $\eta > \eta_2$ the knowledge of the function $\phi(\eta)$ is required. This is why we are not concerned with U_b in the present text. It should be mentioned, however, that in some of the (comparatively recent) works U_b is given as a function of Fr (rather than η), although it is well known that ripples and dunes occur, and migrate, also in deserts and closed conduits.

---

[16] In Fig. 3.14, $(q_{sb})_a$ appears to be negative only because its direction opposes the flow direction x; not because it creates erosion instead of deposition. Hence $(q_{sb})_a$ (which generates the deposition ω_1) is treated as positive.

3.4 Existence Regions of Bed Forms

3.4.1 *Formulation of existence regions*

i- Since the bed form length Λ is always finite, a bed form "exists" if it has a non-zero height Δ, and thus a non-zero steepness Δ/Λ. This is the case when η satisfies $1 < \eta < \eta_3$ (see 3.3.3). However, as should be clear from 3.1.2, the existence of a bed form depends also on the values of (some of) the parameters

$$\frac{B}{h} \;;\quad Z = \frac{h}{D} \;;\quad X = \frac{Dv_*}{\nu} \;;\quad Fr = \frac{v^2}{gh}\,. \qquad (3.43)$$

In the present context, the value of the Froude number Fr merely indicates whether the bed forms whose length is proportional to h are dunes or antidunes; therefore, in the following, attention will be focused on the role of the first three variables in (3.43).

ii- Consider the three-dimensional space formed by the coordinates B/h, Z and X (Fig. 3.15). The characteristics of the $(B/h; Z)$-plane P must, in general, be expected to vary depending on its location on the X-axis. From the content of preceding sections, it should be clear that the role of X on the formation of steepness of the larger bed forms (dunes and bars) is of an indirect nature: it is via ripples which occur for small $X \sim \nu^{-1}$. Ripples, which come into being soon after $t = 0$, convert the initial bed roughness $k_s \sim D$ into the ripple roughness $K_s \sim \Delta_r$ (3.1.3 (iv)). Thus, the (subsequent) development of larger bed forms takes place virtually past the bed covered by ripple roughness. Clearly, the increment of $K_s \sim \Delta_r$ must lead to the reduction of prominence of the larger bed forms (see 3.1.3 (iv)). One can say that the (earlier developed) ripples "protect" the bed against any further deformation. Experiment shows [8], [38] that when $X < \approx 2.5$ (hydraulically smooth flow past the flat bed at $t = 0$), then it is mainly ripples which are present on the bed surface, the steepness of larger bed forms being comparatively small (plane P in the position P_1). When $\approx 2.5 < X < \approx 35$ (transitional regime at $t = 0$), we have ripples superimposed on larger bed forms (plane P in the position P_2) – with the increment of X the prominence of ripples decreases; that of the larger bed forms increases. When $X > \approx 35$ (rough turbulent flow at $t = 0$), ripples are no longer present and the larger bed forms have their maximum prominence (plane P in the position P_3).

iii- Consider now the larger bed forms in the (rough turbulent) plane P_3 (Fig. 3.15) – we begin with bars. It has been shown in 2.3.5 that if the flow is rough turbulent, then the regions of N- and $(N + 1)$-row horizontal bursts in

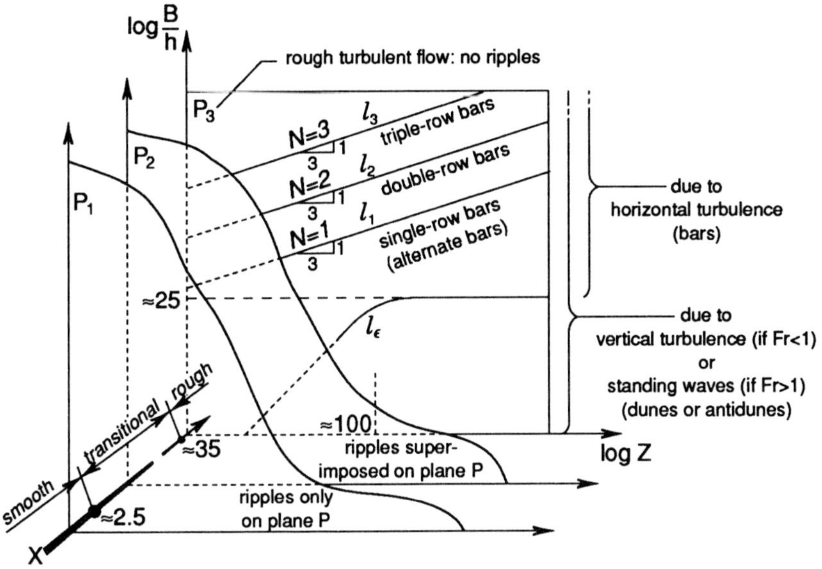

Fig. 3.15

the log-log (B/h; Z)-plane are separated from each other by the 1/3-inclined straight lines l_N representing the function (2.38), i.e.

$$\frac{B}{h} = \overline{\psi}(N)Z^{1/3}.^{17} \tag{3.44}$$

Clearly, the lines l_N are, at the same time, the boundaries separating the regions of N- and $(N+1)$-row bars. The line l_ϵ, which represents the function (2.27) (viz $\overline{\psi}_\epsilon(Z)$), is the lower boundary of the alternate bar region.

The lengths of dunes and antidunes are comparable, and therefore they are mutually exclusive. Yet neither of them is mutually exclusive with alternate bars (for $\Lambda_a \sim B \gg h \sim \Lambda_d$ or Λ_g), and experiment shows [18] that alternate bars can occur indeed with dunes *or* antidunes superimposed on them. The explanation below is given in terms of dunes.

Since the earlier occurring ripples "protect" the bed against any further deformation, they render the bars and dunes less prominent (and the more so, the steeper are ripples, i.e. the smaller is X). No ripples are present, of course, in the rough turbulent plane P_3. Yet the bed which will subsequently be covered by bars may be covered first by the (earlier developed) dunes. Indeed, since $B/h \approx \Lambda_a/\Lambda_d$ is large, then T_a/T_d is large too (in analogy to (3.22)), and the bed may already be covered by the nearly developed dunes when the formation of bars has only just started. Clearly, the steeper the earlier developing dunes, the flatter are the subsequently forming bars. On the other hand, the steepness of dunes progressively decreases with $D \sim Z^{-1}$ (3.3.2 (ii)) and when $Z < \approx 30$, say, they are barely detectable (see Fig. 3.19).

---

17 h/k_s is replaced by $h/D = Z$, hence $\overline{\psi}$ instead of ψ.

Consequently, the steepness of bars must increase with $D \sim Z^{-1}$; and, as is well known, bars forming on a coarse material usually are more prominent.[18] The zone below l_ϵ is the zone of comparatively large flow depths: horizontal eddies are not rubbing the bed and the action of vertical bursts dominates. In this zone, like anywhere else on P_3 where Z is sufficiently large, dunes are present if $Fr < \approx 1$; and antidunes if $Fr > \approx 1$.

If B/h is very large, then we have a non-zero region b_c where only vertical bursts are present (see Fig. 2.24). Clearly, no bars can occur within b_c where no large-scale horizontal turbulence is present: the bed forms are either dunes or antidunes (as in the region below l_ϵ).

3.4.2 *Experimental determination of existence regions*

All the available data (a total of 507 data-points from 27 sources) are plotted on the log-log $(B/h, Z)$-plane, i.e. $(B/h; h/D)$-plane, in Fig. 3.16. The points A, C and D in Fig. 3.16 represent alternate bars, multiple bars and dunes respectively.[19] No distinction was made by the authors of the data as to whether the multiple bars were double-, or triple-row bars, etc., and therefore they are all marked by C.[20]

As can be seen from Fig. 3.16, the points A, C and D exhibit a clear tendency to form "their own" regions, although their "diffusion" from one region to another is substantial. In order to reveal the upper and lower limits of the alternate bar region more accurately, a series of special experiments was carried out (Ref. [40]). The encircled additional points, as well as the transitional points T, obtained from these experiments were of considerable

---

18 The quantitative relation representing the decrement of alternate bar steepness with Z is determined (from the data) in 3.5.3.

19 The sources are too numerous to assign a special symbol for each of them. On the other hand, the utilization of only A, C and D makes it possible to see more clearly how alternate bars, multiple bars and dunes group themselves on the $(B/h; h/D)$-plane.
The data sources are given in "References A" and "B":
Flume data:
Point-symbol A ($0.38mm \leq D_{50} \leq 6.40mm$): [1b], [2b], [3b], [4b], [5b], [6b], [7b], [9b], [11b], [12b] (=[40]), [13b].
Point-symbol C ($0.61mm \leq D_{50} \leq 1.71mm$): [5b], [9b], [11b], [12b] (=[40]).
Point-symbol D ($0.25mm \leq D_{50} \leq 1.71mm$): [2a], [5a], [10a], [12a], [17a], [31a], [32a], [36a], [37a].
River data:
Point-symbols A and C: From Ref. [8b] (B/h and h/D given, D-range unspecified). The plot contains also the data from various Japanese rivers – kindly provided by Dr. S. Ikeda (Tokyo Institute of Technology): D-range is too wide, hence not given here.
Point-symbol D ($0.20mm \leq D_{50} \leq 0.56mm$): [1a], [13a], [14a], [19a], [27a].

20 Hence the available data do not permit the determination of the existence regions of bars corresponding to $N > 1$.

help in locating the "levels" of the lines forming the limits mentioned. Thus, the following relations have been established:

i) the line l_1 (upper limit of the alternate bar region),

$$\frac{B}{h} = 25.0\left(\frac{h}{D}\right)^{1/3} ; \tag{3.45}$$

ii) the line l_ϵ (lower limit of the alternate bar region),

$$\text{if } \left(\frac{h}{D}\right) < \approx 100 \text{ then } \frac{B}{h} = 0.25\,\frac{h}{D}$$

and (3.46)

$$\text{if } \left(\frac{h}{D}\right) > \approx 100 \text{ then } \frac{B}{h} = 25.$$

The lower limit of the region of alternate bars is much vaguer than its upper limit – it is more a "transitional ribbon" than a line, l_ϵ being but the central line of that ribbon.[21]

Most of the data plotted in Fig. 3.16 correspond to rough turbulent flows ($X > \approx 35$). Yet the points corresponding to $X < \approx 35$ (unmarked in Fig. 3.16) did not tend to locate themselves "differently" on the $(B/h;\, h/D)$ -plane; and any influence of X on the lines l_1 and l_ϵ could not have been

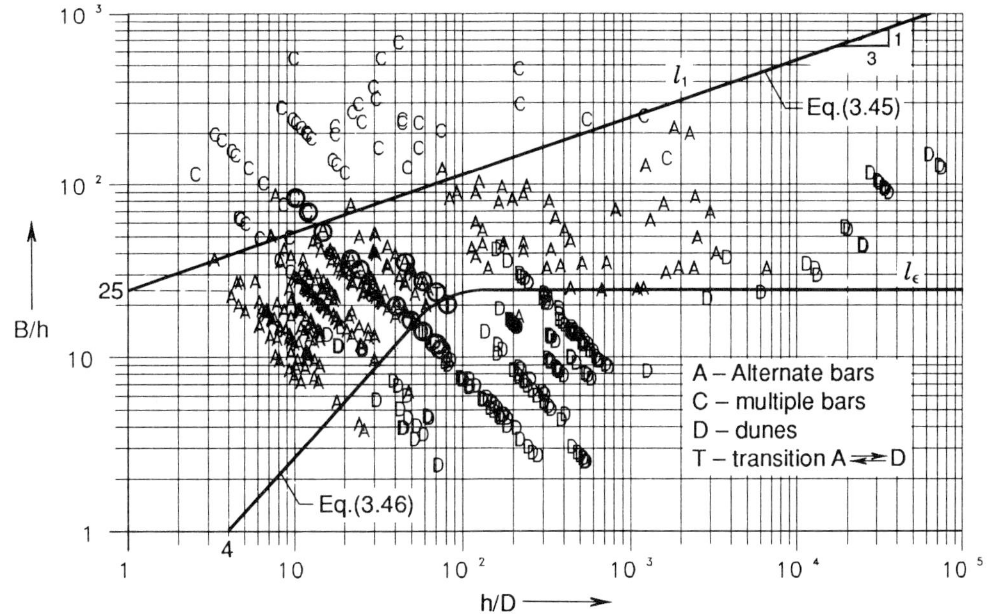

Fig. 3.16 (from Ref. [40])

[21] The theoretical reason for l_ϵ to be a 1/1-inclined straight line when Z is small is given in Ref. [48].

detected. This is not surprising, for it is hardly likely that the plan configuration and thickness of the large-scale horizontal eddies e_H (which determine l_N and l_ϵ) should be affected much by the earlier developing dunes and/or ripples. (More information on the plot in Fig. 3.16 and its characteristics can be found in Ref. [40]).

The existence regions of bed forms has been a popular topic for a long time, the recent contributions to it being mostly by Japanese researchers. Apparently, it was first Sukegawa 1972 [41] who introduced the combination $S(B/h)$, which was subsequently referred to as "channel form index" (Ikeda 1973 [19]). Using this combination, in conjunction with $\eta = (v_*/v_{*cr})^2$, Sukegawa attempted to reveal the existence regions of bed forms by plotting the data on the plane $[\eta\,; [\eta S(B/h)]^{1/2}]$. The resulting plot was only moderately successful (see Fig. 2 in [41]) − in spite of the fact that the spurious correlation due to the presence of η in both abcissa and ordinate has contributed considerably to its visual appearence. Ikeda 1973 [19], and Kishi and Kuroki 1975 [28], have also used η and $S(B/h)$ (without "mixing" them with each other, however) to determine the existence regions. The zone-boundaries obtained in the aforementioned three works are not quite the same: they vary depending on the data used.[22] But this means that the boundaries separating various types of bed forms cannot be "*immobilized*" on the plane $[\eta; S(B/h)]$. And this is because the transport-related parameter η (which, for a given flow, varies depending on material ($\gamma_s D$)) has hardly any influence on the formation of (flow-induced) bursts, standing waves, etc., which determine the bed form types. The value of η merely indicates whether a given flow is capable of transporting the sediment and thus of imprinting a certain bed form; and what the steepness of the imprint might be − it cannot, however, throw any light on the *type* of that "certain" bed form.

It seems that the parameter η was first discarded by Parker 1976 [35], who introduced the (flow-related) Froude number instead of η. According to Ref. [35], the zone-separating lines (l_N) should be given by the proportionalities $Fr \sim [S(B/h)]^2$, which are equivalent to $c \sim S^{1/2}(B/h)$ (for $Fr = c^2 S$ (see (1.31)). But, according to [41], [19] and [28] we have $\eta \sim gSh/v_{*cr}^2 \sim [S(B/h)]^m$ and thus $(gh/v_{*cr}^2)^{1/m} \sim (B/h)/S^{(1/m-1)}$ (where $1/m > 1$ and consequently $((1/m) - 1) > 0$). Hence, according to [35] the ratio B/h must be multiplied by a positive power of S, whereas according to [41], [19] and [28] it must be divided by it. Since all of the proportionalities above were "supported" by the data, the discrepancy between them should be due mainly to the utilization of S which, in fact, is *not* a parameter determining the type of bed forms. This was realized by Muramoto and Fujita 1978 [32] who have used only the dimensionless variables B/h and $Z = h/D$, and who have expressed the zone-separating lines by the proportionalities $(B/h) \sim Z^{1/2}$. (Note that their empirical exponent 1/2 is comparable with the present 1/3). More recently, Hayashi and Ozaki 1980 [15] have proposed, for the zone-separating lines, the proportionalities $Fr \sim S(B/h)$. Here S is an "apparent parameter"; for $Fr = c^2 S$, and $Fr \sim S(B/h)$ implies $c^2 \sim (B/h)$. In the case of a rough turbulent flow, where $c \sim (h/k_s)^{1/6}$, the proportionalities of Ref. [15] acquire the form $B/h \sim (h/k_s)^{1/3}$, which is the same as the present expression (2.43) of the zone-separating lines. Fig. 3.17 is the reproduction of the original plot published by the Public Works Research Institute, Ministry of Construction of Japan 1982 [36] (see also [25]). Observe that this plot is very similar to the present plot in Fig. 3.16: even the zone boundaries resemble each other. Finally, it should be mentioned that in Znamenskaya 1976 [59] the alluvial forms were also classified by using only the flow related parameters, viz $(B/h)c^2$ and h/D.

---

[22] Each of the three works mentioned proposes a power relation $\eta \sim [S(B/h)]^m$ to separate various bed form regions. According to [41], [19] and [28] the value of m is 1/4, 1/3 and 0.4 respectively.

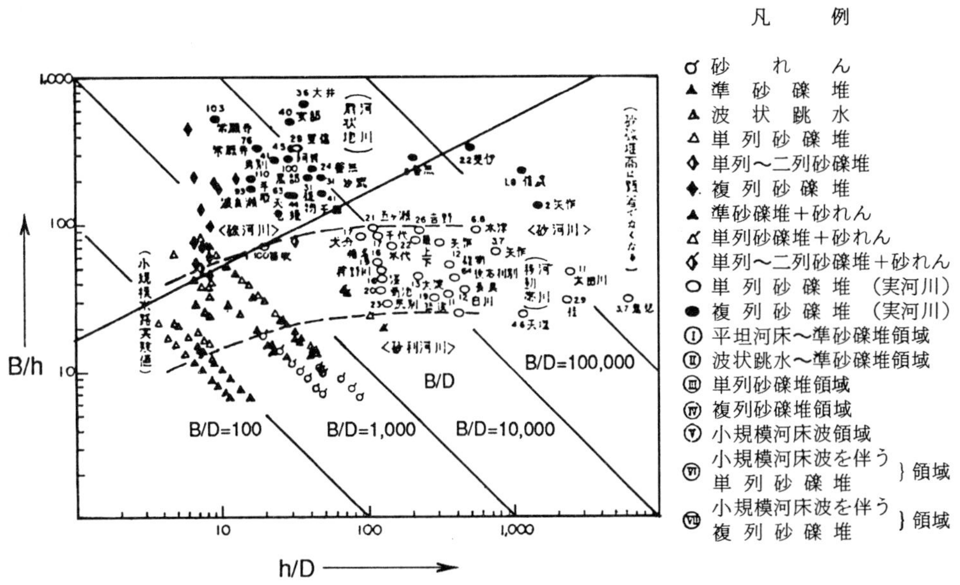

Fig. 3.17 (from Ref. [36])

3.5 Experimental Data and Bed Form Equations

This section concerns the relations which can be used to compute the length and steepness of dunes, ripples and alternate bars. These relations were determined from the available data.

3.5.1 *Dunes*

i- *Dune length*

According to 3.1.2 (ii), the dune length Λ_d is given by (3.11), i.e. by

$$\Lambda_d = \Phi_d(X, Z)\, 6h. \tag{3.47}$$

If the flow past the flat initial bed is rough turbulent, then $\Lambda_d \approx 6h$, i.e. $\Phi_d(X, Z) \approx 1$. The curve family in Fig. 3.18, which was determined from the voluminous data of various sources (given in Ref. [54]), can be considered as the graph of the function $\Phi_d(X, Z)$ (for the ratio of the coordinates in Fig. 3.18, viz $(\Lambda_d/D)/Z$, is equal to $6\Phi_d(X, Z)$). The following comp-eq. can be used for the evaluation of $\Phi_d(X, Z)$:

$$\Phi_d(X, Z) = 1 + 0.01\,\frac{(Z-40)(Z-400)}{Z}\,e^{-m} \tag{3.48}$$

$$\text{(with } m = 0.055\sqrt{Z} + 0.04X\text{)}.$$

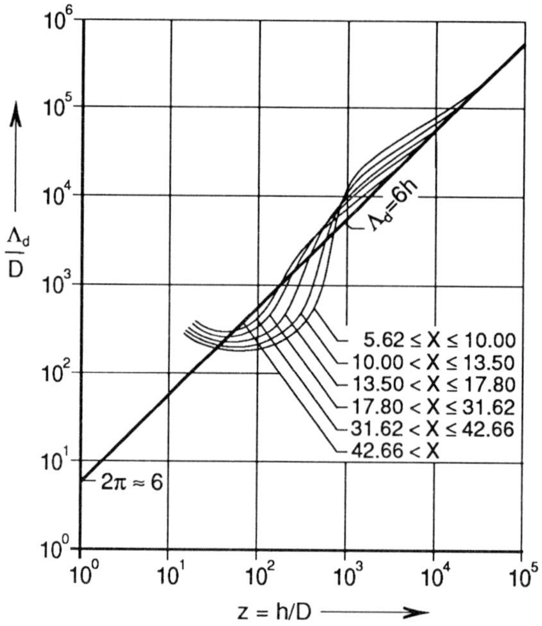

Fig. 3.18 (from Ref. [54])

ii- *Dune steepness*

From 3.3.3 and 3.3.2 it should be clear that the dune steepness $(\Delta/\Lambda)_d = \delta_d$ varies as a function of the dimensionless flow intensity η, and its value should decrease with the increment of the relative skin roughness $k_s/h \sim Z^{-1}$ (3.3.2 (ii)). Furthermore, the value of δ_d must be affected by the grain size Reynolds number X (for with the decrement of X, the steepness of the (earlier occurring) ripples increases, and the prominence (δ_d) of dunes decreases). Hence, in general, the dune steepness must be a function of η, Z and X, its unimpeded (by the ripple roughness) value being a function of η and Z only. Hence, the dune steepness δ_d can be given by the form

$$\delta_d = \psi_d(X, \eta, Z) = \psi_{Xd}(X)\Psi_d(\eta, Z), \tag{3.49}$$

which reduces to the expression of the unimpeded dune steepness

$$\delta_d = \Psi_d(\eta, Z) \tag{3.50}$$

when X is "sufficiently large" ($X > \approx 25$, say)[23] and $\psi_{Xd}(X) \equiv 1$.

1- Fig. 3.19, which corresponds to $X > \approx 25$, shows the family of experimental curves which can be taken to imply (3.50). Note from this graph that the absolute maximum of the dune steepness is ≈ 0.06, and that it occurs (for

---

[23] Strictly speaking, the unimpeded dune steepness must be expected to be present when there are no ripples at all (i.e. when $X > \approx 35$; rough turbulent flow past the flat initial bed (3.4.1)).

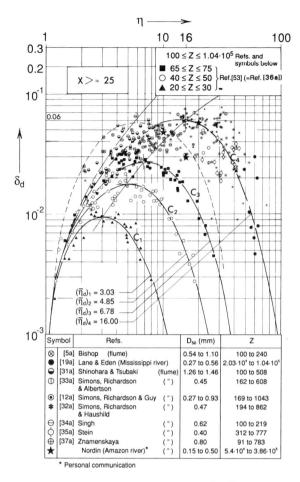

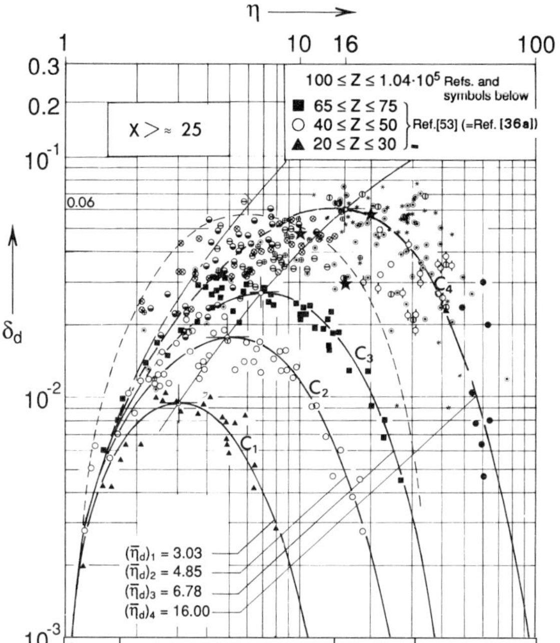

Symbol	Refs.		D_{50} (mm)	Z
⊗	[5a] Bishop (flume)		0.54 to 1.10	100 to 240
●	[19a] Lane & Eden (Mississippi river)		0.27 to 0.56	$2.03\cdot10^4$ to $1.04\cdot10^5$
◒	[31a] Shinohara & Tsubaki	(flume)	1.26 to 1.46	100 to 508
⦶	[33a] Simons, Richardson & Albertson	(")	0.45	162 to 608
⊙	[12a] Simons, Richardson & Guy	(")	0.27 to 0.93	169 to 1043
✱	[32a] Simons, Richardson & Haushild	(")	0.47	194 to 862
⊖	[34a] Singh	(")	0.62	100 to 219
⏀	[35a] Stein	(")	0.40	312 to 777
⊕	[37a] Znamenskaya	(")	0.80	91 to 783
★	Nordin (Amazon river)*	(")	0.15 to 0.50	$5.4\cdot10^4$ to $3.86\cdot10^5$

* Personal communication

Fig. 3.19 (from Ref. [53])

$X > \approx 25$ and $Z > \approx 100$) when $\eta \approx 16$. In Ref. [53] it has been found that the curves C_i in Fig. 3.19 can be normalized by introducing

$$n_d = \frac{\delta_d}{(\delta_d)_{\max}} \quad \text{and} \quad \zeta = \frac{\eta - 1}{\bar{\eta}_d - 1}\,, \tag{3.51}$$

where $(\delta_d)_{\max}$ is the largest δ_d of each C_i-curve and $\bar{\eta}_d$ is the abcissa of $(\delta_d)_{\max}$. Indeed, using (3.51), it was possible to arrive at the unified plot shown in Fig. 3.20. Here the scatter is substantial; yet the points of different symbols do not form "their own" patterns. Hence n_d can be treated as a function of only ζ ($n_d = f_d(\zeta)$), and in Ref. [53] the following form was proposed (on the basis of theoretical considerations) for this function:

$$f_d(\zeta) = \zeta e^{1-\zeta}. \tag{3.52}$$

The solid curve in Fig. 3.20 is the graph of (3.52). Observe that in the limit $\zeta \to 0$, the function $f_d(\zeta)$ reduces into the straight line ζe (asymptote in Fig. 3.20). Thus, when $\zeta \sim (\eta - 1) \to 0$, then δ_d' becomes proportional to $(\eta - 1)$;

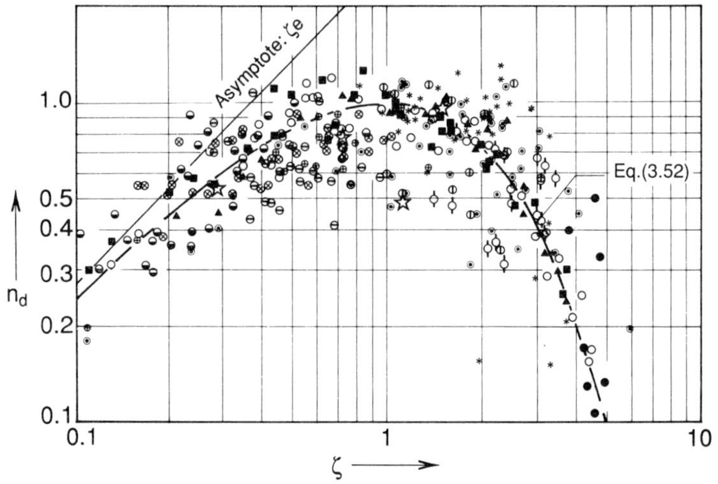

Fig. 3.20 (from Ref. [53])

and this is consistent with the relation (3.33) derived on the basis of Bagnold's formula (straight line σ in Fig. 3.13). Further data,[24] plotted in Fig. 3.21, also appear to be in a reasonable agreement with Eq. (3.52).[25] The quantities $(\delta_d)_{max}$ and $\bar{\eta}_d$ must be expected to vary as functions of Z. Using the data at the tops of the $Z = const_i$ curves (viz, the curves C_i in Fig. 3.19), the point patterns shown in Figs. 3.22 and 3.23 were produced. The solid lines in these plots are the graphs of the comp-eqs.

$$(\delta_d)_{max} = 0.06\,(1 - e^{-0.008Z}) \tag{3.53}$$

and

$$\bar{\eta}_d = 14\,(1 - e^{-0.003Z}) + 2\,. \tag{3.54}$$

---

[24] The sources of the data plotted in Figs. 3.21 to 3.23 and 3.25 to 3.27 (all of which are from [39]), as well as in Fig. 3.28, are given in "References A". The point-symbols in these plots are "numbers" which increase with the grain size D as indicated below.
Point-symbol 1 ($0.02mm \le D_{50} \le 0.04mm$): [13a], [22a], [23a].
Point-symbol 2 ($0.10mm \le D_{50} \le 0.30mm$): [1a], [3a], [4a], [8a], [13a], [16a], [17a], [20a], [21a], [22a], [26a], [27a], [28a], [29a].
Point-symbol 3 ($0.32mm \le D_{50} \le 0.47mm$): [11a], [13a], [25a], [26a], [27a].
Point-symbol 4 ($0.51mm \le D_{50} \le 0.93mm$): [2a], [7a], [14a], [18a], [28a], [29a].
Point-symbol 5 ($1.00mm \le D_{50} \le 5.10mm$): [6a], [9a], [10a], [15a], [18a], [24a], [30a].

[25] Future research may reveal that a curve with a longer "plateau" (such as, for example, C in Fig. 3.21), may reflect n_d more realistically than Eq. (3.52).

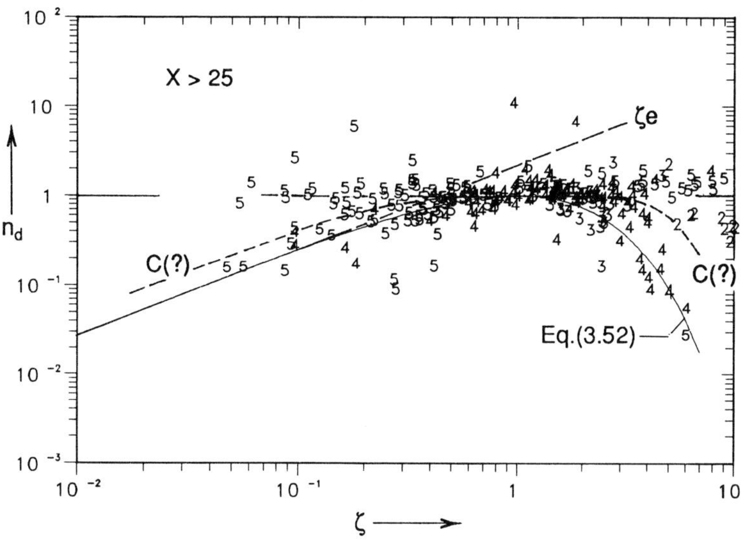

Fig. 3.21

Note that the values of $(\delta_d)_{max}$ and $\bar{\eta}_d$ given by these equations (which correspond to $X > \approx 25$) rapidly approach 0.06 and 16, respectively, with the increasing values of Z.[26] It follows that the unimpeded dune steepness can be expressed as

$$\delta_d = \Psi_d(\eta , Z) = (\delta_d)_{max} f_d(\zeta) , \tag{3.55}$$

where $(\delta_d)_{max}$ and $f_d(\zeta)$ are determined by (3.53) and (3.52) respectively.

2- No special measurements were carried out, to the author's knowledge, to reveal how the dune steepness decreases with the decreasing values of X (only). An indirect determination of this decrement from the available data is virtually impossible, for most of the data is due to experiments where only one (dominant) mode of bed forms was measured. Similarly, the number of available bed profiles obtained by means of bed-level-plotters (which correspond to various X (from $\approx 2.5 < X < \approx 20$) and yet to some constant η and Z) is also too limited to permit the determination of the function $\psi_{Xd}(X)$. Hence, at present, one has no alternative but to identify $\psi_{Xd}(X)$ with a non-decreasing continuous function which is zero when $X = 0$ and which is (nearly) unity when $X > \approx 35$. The comp-eq.

$$\psi_{Xd}(X) = 1 - e^{-(X/10)^2} \tag{3.56}$$

is such a function. The analogous decrement of the ripple steepness δ_r with the increasing X can be reflected by

[26] Since $(\delta_d)_{max}$ and $\bar{\eta}_d$ are functions of only Z, the product $n_d(\delta_d)_{max} = \delta_d$ is thus a function of ζ and Z or, which is the same, of η and Z (as required by $\delta_d = \Psi_d(\eta , Z)$).

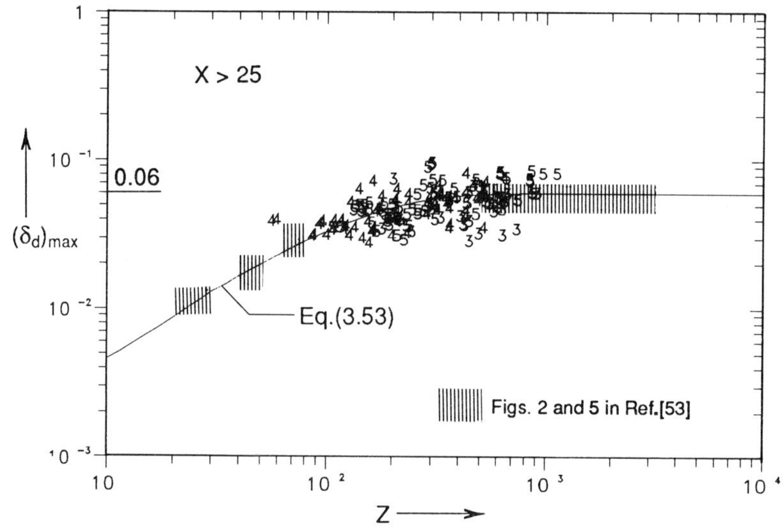

Fig. 3.22

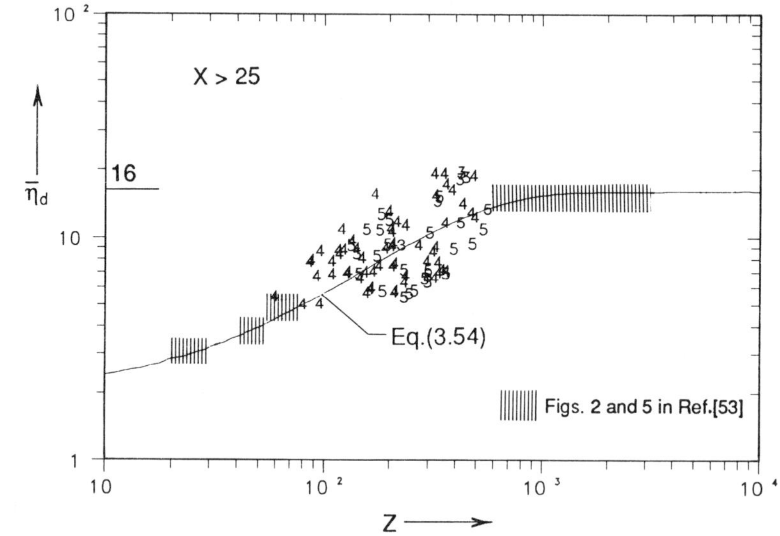

Fig. 3.23

$$\psi_{Xr}(X) = e^{-((X-2.5)/14)^2} \text{ if } X > 2.5\,, \quad \text{and} \quad \psi_{Xr}(X) \equiv 1 \text{ if } X < 2.5\,, \qquad (3.57)$$

where 2.5 is the X-value which signifies the upper limit of the hydraulically smooth regime (of the flow past the flat initial bed at $t = 0$). The expressions (3.56) and (3.57) proved to give realistic results when used alongside other relations of this section in computations of the friction factor c.

3.5.2 *Ripples*

i- *Ripple length*

We have, according to (3.19),

$$\Lambda_r = \Phi_r(X, \xi)D. \tag{3.58}$$

Consider the plot in Fig. 3.24 where the experimental values of Λ_r/D are plotted versus X. This plot is consistent with (3.58): the data points sort themselves out depending on the constant values of ξ. As will be seen in the

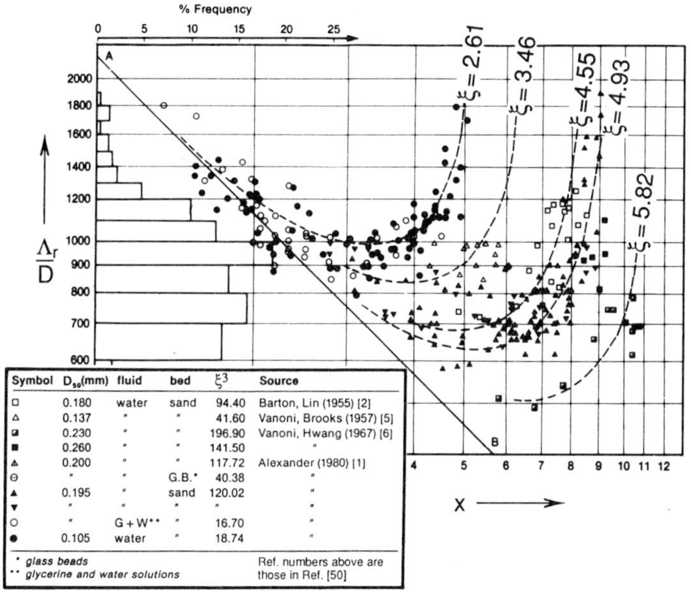

Fig. 3.24 (from Ref. [50])

next paragraph, ripples disappear, their steepness $(\Delta/\Lambda)_r$ reduces to zero, when $(\eta - 1) \approx 20$. This means that the X-values at the (upward bending) right-hand ends of the individual point patterns $\xi = const_i$ must be in coincidence with $\eta \approx 21$. Hence the abcissa of the right-hand ends will be unified if Λ_r/D is plotted versus $\sim \eta^m$ (rather than versus X).[27] Similarly, the left-hand ends of the individual patterns in Fig. 3.24 can be brought together by taking into account the fact that they are situated on the common straight line AB. The plot of the ripple data (mainly symbols 1 and 2 of the data-set in footnote 24) produced along these lines is shown in Fig. 3.25. The solid line passing through the (unified) pattern is the graph of the comp-eq.

---

[27] The consideration of η and ξ instead of X and ξ is obviously permissible, for $X = \sqrt{\xi^3 Y} = \sqrt{\xi^3 \Psi(\xi) \eta}$ (see (1.12) and (1.14)).

$$\frac{\Lambda_r}{D} = \Phi_r(X, \xi) = \overline{\Phi}_r(\eta, \xi) = \frac{(\Lambda_r)_{min}}{D}[4H(1 - H)]^{-1}, \qquad (3.59)$$

where

$$\frac{(\Lambda_r)_{min}}{D} = \frac{2650}{\xi^{0.88}} \quad \text{and} \quad H = \sqrt{\frac{\eta}{\eta_{max}}} = \sqrt{\frac{\eta}{21}}. \qquad (3.60)$$

$(\Lambda_r)_{min}/D$ is the smallest relative length of ripples that occurs for the specified ξ: more detailed information on the determination of (3.59) and (3.60) can be found in Ref. [39].

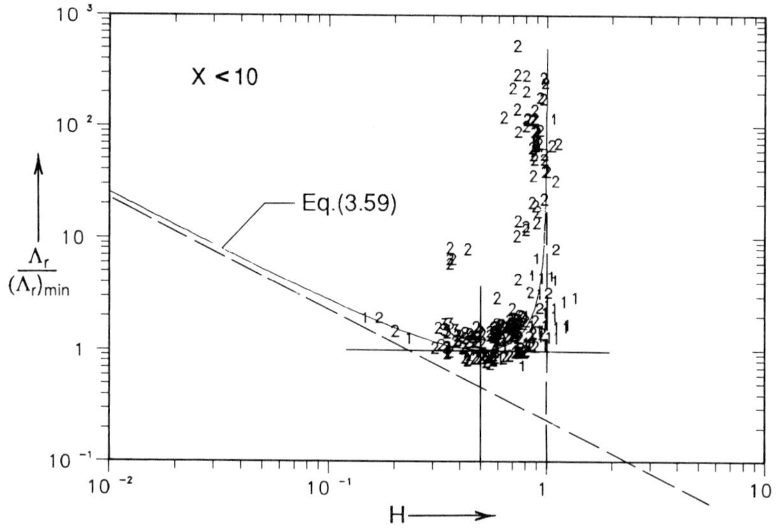

Fig. 3.25

ii- *Ripple steepness*

The geometry of ripples does not depend on the flow depth (see 3.1.2 (iii)).[28] Hence, $Z \sim h$ cannot be a variable in the expression of $(\Delta/\Lambda)_r = \delta_r$:

$$\delta_r = \psi_r(X, \eta) = \psi_{Xr}(X)\,\Psi_r(\eta). \qquad (3.61)$$

The form of the function $\psi_{Xr}(X)$ has already been discussed (see (3.57)), and it remains to reveal $\Psi_r(\eta)$ only. The experimental values of the unimpeded ripple-steepness

$$\delta_r = \Psi_r(\eta) \qquad (3.62)$$

[28] The only (insignificant) case when h might be of relevance will be discussed at the end of the paragraph.

(which can be taken as to correspond to $X < \approx 10$, say) are plotted versus $(\eta - 1)$ in Fig. 3.26. Note from this graph that

$$(\delta_r)_{max} \approx 0.14 \ , \quad \bar{\eta}_r - 1 \approx 10 \quad \text{and} \quad (\eta_r)_{max} - 1 \approx 20 \, , \tag{3.63}$$

where $\bar{\eta}_r$ and $(\eta_r)_{max}$ are the η-values which correspond to $\delta_r = (\delta_r)_{max}$ and

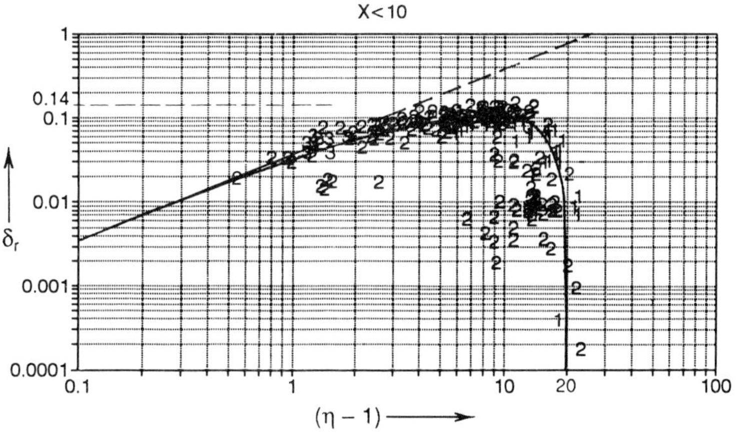

Fig. 3.26

$\delta_r \approx 0$ (at advanced stages) respectively. Introducing the normalized coordinates

$$n_r = \frac{\delta_r}{(\delta_r)_{max}} \quad \text{and} \quad \zeta' = \frac{\eta - 1}{\bar{\eta}_r - 1} \tag{3.64}$$

(where $\bar{\eta}_r - 1 \approx 10$), one may attempt, in analogy to dunes, to establish the normalized ripple function $n_r = f_r(\zeta')$. The plot of n_r versus ζ' is shown in Fig. 3.27. Here the broken line (Eq. (3.52)) is the normalized dune function $n_d = f_d(\zeta)$ with $\zeta = \zeta'$. Observe that $f_r(\zeta')$ (solid line) can be adequately represented by $f_d(\zeta)$ (broken line) as long as $\zeta' < 1$. In the region $\zeta' > 1$, however, the dune function must be "corrected" – as to yield $\delta_r = 0$ for $(\eta_r)_{max} - 1 \approx 20$, i.e. for $\zeta' = 2$. Thus we adopt

$$f_r(\zeta') = r f_d(\zeta) = r\zeta' e^{1 - \zeta'} \, , \tag{3.65}$$

where the "correction factor" r can be given e.g. by

$$r \equiv 1 \quad \text{if } \zeta' \leq 1 \quad ; \quad r \equiv \zeta'(2 - \zeta') \quad \text{if } 1 < \zeta' < 2 \, . \tag{3.66}$$

The solid line in Fig. 3.27 is the graph of (3.65) (with r computed from (3.66)). Hence

$$\delta_r = \Psi_r(\eta) = (\delta_r)_{max} \, f_r(\zeta') \approx 0.14 f_r(\zeta'), \tag{3.67}$$

where $f_r(\zeta')$ is determined by (3.65) and (3.66).

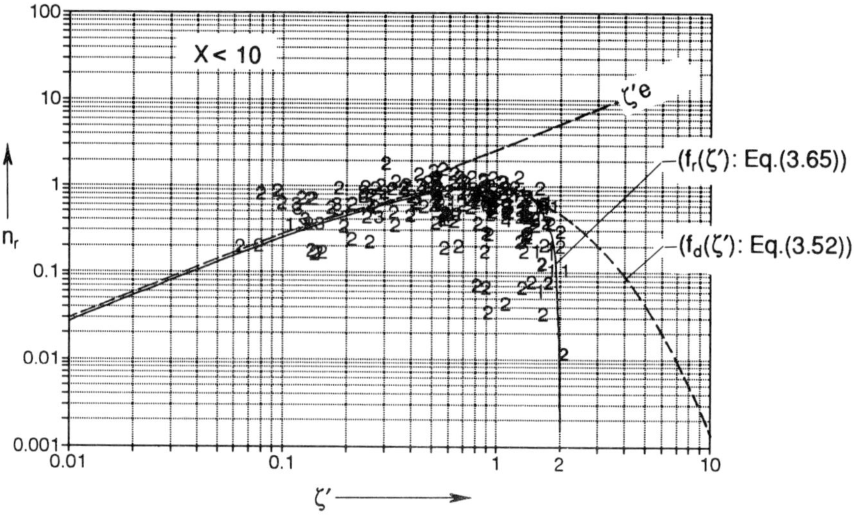

Fig. 3.27

From (3.59) and (3.60) it is clear that Λ_r is a function of ξ and η only, while the unimpeded δ_r is determined by η alone (see (3.62)). Hence in the usual case of $h \gg \Delta_r$, the value of $\Delta_r = \delta_r \Lambda_r$ depends on ξ and $\eta \sim Sh$ but not on h separately. Suppose, however, that the value of Δ_r (determined by ξ and $\eta \sim Sh$) turns out to be comparable with h. In such cases (very rare in river engineering), the ripple height Δ_r and thus δ_r can obviously not grow to their "full extent", i.e. to the values determined by the curve in Fig. 3.26: their growth will be damped by the (too small) flow thickness h. Let α be the "damping factor", implying the ratio of the damped ripple height Δ_r' to the regular ripple height Δ_r; the ratio $\alpha = \Delta_r'/\Delta_r = \delta_r'/\delta_r$ can be given by

$$\alpha = 1 - 0.95e^{-0.17(h/\Delta_r)}. \tag{3.68}$$

Observe from Fig. 3.28 that α is nearly unity for almost all h/Δ_r encountered in practice. [First

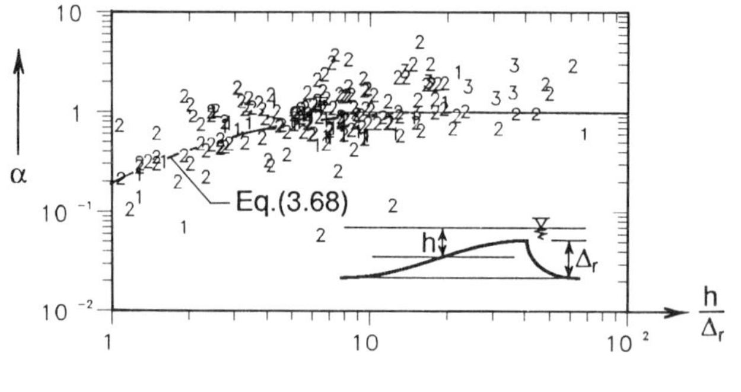

Fig. 3.28

determine δ_r and thus Δ_r (from the expressions of this subsection), then compute α from (3.68) and use $\alpha\delta_r$ instead of δ_r]. An analogous damping of dunes cannot arise, for the (statistical) height of dunes is always less than approximately one third of the flow depth. Indeed, $\Delta_d < (\Delta_d)_{max} = (\delta_d)_{max} \Lambda_d \approx (0.06)(6h) = 0.36h$.

3.5.3 *Alternate bars*

i- *Bar length*

Figs. 3.29a and b are two examples of the currents plots of the alternate bar data. As has already been mentioned earlier, the gross scatter in these plots does not permit the determination of the form

$$\Lambda_a = \Phi_a(X, Z)\, 6B, \qquad (3.69)$$

which is analogous to (3.47). Thus we will consider Λ_a as given by (3.12), viz

$$\Lambda_a \approx 6B,$$

for all regimes of turbulent flow.

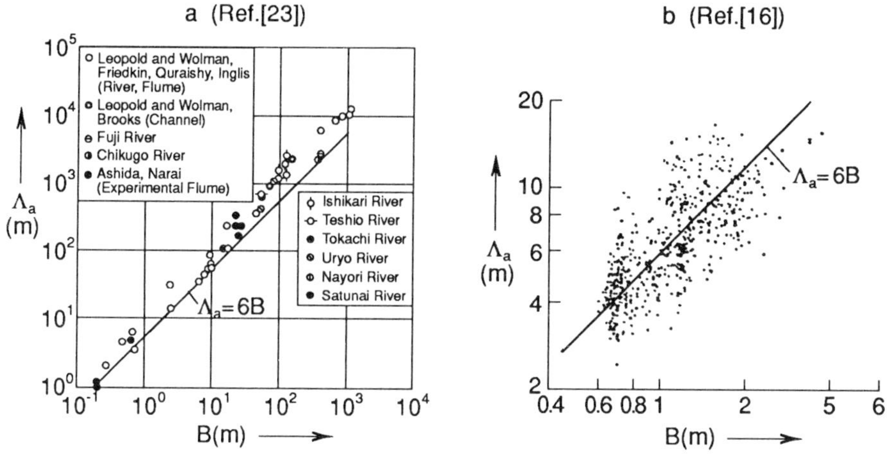

Fig. 3.29

ii- *Bar steepness*

Applying formally the Π-theorem to the set of parameters

$$\rho,\ \nu,\ \gamma_s,\ D, \upsilon_*,\ h,\ B, \qquad (3.70)$$

we obtain for the alternate bar steepness

$$\delta_a = (\Delta/\Lambda)_a = \psi_a(X, \eta, Z, B/h) = \psi_{Xa}(X)\Psi_a(\eta, Z, B/h). \qquad (3.71)$$

We will confine our considerations to the bars caused by rough turbulent flows $(\psi_{Xa}(X) \equiv 1)$, and thus we will identify δ_a with the unimpeded bar steepness

$$\delta_a = \Psi_a(\eta, Z, B/h). \qquad (3.72)$$

In order to reveal the nature of this function, the plots shown in Figs. 3.30

to 3.32 were produced.[29]

When $\eta \leq 1$, then δ_a is zero, and so it must be when η exceeds a certain value $(\eta_3)_a$ (see 3.3.3). Consequently, δ_a must (necessarily) vary with η, in a not yet known manner as indicated symbolically in Fig. 3.30 by the broken line (?). Yet none of the recent expressions for $\delta_a \sim \Delta_a$ contains η (see e.g.

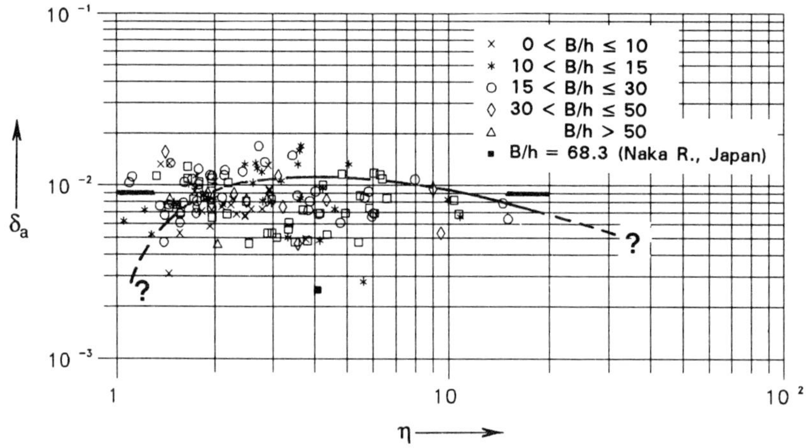

Fig. 3.30

Δ_a-formulae in Refs. [20], [22], [27], [10]): there appears to be a general agreement that δ_a does not depend on η. The reason for this, apparently, lies in the fact that for a considerable interval of η ($\approx 1.5 < \eta < \approx 15$, say) the experimental points are grossly, but evenly, scattered around the horizontal line $\delta_a = 0.009$. Complying with the "general agreement", we will consider (3.72) in the reduced form

$$\delta_a = 0.009 \overline{\Psi}_a(Z, B/h) \tag{3.73}$$

(which is, in fact, valid for the aforementioned η-interval only). No agreement appears to have been reached yet, however, with regard to the role of B/h. Consider, for example, the Δ_a-formulae of S. Ikeda [20] and M. Jaeggi [22], viz

$$\frac{\Delta_a}{h} = 0.044\left(\frac{B}{h}\right)^{1.45} Z^{-0.45} \quad \text{and} \quad \frac{\Delta_a}{B} = \frac{0.219}{(B/D)^{0.15}} \tag{3.74}$$

respectively. Using $\Lambda_a \approx 6B$, one determines from these expressions

$$\delta_a = 0.0073\left(\frac{B}{h}\right)^{0.45} Z^{-0.45} \quad \text{and} \quad \delta_a = 0.0365\left(\frac{B}{h}\right)^{-0.15} Z^{-0.15} \tag{3.75}$$

---

[29] The Refs. to the data sources are given in "References B": [1b], [3b], [6b], [7b], [9b], [11b], [13b].

respectively. Note that, according to S. Ikeda, δ_a increases with B/h, whereas according to M. Jaeggi δ_a decreases with it. The presence of these opposing B/h-powers can be taken as an indication that the influence of B/h on δ_a (if it exists) is not readily detectable from the (rather scattered) plot of the available data in Fig. 3.31. Hence, with an accuracy sufficient for practical

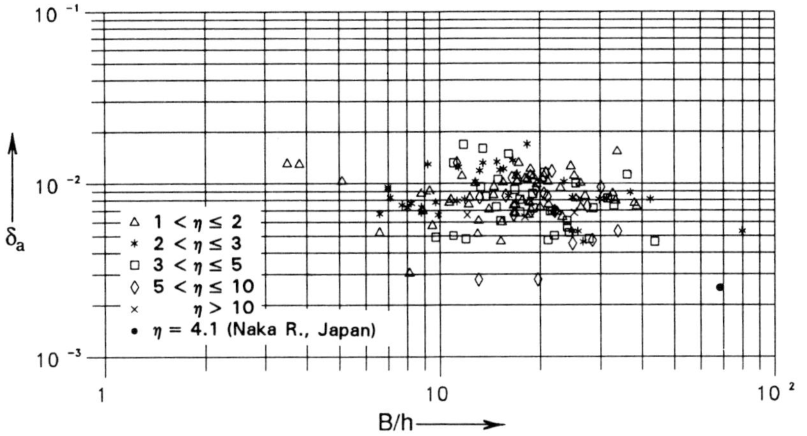

Fig. 3.31

purposes, δ_a can be treated as a function of Z only:

$$\delta_a = 0.009\, \overline{\overline{\Psi}}_a(Z)\,. \tag{3.76}$$

Both expressions in (3.75) indicate that δ_a decreases with the increment of Z; and so does the point pattern in Fig. 3.32. The solid straight line having S. Ikeda's Z-exponent, viz -0.45, adequately characterizes the inclination of this pattern. Note also that the center of the pattern (its approximate median)

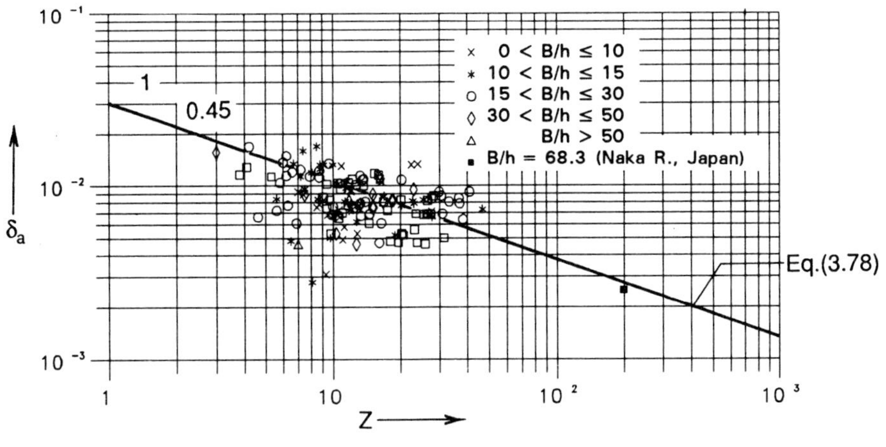

Fig. 3.32

is at the level[30] ≈ 0.009, its abcissa being $Z = 15$. Hence,

$$\bar{\bar{\Psi}}_a(Z) = aZ^{-0.45} \qquad (\text{with } a = (15)^{0.45} = 3.38), \tag{3.77}$$

which yields for the alternate bar steepness

$$\delta_a = 0.03 Z^{-0.45}. \tag{3.78}$$

This relation indicates that the alternate bar height Δ_a ($= \delta_a \Lambda_a$) also scales with the flow width:

$$\frac{\Lambda_a}{B} \approx 6 \quad ; \quad \frac{\Delta_a}{B} \approx 0.18 Z^{-0.45}. \tag{3.79}$$

From the content of this paragraph, one may conclude that although the variables η and B/h determine the *existence* of alternate bars, as

$$\delta_a > 0, \text{ when } 1 < \eta < (\eta_3)_a \quad \text{and} \quad \psi_\epsilon(Z) < B/h < \psi_1(Z),$$

their effect on the steepness of *existing* bars, if any, is not detectable in the interval $\approx 1.5 < \eta < \approx 15$.

Fig. 3.33 shows the plot of S. Ikeda [20], to which the data from Refs. [1b], [7b], [11b] have also been added. The above equation of S. Ikeda, i.e. the first equation of (3.75), is the power-approximation of his transcendental function $I = \psi(B/h)$, say.[31] The curve $\psi(B/h)$ in Fig. 3.33 is the graph of this function, the lines L_1 and L_2 being the graphs of the first equation of (3.75) and of the present equation (3.78), respectively.

3.6 Friction Factor

3.6.1 *Formulation of c*

Consider the friction factor relations (1.36) to (1.39). The (simultaneously present) bed forms 1 and 2 (see Fig. 1.6b) will be identified in the following with ripples and dunes, respectively. In this subsection, c, c_f and $\bar{c}$ have the same meaning as in 1.4.2 (ii) (where they were introduced).

i- Since c_f can be identified with $\bar{c}$ (see (1.40)), the relation (1.36) can be expressed as

[30] It should be noted that according to T. Kishi [27] the height of alternate bars is given by $\Delta_a = 0.05B$. Substituting this value together with $\Lambda_a = 6B$ in $\delta_a = \Delta_a/\Lambda$, one obtains $\delta_a = 0.05/6 = 0.0083$, which is remarkably near to the level ≈ 0.009.

[31] $I = (B/D)^{0.45}(\Delta_a/h)$; $\psi(B/h) = 9.34 \exp(2.53 erf[(\log(B/h) - 1.22)/0.594])$ (see Eq. (19) in [20] or Eq. (13) in [10]).

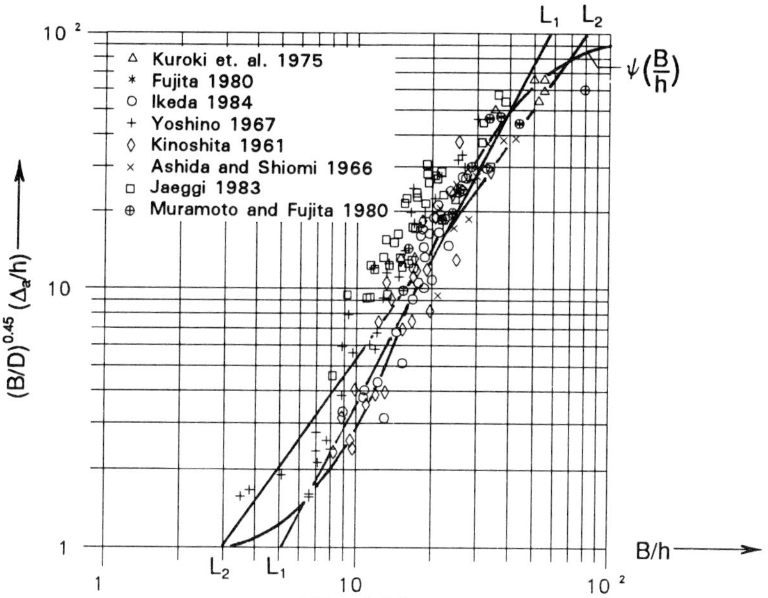

Fig. 3.33

$$\frac{1}{c^2} = \frac{1}{\bar{c}^2} + \frac{1}{2}\left[\delta_r^2 \frac{\Lambda_r}{h} + \delta_d^2 \frac{\Lambda_d}{h}\right] \tag{3.80}$$

(where δ_r and δ_d stand for $(\Delta/\Lambda)_r$ and $(\Delta/\Lambda)_d$ respectively).[32] From the preceding section, we have

$$\left.\begin{aligned} \Lambda_d/h &= \Phi_d(X, Z)6 && \text{(Eq. (3.47))} \\ \Lambda_r/h &= \Phi_r(X, \xi)/Z && \text{(Eq. (3.58))} \\ \delta_d &= \psi_{Xd}(X)\Psi_d(\eta, Z) && \text{(Eq. (3.49))} \\ \delta_r &= \psi_{Xr}(X)\Psi_r(\eta) && \text{(Eq. (3.61))}\ . \end{aligned}\right\} \tag{3.81}$$

The six functions which appear in these relations, viz $\Phi_d(X, Z)$, $\Phi_r(X, \xi)$, $\Psi_d(\eta, Z)$, $\Psi_r(\eta)$, $\psi_{Xd}(X)$ and $\psi_{Xr}(X)$, are given by (3.48), (3.59), (3.55), (3.67), (3.56) and (3.57) respectively. The value of $\bar{c}$ is determined (by (1.26)) as

$$\bar{c} = 2.5 \ln\left(b_s \frac{h}{k_s}\right) = 2.5 \ln\left[e^{0.4B_s - 1} Z \frac{D}{k_s}\right]. \tag{3.82}$$

Here, B_s is a function of $Re_* = X(k_s/D)$ (see 1.4.1) and k_s/D is a function of Y (see (1.72)). Hence,

$$\bar{c} = \overline{\Psi}_c^{**}(X, Z, k_s/D) = \overline{\Psi}_c^{*}(X, Z, Y), \tag{3.83}$$

---

32 The reduced version of this relation, corresponding to one mode of bed forms, has already been used to correlate the friction factor data in Refs. [7], [51], [52] and [58].

which in the usual case $Y < \approx 1$, when $k_s/D \approx 2$, reduces into

$$\bar{c} = \bar{\Psi}_c(X, Z) \,. \tag{3.84}$$

Observe that the variables which appear on the right-hand side of (3.80) (i.e. in (3.81) and in the expression of $\bar{c}$ (3.83)) are

$$X,\ Z,\ \xi,\eta,\ \text{and}\ Y.$$

But $Y = \eta\Psi(\xi)$ while $X = \sqrt{\xi^3 Y}$ (see (1.14) and (1.11)), and therefore c is, in fact, determined by three dimensionless variables only. Thus, we can write e.g.

$$c = \Psi_c(X, Y, Z) = \psi_c(\xi, \eta, Z) \tag{3.85}$$

which is consistent with (1.9) and (1.15).

ii- The flow in a wide channel can be given by its specific flow rate $q = Q/B$ (rather than by its depth h). Dividing both sides of the Chézy resistance equation (1.31), viz

$$q = vh = hc\sqrt{gSh} = hcv_* \,, \tag{3.86}$$

by the product Dv_{*cr}, and taking into account that $v_*/v_{*cr} = \sqrt{\eta}$, one arrives at the following definition (and expression) of the dimensionless specific flow rate N

$$N = \frac{q}{Dv_{*cr}} = Zc\sqrt{\eta} \,. \tag{3.87}$$

Eliminating Z between (3.87) and (3.85), one can determine c in terms of $N \sim q$ (rather than in terms of $Z \sim h$) as

$$c = \psi_c(\xi, \eta, N/(c\sqrt{\eta})) \,. \tag{3.88}$$

This relation is transcendental, and thus it cannot be solved algebraically with respect to c in the form

$$c = \phi_c(\xi, \eta, N) \,. \tag{3.89}$$

Nonetheless, it can be solved numerically, and therefore (3.88) *implies* (3.89). In the present text we will always designate (3.88) by (the clearer) (3.89).

The form (3.89) is more suitable for the study of self-forming channels in the next chapter, and therefore the values of c are plotted in the following on the basis of (3.89) (rather than (3.85)). The function of three variables (3.89) is a family of surfaces or a set of the family of curves. In this text, the values of c (ordinate) are always plotted versus η (abcissa). Each $(c;\eta)$-plane, containing one curve family, is specified by a constant value of ξ ($(const)_i$), each curve on that plane being distinguished by "its" constant value of N ($(const)_j$). An individual curve

$$c = \phi_c((const)_i, \eta, (const)_j) = \phi_{c,\,ij}(\eta) \tag{3.90}$$

will be referred to as a *c-curve.*

iii- The combination $(1/2)(\Delta/\Lambda)^2(\Lambda/h)$, which reflects the contribution of bed forms on c, has emerged (in Refs. [7] and [58]) from the consideration of two-dimensional bed forms. There is no reason to expect that this combination should still remain as it stands if bed forms are no longer two-dimensional. If bed forms are three-dimensional, then part of the flow passes around them, and the energy losses (which are mainly due to the sudden expansion of flow at the abrupt downstream faces) are smaller. For given Δ/Λ and Λ/h, this reduction of losses can be achieved by generalizing the aforementioned combination into the form

$$a\left(\frac{\Delta}{\Lambda}\right)^b \frac{\Lambda}{h} \qquad (3.91)$$

where $a \leq 1/2$ and $b \geq 2$ (with $a = 1/2$ and $b = 2$, if the bed forms are two-dimensional). Clearly, for a specified ξ, the three-dimensionality of bed forms must increase with the increment of B/h and h/D ($= Z$); and since the channels under study are "wide", it should increase with Z (or $N \sim Z$) only. At present, it is not known how a and b vary with ξ and Z (or N); and in order to gain some information on this score, numerous $(a; b)$-pairs were tested for a series of pertinent ξ and N: only the dune combination (3.91) ($a = a_d$; $b = b_d$) was varied during these trials (the ripple combination was used with $a_r = 0.5$; $b_r = 2$). The best fit was present when the $(a_d; b_d)$-pairs had the values shown in Table 3.1: the c-curves presented in the next sub-

Table 3.1

		log N				
D (mm)	ξ	3	4	5	6	7
0.2	5.1	0.50;2.00	0.50;2.00	0.50;2.00	0.35;2.00	0.35;2.30
0.3	7.6	0.50;2.00	0.50;2.00	0.45;2.00	0.45;2.30	0.45;2.50
0.5	12.6	0.50;2.00	0.50;2.00	0.45;2.00	0.45;2.25	0.45;2.50
0.7	17.7	0.50;2.00	0.50;2.00	0.50;2.00	0.45;2.25	0.45;2.50
1.0	25.3	0.50;2.00	0.50;2.00	0.50;2.00	0.50;2.00	0.50;2.00

section were computed according to this table. Observe that the $(a_d; b_d)$-pairs corresponding to the three-dimensional dunes do not deviate much (e.g. by multiple times) from (0.50;2.00); yet, this deviation often is sufficient to cause a noticeable difference in the computed c-values.

3.6.2 *Comparison with experiment*

The data from 84 sources given in "References C" (all together 6982 "data-points", D-range: $0.01\,mm < D < 52\,mm$) were compiled by T. Hayashi, who used them in Ref. [14]: the data corresponds to sand-or-gravel and water only ($\xi = 25.3D$, where D is in mm). These data were used to test how the values of c determined from the expressions above compare with experiment.[33] Four examples of these test-plots are shown in Figs. 3.34 to 3.37: only the data corresponding to $B/h \geq 5$ were used. Each plot corresponds to a certain $\xi_i \sim D_i$. However, one can hardly find a reasonable amount of data corresponding to a *value* of D, and therefore each graph contains the data from a *range* of D (which includes D_i).

The data-points in Figs. 3.34 to 3.37 are symbolized by digits. Each digit is the exponent m of that $N = 10^m$ which is the nearest to the N-value of the point plotted. The scatter in the graphs is gross. Nonetheless, one can realize that the "point clouds" are, in general, in coincidence with the areas occupied by the computed c-curves. Furthermore, the larger the digits, the higher tend the points to be situated on the $(c;\eta)$-plane (as the m-values of the c-curves themselves). A comparable behaviour is observable in each of the remaining test-plots.

From Figs. 3.34 to 3.37, one infers that with the increment of D and decrement of N the computed c-curves tend to become monotonous, whereas with the decrement of D and increment of N they usually exhibit "dips" (where $\partial c/\partial\eta = 0$). As should be clear from preceding sections, with the increment of η from unity onwards, the steepness (δ_i) of ripples and dunes first increases, then it reaches its maximum value and then it decreases again; as to yield the flat bed at advanced stages. Correspondingly, as can be inferred from (3.80), the value of c must first decrease (in comparison to $\bar{c}$),

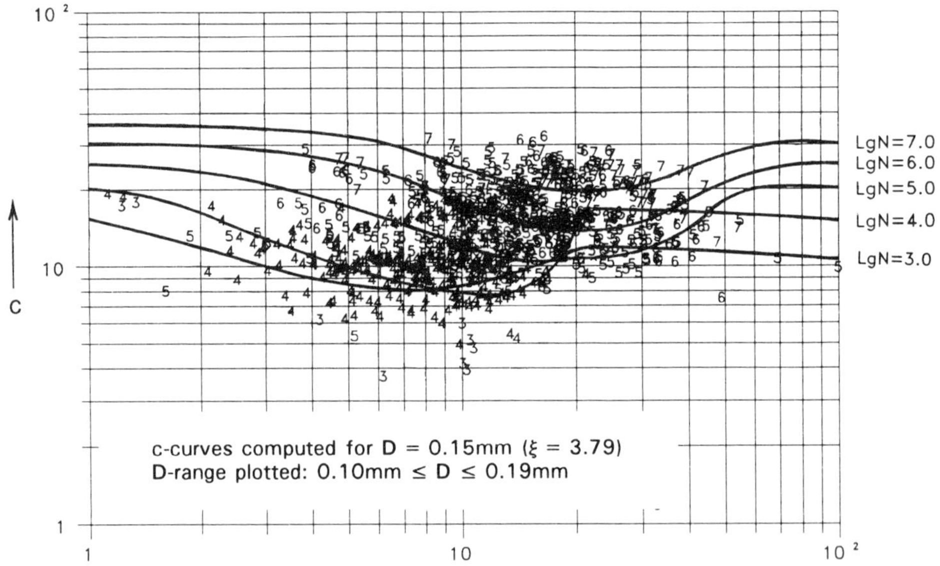

Fig. 3.34

[33] The author is grateful to Professor T. Hayashi for kindly providing him with these comprehensive data.

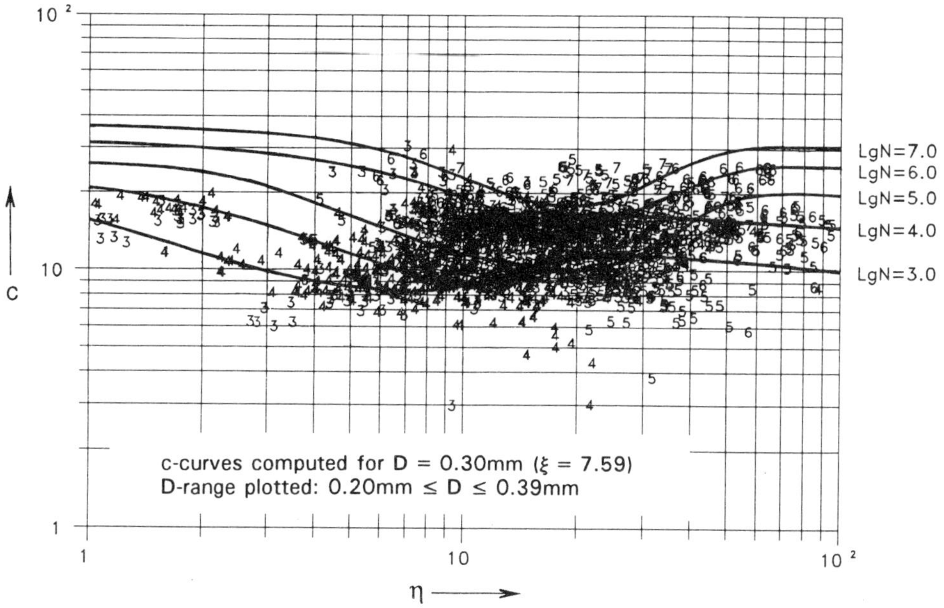

Fig. 3.35

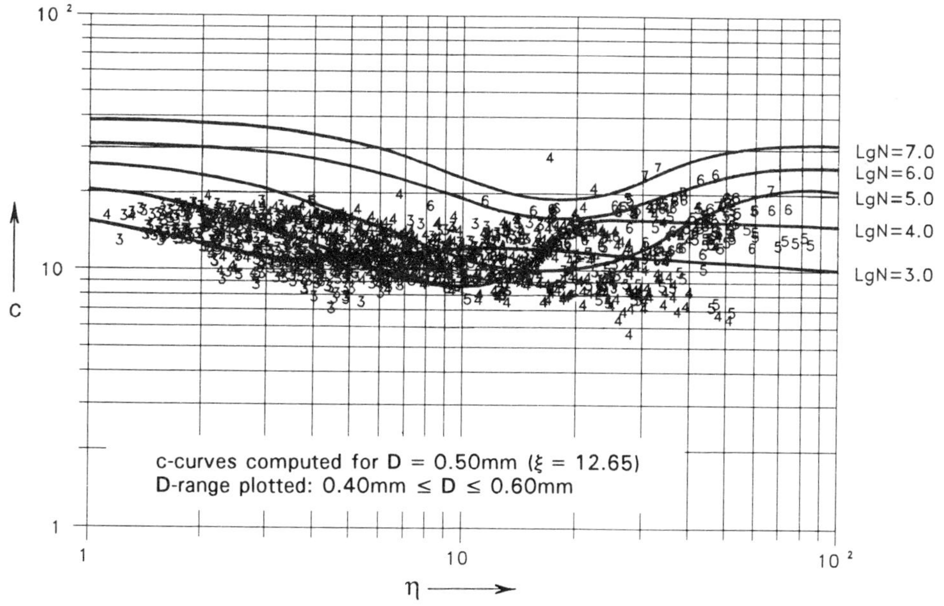

Fig. 3.36

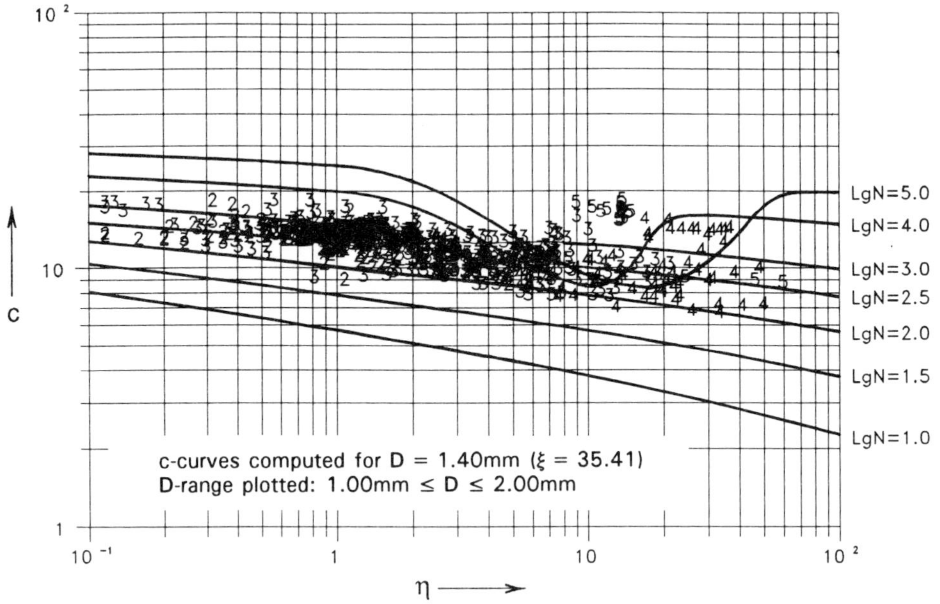

Fig. 3.37

then reach its minimum value and then increase again as to become comparable with $\bar{c}$. However, the increment of D reduces the steepness of bed forms (see 3.3.2 (ii)), and consequently the deviation of c from $\bar{c}$ (for all η). This is why the c-curves progressively become undistinguishable from the monotonous $\bar{c}$-curves with the increment of D.

3.6.3 *Total bed roughness K_s*

The consideration of the friction factor c can always be replaced by that of the total bed roughness K_s/D $(=\omega)$. Indeed if ω is given, then c can be computed from

$$c = 2.5 \ln[e^{0.4B_s - 1} Z/\omega] , \tag{3.92}$$

where $B_s = \phi_{B_s}(Re_*) = \phi_{B_s}(X\omega)$ is a known function (Fig. 1.5).

A significant contribution to the determination of $\omega = K_s/D$ is due to the recent work [14] of T. Hayashi. In Ref. [14] the values of ω were revealed from the same set of data which was referred to in the preceding subsection. The data were classified according to the ranges of the slope S, and for each range the ω-values were plotted versus the corresponding values of the friction mobility number Y_f $(= Y(c/\bar{c})^2)$. Thus a set of graphs such as that shown in Fig. 3.38 was obtained. The point pattern in each graph was characterized by a line (L in Fig. 3.38). The totality of these lines, which is shown in Fig. 3.39, reflects the functional relation

$$\omega = \phi_\omega(Y_f, S). \tag{3.93}$$

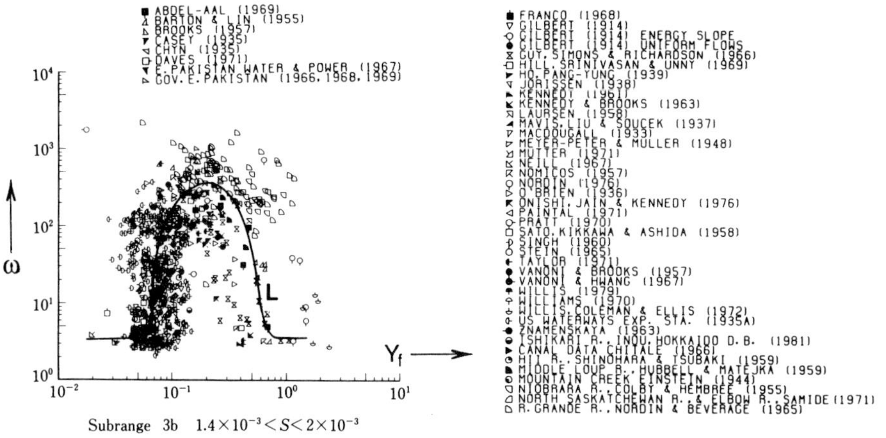

Fig. 3.38 (from Ref. [14])

From (3.93) it does not follow that ω is a function of two variables only; for

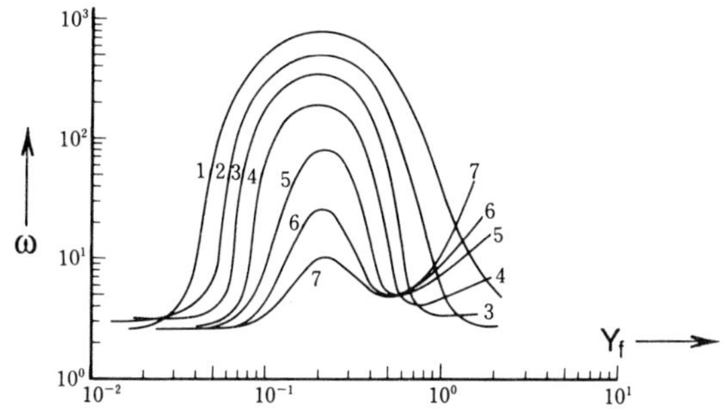

Fig. 3.39 (from Ref. [14])

the "variables" in the function (3.93) are themselves functions of the dimensionless variables of the two-phase motion. Indeed, if γ_s/γ is specified, then $S = (\gamma_s/\gamma)Y/Z \sim Y/Z$, while $Y_f = Y(c/\bar{c})^2$ where $c/\bar{c}$ is determined by X, Y and Z (see (3.84) and (3.85)). Consequently, $\omega = \bar{\phi}_\omega(X, Y, Z)$. No restriction is present in the theory of dimensions as to how one function of a phenomenon should be given in terms of its other functions — nor by how many of them.

References

1. Alexander, L.J.D.: *On the geometry of ripples generated by unidirectional open channel flows.* M.Sc. Thesis, Dept. of Civil Engrg., Queen's Univ., Kingston, Canada, 1980.
2. Allen, J.: *Bed forms due to mass transfer in turbulent flows: a kaleidoscope of phenomena.* J. Fluid Mech., Vol. 49, Part 1, 1971.
3. Anderson, A.G.: *The characteristics of sediment waves by flow in open channels.* Proc. 3rd Midwestern Conf. on Fluid Mechanics, Univ. of Minnesota, Minneapolis, 1953.
4. Binnie, A.M., Williams, E.E.: *Self-induced waves in a moving open channel.* J. Hydr. Res., Vol. 4, No. 1, 1966.
5. Bishop, C.T.: *On the time-growth of dunes.* M.Sc. Thesis, Dept. of Civil Engrg., Queen's Univ., Kingston, Canada, 1977.
6. Coleman, J.M.: *Brahmaputra river: channel processes and sedimentation.* Sediment. Geol., Vol. 3, 1969.
7. Engelund, F.: *Hydraulic resistance of alluvial streams.* J. Hydr. Div., ASCE, Vol. 92, No. HY2, March 1966.
8. Fok, A.T.: *On the development of ripples by an open channel flow.* M.Sc. Thesis, Dept. of Civil Engrg., Queen's Univ., Kingston, Canada, 1975.
9. Fujita, Y., Muramoto, Y.: *Multiple bars and stream braiding.* Int. Conf. on River Regime, W.R. White ed., John Wiley and Sons, 1989.
10. Fukuoka, S.: *Finite amplitude development of alternate bars.* Flow and Waves, No. 4, Tokyo Inst. of Technology, 1989.
11. Grass, A.: *Structural features of turbulent flow over smooth and rough boundaries.* J. Fluid Mech., Vol. 50, Part 2, 1971.
12. Grass, A.: *Initial instability of fine bed sand.* J. Hydr. Div., ASCE, Vol. 96, No. HY3, March 1970.
13. Hardtke, P.: *Turbulenzerzeugte Sedimentriffeln.* Mitt. Inst. Wasserbau III, Universitat Karlsruhe, Heft 14, 1979.
14. Hayashi, T.: *Alluvial bedforms of small scale and roughness.* (In Japanese) Proc. JSCE, No. 411/II-12, 1989.
15. Hayashi, T., Ozaki, S.: *Alluvial bed form analysis – formation of alternating bars and braids.* in *Application of Stochastic Processes in Sediment Transport*, H.W. Shen and H. Kikkawa eds., Water Resources Publications, Litleton, Colo., 1980.
16. Hayashi, T.: *Study of the cause of meandering rivers.* Trans. JSCE, Vol. 2, Part 2, 1971.
17. Hino, M.: *Equilibrium spectre of sand waves forming by running water.* J. Fluid Mech., Vol. 34, Part 3, 1969.
18. Ikeda, H.: *Experiments on bed load transport, bed forms and sedimentary structures using fine gravel in the 4-meter-wide flume.* Environmental Research Center Papers, No. 2, The University of Tsukuba, 1983.
19. Ikeda, H.: *A study on the formation of sand bars in an experimental flume.* (In Japanese) Geographi. Rev. Japan, 46-7, 1973.
20. Ikeda, S.: *Prediction of alternate bar wavelength and height.* J. Hydr. Engrg., ASCE, Vol. 110, No. 4, April 1984.
21. Jackson, R.G.: *Sedimentological and fluid-dynamics implications of the turbulent bursting phenomenon in geophysical flows.* J. Fluid Mech., Vol. 77, 1976.
22. Jaeggi, M.: *Formation and effects of alternate bars.* J. Hydr. Engrg., ASCE, Vol. 110, No. 2, Feb. 1984.
23. JSCE Task Committee on the Bed Configuration and Hydraulic Resistance of Alluvial Streams: *The bed configuration and roughness of alluvial streams.* (In Japanese) Proc. JSCE., No. 210, Feb. 1973.
24. Kennedy, J.F.: *The mechanics of dunes and antidunes in erodible-bed channels.* J. Fluid Mech., Vol. 16, Part 4, Aug. 1963.
25. Kinoshita, R.: *Model experiments based on the dynamic similarity of alternate bars.* Research Report, Ministry of Construction, Aug. 1980.

26. Kinoshita, R.: *An analysis of the movement of flood waters by aerial photography, concerning characteristics of turbulence and surface flow.* (In Japanese) Photographic Surveying, Vol. 6, 1967.
27. Kishi, T.: *Bed forms and hydraulic relations for alluvial streams.* in *Application of Stochastic Processes in Sediment Transport,* H.W. Shen and H. Kikkawa eds., Water Resources Publications, Litleton, Colo., 1980.
28. Kishi, T., Kuroki, M.: *Hydraulic characteristics of alternating bars.* (In Japanese) Hydraulics Laboratory, Hokkaido Univ., 1975.
29. Kondratiev, N., Popov, I., Snishchenko, B.: *Foundations of hydromorphological theory of fluvial processes.* (In Russian) Gidrometeoizdat, Leningrad, 1982.
30. Muller, A., Gyr, A.: *On the vortex formation in the mixing layer behind dunes.* J. Hydr. Res., Vol. 24, No. 5, 1986.
31. Muller, A., Gyr, A.: *Visualization of the mixing layer behind dunes.* Mech. of Sediment Transport, Balkema, B.M. Sumer and A. Muller eds., 1982.
32. Muramoto, Y., Fujita, Y.: *The configuration of meso-scale river bed configuration and the criteria of its formation.* 2nd Meeting of Hydr. Res. in Japan, 1978.
33. Nezu, I., Nakagawa, H.: *Self forming mechanism of longitudinal sand ridges and troughs in fluvial open-channel flows.* Proc. 23rd IAHR Congress, Vol. 2, Ottawa, Canada 1989.
34. Nezu, I., Nakagawa, H., Kawashima, N.: *Cellular secondary currents and sand ribbons in fluvial channel flows.* Proc. 6th Congress APD-IAHR, Kyoto, Japan, Vol. II, 1988.
35. Parker, G.: *On the cause and characteristic scales of meandering and braiding in rivers.* J. Fluid Mech., Vol. 76, 1976.
36. Public Works Research Institute, Ministry of Construction of Japan: *Meandering phenomenon and design of river channels.* (In Japanese) 1982.
37. Schlichting, H.: *Boundary layer theory.* McGraw-Hill Book Co. Inc., Verlag G. Braun (6th edition), 1968.
38. Schmid, A.: *Wandnache turbulente Bewegungsablaufe und ihre Bedeutung fur die Riffelbildung.* Institut Fuer Hydromechanik und Wasserwirtschaft, Zurich, Jan. 1985.
39. Shizong, L.: *A regime theory based on the minimization of the Froude number.* Ph.D. Thesis, Dept. of Civil Engrg., Queen's Univ., Kingston, Canada (in preparation).
40. Silva, A.M.F.: *Alternate bars and related alluvial processes.* M.Sc. Thesis, Dept. of Civil Engrg., Queen's Univ., Kingston, Canada, 1991.
41. Sukegawa, N.: *Criterion for alternate bar formation.* Memoirs of the School of Science and Engrg., Waseda Univ., No. 36, 1972.
42. Tsujimoto, T.: *Longitudinal stripes of sorting due to cellular secondary currents.* J. Hydroscience and Hydraulic Engineering, Vol. 7, No. 1, Nov. 1989.
43. Tsujimoto, T.: *Formation of longitudinal stripes due to lateral sorting by cellular secondary currents.* (In Japanese) Proc. 33rd Japanese Conf. on Hydraulics, JSCE, 1989.
44. Tsujimoto, T.: *Longitudinal stripes of alternate lateral sorting due to cellular secondary currents.* Proc. 23rd IAHR Congress, Vol. 2, Ottawa, Canada 1989.
45. Velikanov, M.A.: *Alluvial processes: fundamental principles.* State Publishing House for Physical and Mathematical Literature, Moscow, 1958.
46. Velikanov, M.A.: *Dynamics of alluvial streams. Vol. II. Sediment and Bed Flow.* State Publishing House for Theoretical and Technical Literature, Moscow, 1955.
47. Williams, P., Kemp, P.: *Initiation of ripples on flat sediment bed.* J. Hydr. Div., ASCE, Vol. 97, No. HY4, April 1971.
48. Yalin, M.S., Silva, A.M.F.: *On the formation of alternate bars.* Euromech 262, Wallingford, 1990.
49. Yalin, M.S.: *On the formation mechanism of dunes and ripples.* Euromech 261, Genoa, Sept. 1987.
50. Yalin, M.S.: *On the determination of ripple geometry.* J. Hydr. Engrg., ASCE, Vol. 111, No. 8, Aug. 1985.
51. Yalin, M.S., Lai, G.: *On the form drag caused by sand waves.* Proc. JSCE, No. 363/II-4, Nov. 1985.

52. Yalin, M.S.: *On the friction factor of alluvial streams.* Proc. JSCE, No. 335, July 1983.
53. Yalin, M.S., Karahan, E.: *Steepness of sedimentary dunes.* J. Hydr. Div., ASCE, Vol. 105, No. HY4, April 1979.
54. Yalin, M.S.: *Mechanics of sediment transport.* Pergamon Press, Oxford, 1977.
55. Yalin, M.S., Price, W.A.: *On the origin of submarine dunes.* Proc. XV Int. Coastal Engrg. Conf., Honolulu, Hawaii, July 1976.
56. Yalin, M.S.: *On the development of sand waves in time.* Proc. XVI IAHR Cong., Sao Paulo, Brasil, July 1975.
57. Yalin, M.S.: *Geometric properties of sand waves.* J. Hydr. Div., ASCE, Vol. 90, No. HY5, Sept. 1964.
58. Yalin, M.S.: *On the average velocity of flow over a mobile bed.* La Houille Blanche, No. 1, Jan/Feb. 1964.
59. Znamenskaya, N.S.: *Sediment transport and alluvial processes.* Hydrometeoizdat, Leningrad, 1976.

References A
Sources of Dune and Ripple Data

1a. Annambhotla, V.S., Sayre, W.W., Livesey, R.H.: *Statistic properties of Missouri river bed forms.* J. Waterways, Harbours and Coastal Engrg., ASCE, WW4, 1972.
2a. Ashida, K., Tanaka, Y.: *A statistical study of sand waves.* Proc. XXII Cong. IAHR, Fort Collins, Colo., 1967.
3a. Banks, N.L., Collinson, J.D.: *The size and shape of small scale ripples: an experimental study using medium sand.* Sedimentology, Vol. 12, 1975.
4a. Barton, J.R., Lin, P.N.: *Sediment transport in alluvial channels.* Rept. No. 55JRB2, Civil Engrg. Dept., Colorado A&M College, Fort Collins, Colo., 1955.
5a. Bishop, C.T.: *On the time-growth of dunes.* M.Sc. Thesis, Dept. of Civil Engrg., Queen's Univ., Kingston, Canada, 1977.
6a. Casey, H.J.: *Bed load movement.* Ph.D. Dissertation, Technische Hochschula, Berlin, 1935.
7a. Crickmore, M.J.: *Effect of flume width on bed-form characteristics.* J. Hydr. Div., ASCE, Vol. 96, No. HY2, Feb. 1970.
8a. Fok, A.T.: *On the development of ripples by an open channel flow.* M.Sc. Thesis, Dept. of Civil Engrg., Queen's University, Kingston, Canada, 1975.
9a. Fredsoe, T.: *Unsteady flow in straight alluvial streams.* J. Fluid Mech., Vol. 102 (part 2), 1981.
10a. Gilbert, G.K.: *The transportation of debris by running water.* U.S. Geol. Survey Prof. Paper 86, 1914.
11a. Grigg, N.S.: *Motion of single particle in alluvial channels.* J. Hydr. Div., ASCE, Vol. 96, No. HY12, Dec. 1970.
12a. Guy, H.P., Simons, D.B., Richardson, E.V.: *Summary of alluvial channel data from flume experiments 1956-1961.* U.S. Geol. Survey Prof. Paper 462-I, 1966.
13a. Haque, M.I., Mahmood, K.M.: *Analytical determination of form friction factor.* J. Hydr. Engrg., ASCE, Vol. 109, No. 4, April 1983.
14a. Hubbell, D.W., Sayre, W.H.: *Sand transport studies with radioactive tracers.* J. Hydr. Div., ASCE, Vol. 90, No. HY3, May 1964.
15a. Hung, C.S., Shen, H.W.: *Statistical analysis of sediment motions of dunes.* J. Hydr. Div., ASCE, Vol. 105, No. HY3, March 1979.
16a. Hwang, L.S.: *Flow resistance of dunes in alluvial streams.* Ph.D. Thesis, California Inst. of Thec., Pasadene, California, 1965.
17a. Jain, S.C., Kennedy, J.F.: *The growth of sand waves.* Proc. I Int. Symp. on Stochastic Hydraulics, University of Pittsburgh Press, 1971.
18a. Korchokha, Y.M.: *Investigation of the dune movement of sediments on the Polomet River.* Soviet Hydrology, No. 6, 1968.

19a. Lane, E.W., Eden, E.W.: *Sand waves in Lower Mississippi river.* J. Western Soc. Engrs., No. 6, 1940.
20a. Lau, Y.L., Krishnappan, B.: *Sediment transport under ice cover.* J. Hydr. Engrg., ASCE, Vol. 111, No. 6, June 1985.
21a. Mahmood, K., Amadi, H.: *Analysis of bed profile in sand canals.* III Annual Symp. of Waterways, Harbours and Coastal Eng., Colo. State Univ., Fort Collins, Colo., 1976.
22a. Mantz, P.A.: *Semi-empirical correlations for fine and coarse sediment transport.* Proc. Instn. Civ. Engrs., Vol. 75, Part 2, 1983.
23a. Mantz, P.A.: *Laboratory flume experiment on the transport of cohesionless silica silts by water streams.* Proc. Instn. Civ. Engrs., Vol. 68, Part 2, 1980.
24a. Martinec, J.: *The effect of sand ripples on the increase of river bed roughness.* Proc. XXII Cong. IAHR, Fort Collins, Colo., 1967.
25a. Matsunashi, J.: *On a solution of bed fluctuation in an open channel with a movable bed and steep slopes.* Proc. XXII Cong. IAHR, Fort Collins, Colo., 1967.
26a. Nordin, C.F.: *Statistical properties of dune profiles.* U.S. Geol. Survey Prof. Paper 562-F, 1971.
27a. Nordin, C.F., Algert, J.H.: *Spectral analysis of sand waves.* J. Hydr. Div., ASCE, Vol. 92, No. HY5, Sept. 1966.
28a. Nordin, C.F.: *Aspects of flow resistance and sediment transport – Rio Grande near Bernalillo, New Mexico.* U. S. Geol. Survey Water Supply Paper, 1964.
29a. Raichlan, F., Kennedy, J.F.: *The growth of sediment bed forms from an initially flattened bed.* Proc. XXI IAHR Cong., Leningrad, Vol. 3, 1965.
30a. Shen, H.W., Cheong, H.F.: *Statistical properties of sediment bed profiles.* J. Hydr. Div., ASCE, Vol. 103, No. HY11, Nov. 1977.
31a. Shinohara, K., Tsubaki, T.: *On the characteristics of sand waves formed upon the beds of open channels and rivers.* Rept. of the Research Inst. for Applied Mechanics, Vol. VII, No. 25, 1959.
32a. Simons, D.B., Richardson, E.V., Hausbild, W.L.: *Some effects of fine sediment on flow phenomena.* U.S. Geol. Survey Water Supply Paper 1498-G, Washington, 1963.
33a. Simons, D.B., Richardson, E.V., Albertson, M.L.: *Flume studies of alluvial using medium sand.* U.S. Geol. Survey Water Supply Paper 1498-A, Washington, 1961.
34a. Singh, B.: *Transport of bed load in channels with special reference to gradient and form.* Ph.D. Thesis, London Univ., England, 1960.
35a. Stein, R.A.: *Laboratory studies of total load and bed load.* J. Geoph. Res., Vol. 70, No. 8, April 1965.
36a. Yalin, M.S., Karahan, E.: *Steepness of sedimentary dunes.* J. Hydr. Div., ASCE, Vol. 105, No. HY4, April 1979.
37a. Znamenskaya, N.S.: *Experimental study of the dune movement of sediment.* Soviet Hydrology: Selected papers. Published by American Geophysical Union, No. 3, 1963.

References B
Sources of Bar Data

1b. Ashida, K., Shiomi,Y.: *On the hydraulic of dunes in alluvial channels.* Disaster Prevention Res. Inst., Kyoto Univ., Annual Report No. 9, 1966.
2b. Chang, H.Y., Simons, D., Woolhiser, D.: *Flume experiments on alternate bar formation.* J. Waterways, Harbors and Coastal Engineering Div., ASCE, Vol. 97, No. 1, Feb. 1971.
3b. Fujita, Y.: *Fundamental study on channel changes in alluvial rivers.* Thesis presented to Kyoto University, Kyoto, Japan, 1980, in partial fulfillment of the requirements for the degree of Doctor of Engineering.
4b. Iguchi, M.: *Tests for fine gravel transport in a large laboratory flume* (In Japanese) Report for National Science Foundation, School of Earth Science, Tsukuba Univ., Japan 1980.

5b. Ikeda, H.: *Experiments on bed load transport, bed forms, and sedimentary structures using fine gravel in the 4-meter-wide flume.* Environmental Research Center Papers, No. 2, The University of Tsukuba, 1983.
6b. Ikeda, S.: *Prediction of alternate bar wavelength and height.* J. Hyd. Engrg., ASCE, Vol. 110, No. 4, April 1984.
7b. Jaeggi, M.: *Alternierende Kiesbanke.* Mitteilungen der Versuchsanstal fur Wasserbau, Hydrologie und Glaziologie, Zurich, No. 62, 1983.
8b. Kinoshita, R.: *Model experiments based on the dynamic similarity of alternate bars.* (In Japanese) Research Report, Ministry of Construction, Aug. 1980.
9b. Kinoshita, R.: *Investigation of the channel deformation of the Ishikari River.* (In Japanese) Science and Technology Agency, Bureau of Resources, Memorandum No. 36, 1961.
10b. Kuroki, M., Kishi, T., Itakura, T.: *Hydraulic characteristics of alternate bars.* (In Japanese) Report for National Science Foundation, Dept. of Civil Engrg., Kokkaido Univ., Hokkaido, Japan, 1975.
11b. Muramoto, Y., Fujita, Y.: *The classification of meso-scale bed configuration and the criteria of its formation.* 2nd Meeting of Hydr. Res. in Japan, 1978.
12b. Silva, A.M.F.: *Alternate bars and related alluvial processes.* M.Sc. Thesis, Dept. of Civil Engrg., Queen' Univ., Kingston, Canada 1991.
13b. Yoshino, F.: *Study on bed forms.* (In Japanese) Collected Papers, Dept. of Civil Engrg., Univ. of Tokyo, Vol. 4, 1967.

References C
Sources of Friction Factor Data

1c. Abdel-aal 1969; 2c. Barton and Lin 1955; 3c. Government West Bengal 1965; 4c. Bogardi and Yen 1939; 5c. Brooks 1957; 6c. Casey 1935; 7c. Chaudhry et. al. 1970; 8c. Chaudhry et. al. 1970; 9c. Chitale 1966 (canal data); 10c. Chyn 1935; 11c. Colby and Hembree 1955 (Niobrara R.); 12c. Costello 1974; 13c. Culbertson, Scott and Bennett 1976 (Rio Grande); 14c. Da Cunha 1973 (Portuguese R.); 15c. Daves 1971; 16c. Einstein 1944 (Mountain Creek); 17c. Einstein and Chien 1955; 18c. E. Pakistan Water and Power 1967; 19c. Gov. E. Pakistan 1966, 1968, 1969; 20c. Foley 1975; 21c. Franco 1968; 22c. Gibbs and Neill 1972; 23c. Gilbert 1914; 24c. Gilbert 1914 (Energy Slope); 25c. Gilbert 1914 (Uniform Flow); 26c. Guy, Simons and Richardson 1966; 27c. Hill, Scrinivasan and Unny 1969; 28c. Ho, Pang-Yung 1939; 29c. Hubell and Matejka 1959 (Middle Loup R.); 30c. Inou, Hokkaido D.B. 1981 (Ishikari R.); 31c. I-o-hashi, Hokkaido D.B. 1981 (Ihsikari R.); 32c. I-kakobashi, Kinoshita 1989 (Ishikari R.); 33c. Jorissen 1938; 34c. Kalinske and Hsia 1945; 35c. Kalkanis 1957; 36c. Kawamoto and Yamamoto 1976 (Tone R.); 37c. Kennedy 1961; 38c. Kennedy and Brooks 1963; 39c. Knott 1974 (Trinity R.); 40c. Mahmood et. al. 1979 (Apoc Canal); 41c. Laursen 1958; 42c. Leopold 1969 (river data); 43c. Mavis, Lin and Soucek 1937; 44c. MacDougall 1933; 45c. Meyer Peter and Muller 1948; 46c. Milhous 1969 (Oak Creek); 47c. Mutter 1971; 48c. Niagara River 1976; 49c. Nedeco 1973 (Rio Magdalena and Canal Del Dique); 50c. Neill 1967; 51c. Nordicos 1957; 52c. Nordin and Beverage 1965 (Rio Grande) 53c. Nordin 1976; 54c. O'Brien 1936; 55c. Onishi, Jain and Kennedy 1976; 56c. Paintal 1971; 57c. Pratt 1970; 58c. Samide 1971 (North Saskatchewan R. and Elbow R.); 59c. Sato, Kikkawa and Ashida 1958; 60c. Seitz 1976 (Snake and Clearwater R.); 61c. Shen et. al. 1978 (Missouri R.); 62c. Shinohara, K and Tsubaki, T. 1959 (Hii R.); 63c. Simons 1957 (American Canals); 64c. Singh 1960; 65c. Soni 1980; 66c. Stein 1965; 67c. Straub 1954, 1958; 68c. Tayler 1971; 69c. Toffaleti 1968 (Atchafalaya R.); 70c. Toffaleti 1968 (Mississippi R.); 71c. Toffaleti 1968 (Red R.); 72c. Toffaleti 1968 (Rio Grande near Bernadillo); 73c. Vanoni and Brooks 1957; 74c. Vanoni and Hwang 1967; 75c. Willis, Coleman and Ellis 1972; 76c. Willis 1979; 77c. Williams 1970; 78c. U.S. Bureau of Reclamation 1958 (Colorado R.); 79c. U.S. Waterways Exp. Station 1935A; 80c. U.S. Waterways Exp. Station 1936A; 81c. U.S. Waterways Exp. Station 1936B; 82c. U.S. Waterways Exp. Station 1963B; 83c. U.S. Waterways Exp. Station 1936C; 84c. Znamenskaya 1963.

CHAPTER 4

REGIME CHANNELS

PART I

4.1 Regime Concept

4.1.1 *Regime channel R and its determining parameters*

Consider a very long, straight open-channel excavated in a cohesionless alluvium; and suppose that, at the time $t = 0$, a constant flow rate Q begins to flow in it. It is assumed that the flow is nearly bankfull, and that it is transporting the sediment (being thus capable of inducing the channel deformation). Experiment shows that, in general, the flow rate Q will not "accept" the *initial* channel: it will gradually deform it so as to establish, eventually, a certain definite channel "of its own", which is referred to as the *regime channel*. Usually the cross-section of a regime channel is trapezoidal and its width-to-depth ratio is "large". Hence, the cross-section geometry of a regime channel can be characterized by two lengths only: flow depth and (the average) flow width. The flow in the regime channel is steady, and it is treated as uniform (or quasi-uniform).

Let T_R be the duration of development of the regime channel, and $[B_0, h_0, S_0]$ and $[B_R, h_R, S_R]$ be the characteristics of the initial channel (at $t = 0$) and of the regime channel (at $t \geq T_R$) respectively. From observations and measurements, it appears that the regime characteristics $[B_R, h_R, S_R]$ do not depend on the initial channel characteristics $[B_0, h_0, S_0]$ – the degree of discrepancy between them can affect only the magnitude of T_R. The regime channel, R say, which comes into being as described above is completely determined by the (constant) flow rate Q, by the physical nature of the liquid and solid phases involved, and by the acceleration due to gravity g. Since the nature of fluid is given by the parameters ρ and ν, while the nature of a cohesionless granular material of a specified internal geometry (i.e. of a specified shape of grains and of the dimensionless grain size distribution curve) can be reflected by γ_s and D [33], [34], the regime channel R can be defined by the following six characteristic parameters:

$$Q, \rho, \nu, \gamma_s, D, g. \tag{4.1}$$

(From the fact that the parameter g determines the uniform flow in an open-channel in the form gS (Chapter 1), it does not follow, and it is in fact false, that it must determine the formation of a channel also in the same manner).

Any quantitative property A_R of the regime channel R (such as B_R, h_R, S_R, $(Q_s)_R$, c_R, etc.) must thus be a certain function of the parameters (4.1):

$$A_R = f_{A_R}(Q, \rho, \nu, \gamma_s, D, g).^{1} \tag{4.2}$$

The variation of a property A, during the formation of the regime channel R, can be symbolized by the time-dependent relation

$$A = F_A(Q, \rho, \nu, \gamma_s, D, g, \theta) \tag{4.3}$$

where θ is the dimensionless time t/T_R. For the special cases $\theta = 0$ and $\theta \geq 1$, this relation acquires its special forms

$$A_0 = F_A(Q, \rho, \nu, \gamma_s, D, g, 0) = f_{A_0}(Q, \rho, \nu, \gamma_s, D, g)$$

and

$$A_R = F_A(Q, \rho, \nu, \gamma_s, D, g, 1) = f_{A_R}(Q, \rho, \nu, \gamma_s, D, g),$$

which correspond to the initial and the regime channels respectively.

Only the characteristic parameters determining a phenomenon must remain constant during its development. Since the (volumetric) transport rate Q_s is not one of the parameters (4.1) determining the regime channel R, it must be expected to vary (as any other A) during the channel formation process (from its initial value $(Q_s)_0$ to its final regime value $(Q_s)_R$). This is only natural, for Q_s is determined, among others, by B, h and S, which vary during T_R.[2]

4.1.2 *Empirical regime formulae*

A reliable prediction of the stable (invariant in time) flow cross-section and slope of an alluvial channel, as implied by B_R, h_R and S_R, is the key to success in river training, and therefore it is not surprising that regime chan-

---

[1] From 3.6.1 it follows that the friction factor is determined by h, gS, γ_s, ρ, ν and D, and thus the regime value of c_R should be given by

$$c_R = f_c(h_R, gS_R, \gamma_s, \rho, \nu, D). \tag{i}$$

Observe that (i) is consistent with (4.2), for if the values of h_R and S_R given by (4.2) are substituted (in principle) in (i), then its right-hand side will contain only the parameters (4.1); i.e. it will become what is meant by (4.2).

[2] In the case of a laboratory channel (having a finite length), it must thus be ensured that the (time variable) rate of sediment fed into the channel at its entrance is equal to the transport rate at its end, for any t (recirculation).

nels have become one of the most popular research topics in the field. There is no shortage of experimental data, nor of empirical relations produced from them: Table 4.1 shows but a few of the existing "regime formulae" (Refs. [1a] to [20a] in "References A" at the end of the chapter). Note, however, that most of these equations are the dimensionally inhomogeneous power functions of Q, viz

$$A_R = a_A Q^{n_A}, \tag{4.4}$$

where the influence of the remaining characteristic parameters (ρ, ν, γ_s, D and g) is "hidden" in the coefficients a_A and the exponents n_A. Nonetheless, these regime equations convey the following *very relevant* messages:

(i) B_R is practically proportional to the square root of Q in all cases:

$$B_R \sim Q^{1/2}. \tag{4.5}$$

(ii) The Q-exponent in $h_R \sim Q^{n_h}$ is affected by the grain size D:

$$n_h \approx 1/3 \quad \text{for fine sand} \quad ; \quad n_h \approx 0.43 \quad \text{for gravel}. \tag{4.6}$$

(iii) The Q-exponent in $S_R \sim Q^{n_S}$ is also affected by D:

$$n_S \approx -0.1 \quad \text{for fine sand} \; ; \quad n_S \approx -0.43 \quad \text{for gravel}. \tag{4.7}$$

Note that the absolute values of n_h and n_S are identical for gravel bed channels.

Some attempts to reveal the influence of ρ, ν, γ_s, D and g (in addition to Q), and to bring the regime equations into the dimensionally homogeneous forms have already been made in the past (see e.g. [12], [14], [19], [25]). However, in most of these works the π-theorem was used formally (that is, without considering the physical side of the phenomenon), and therefore the results produced are of a limited value. For example, the utilization of D as a "typical length" (which in a formal approach is only natural, because D is the only length among the parameters involved) led to the emergence of the dimensionless forms B_R/D and h_R/D. Clearly these forms are not realistic, for the external dimensions (B_R and h_R) of a regime channel corresponding to some specified values of Q, ρ, ν and γ_s do not increase in proportion to the grain size D.

4.2 Extremal (or Rational) Methods

Owing to the pioneering works of C.T. Yang, C.S. Song, H. Chang and other researchers (see e.g. [39], [40], [27], [4], [5]), the so-called extremal, or rational, methods of determination of regime channels have been developed. These methods are motivated by the conviction that a regime channel is forming because a certain physical quantity, A_* say, tends to acquire its

Table 4.1

Source	D_{50} (mm)	B	h	S
[1a] Leopold, et al., 1953	-	$\sim Q_{ma}^{0.45 \text{ to } 0.56}$	$\sim Q_{ma}^{0.37 \text{ to } 0.45}$	$\sim Q_{ma}^{(-0.19) \text{ to } (-0.50)}$
[2a] Leopold, et al., 1956	0.7 to 5.	$5.0Q_{uf}^{0.50}$	$0.10Q_{uf}^{0.28}$	$\sim Q_{uf}^{0.0667}$
[3a] Nixon, 1959	0.1 to 0.6	$1.67Q_{bf}^{0.50}$	$0.55Q_{bf}^{0.33}$	$\sim Q_{bf}^{-0.1}$
[4a] Nash, 1959	clay	$1.32Q_{bf}^{0.54}$	$0.93Q_{bf}^{0.27}$	$\sim Q_{bf}^{0.12}$
[5a] Lacey, 1929	0.1 to 0.4	$2.67Q_d^{0.50}$	$0.47Q_d^{0.33}$	$0.00039f^{1.5}Q_d^{-0.11}$
[6a] Lapturev, 1969	-	$2.58Q_d^{0.50}$	$0.52Q_d^{0.33}$	$\sim Q_d^{-0.10}$
[7a] Ackers, 1964	0.16 & 0.34	$3.6Q_{expt}^{0.42}$	$0.28Q_{expt}^{0.43}$	no good correlation
[8a] Blench, 1957 (*) dunes, sandbed	0.1to0.6 0.3to7.	$\left[\frac{F_b}{F_s}\right]^{0.50} Q_{bf}^{0.5}$	$\left[\frac{F_s}{F_b^2}\right]^{0.33} Q_{bf}^{0.33}$	$\beta_s Q_{bf}^{-0.167}$
[9a] Blench, 1957, no dunes, gravel bed	> 7.	$\left[\frac{F_b}{F_s}\right]^{0.50} Q_{bf}^{0.50}$	$\left[\frac{F_s}{F_b D}\right]^{0.20} Q_{bf}^{0.40}$	$\sim D^{6/5}Q_{bf}^{-0.40}$
[10a] Simons & Albertson, 1960	0.03 to 0.8	$2.5Q_{bf}^{0.51}$	$R = 0.43Q_{bf}^{0.36}$	$0.00675Q_{bf}^{-0.40}$
[11a] Bose, 1936	-	$2.8Q_d^{0.50}$	$0.47Q_d^{0.33}$	$0.209D^{0.86}Q_d^{-0.21}$
[12a] Inglis, 1957, 1949	≈0.2	$\sim Q^{0.50}$	$\sim Q^{0.33}$	$\sim Q^{-0.167}$
[13a] Hey, 1982	21. to 190.	$2.2Q_s^{-0.05}Q^{0.54}$	$0.161D^{-0.15}Q^{0.41}$	$0.68Q_s^{0.13}D^{0.97}Q^{-0.53}$
[14a] Bray, 1982 regression	19. to 145.	$2.08D^{-0.07}Q_2^{0.528}$	$0.256D^{-0.025}Q_2^{0.331}$	$0.0965D^{0.586}Q_2^{-0.334}$
[15a] Bray, 1982 threshold method	19. to 145.	$2.67Q_2^{0.50}$	$0.0585D^{-0.29}Q_2^{0.428}$	$0.968D^{1.285}Q_2^{-0.428}$
[16a] Bray, 1982 Kellerhals meth.	19. to 145.	$1.80Q_2^{0.50}$	$0.166D^{-0.12}Q_2^{0.40}$	$0.12D^{0.92}Q_2^{-0.40}$
[17a] Bray, 1982 dimensional appr.	19. to 145.	$2.0D^{-0.24}Q_2^{0.496}$	$0.157D^{0.008}Q_2^{0.397}$	$0.259D^{0.937}Q_2^{-0.375}$
[18a] Glover & Florey, 1951	-	$0.93D^{-0.15}Q^{0.46}$	$0.12D^{-0.15}Q^{0.46}$	$0.44D^{1.15}Q^{-0.46}$
[19a] Ghosh, 1983	> 6.	$0.87D^{-0.15}Q^{0.46}$	$0.11D^{-0.15}Q^{0.46}$	$0.68D^{1.15}Q^{-0.46}$
[20a] Hey & Thorne, 1983	14. to 176.	$(2.3 \text{ to } 4.3)Q_{bf}^{0.50}$	$(0.16 \text{ to } 0.20) \times$ $\times Q_s^{-0.03}D^{-0.14}Q_{bf}^{0.39}$	$0.42Q_s^{0.17}D^{0.83}Q_{bf}^{-0.57}$

minimum (or maximum) value.[3] Once the minimum value of A_* is reached, the channel formation stops – the regime channel is achieved.

Different authors propose different quantities as A_*. Thus we have e.g.

$A_* = S$ (minimum stream power hypothesis: Refs. [5], [6], [7], [8], [27], etc.)[4]

$A_* = Sv$ (minimum unit stream power hypothesis: Refs. [36], [37], [40], [41], etc.)

$A_* = Q_s^{-1}$ (maximum transport rate (Q_s) hypothesis: Refs. [31], [32])

$A_* = c$ (maximum friction factor ($\sim 1/c^2$) hypothesis: Refs. [9], [10])

$A_* = SL$ (minimum energy dissipation rate hypothesis: Refs. [35], [38], [39])[5]

... etc.

No agreement has been reached yet, however, as to exactly what A_* should be, for none of the trends $A_* \rightarrow \min$ proposed to date rests on a convincing theoretical reasoning. An extensive comparative review of the current extremal hypotheses is given in Ref. [9] (see also [3], [4] and [35]).

Three equations are needed to determine analytically the three properties (viz B_R, h_R and S_R) of a regime channel. All authors use the flow resistance formula as the "first" of these equations, the "third" equation being the condition reflecting the minimization of A_* (i.e. $dA_* = 0$). The choice of A_* and of the "second" equation varies from one author to another.

Very often a transport rate formula is used as the second equation ([4], [18], [31], [38], etc.) and, in this case, one arrives at a system of three equations which (for a given granular material and fluid) can be shown symbolically as[6]

---

3 For the sake of uniformity we will refer in the following only to the minimization of A_*, and if a method is based on $B_* \rightarrow \max$, we will interpret it as if it is based on $A_* = B_*^{-1} \rightarrow \min$.

4 The original quantity subjected to minimization is γQS, which is reduced here to S because $\gamma Q = const.$ The same applies to the next hypothesis: Sv is originally γQSv.

5 Here, L is a length along the channel; the original quantity subjected to minimization is γQSL. For the sake of scientific precision, some authors use $(\gamma Q + \gamma_s Q_s)SL$ instead of γQSL (see e.g. [38]. See also A. Brebner and K.C. Wilson: *Derivation of the regime equations from relationships for the pressurized flow by use of the principle of minimum energy-degradation rate.* Proc. I.C.E., Vol. 36, January 1967). It has been shown in Ref. [9] that if $Q_s/Q < 1000ppm$ (per weight), then the error in neglecting $\gamma_s Q_s$ is less than 0.1%.

6 The reason for the subscript 1 in the set of Eqs. (4.8) will be clarified presently.

$$\left.\begin{aligned} Q &= f_Q(B_{R_1}, h_{R_1}, S_{R_1}, c_{R_1}) \quad &&\text{(resistance equation)} \\ Q_s &= f_{Q_s}(B_{R_1}, h_{R_1}, S_{R_1}, c_{R_1}) \quad &&\text{(transport equation)} \\ dA_* &= 0 &&\text{(minimum } A_*\text{)} \end{aligned}\right\} \tag{4.8}$$

where c_{R_1}, h_{R_1} and S_{R_1} are interrelated:

$$c_{R_1} = f(h_{R_1}, S_{R_1}). \tag{4.9}$$

Clearly, for a specified granular material and fluid (i.e. for given ρ, ν, γ_s and D) the three regime characteristics B_{R_1}, h_{R_1} and S_{R_1} can then be solved from the three equations (4.8) (in conjunction with (4.9)) if, in addition to the flow rate Q, the value of the transport rate Q_s is *also* given. But this means that the regime channel determined by (4.8) is not specified by the *six* characteristic parameters (4.1) as the regime channel R considered in the preceding section: it is a *different* regime channel R_1 (hence the reason for the subscript 1) which is specified by *seven* parameters:

$$Q_s, Q, \rho, \nu, \gamma_s, D, g. \tag{4.10}$$

A quantitative property A_{R_1} of this "new" regime channel R_1 must thus be given by

$$A_{R_1} = f_{A_{R_1}}(Q_s, Q, \rho, \nu, \gamma_s, D, g). \tag{4.11}$$

Since a physical phenomenon is determined by the constant values of its characteristic parameters, one has to assume that both Q and Q_s remain constant throughout the duration of development of the regime channel R_1.[7] The form ($f_{A_{R_1}}$) of the function (4.11), and thus the value of the property A_{R_1} supplied by it depends on the choice of A_* (and, of course, on the choice of the resistance and transport rate equations).

The regime channels of extremal theories must not necessarily be defined by the set of parameters (4.10): other definitions are also conceivable. For example, in Ref. [31] the regime channel is defined by the following seven parameters:

$$S, Q, \rho, \nu, \gamma_s, D, g, \tag{4.12}$$

i.e. it is assumed that this channel, R_2 say, develops for the constant slope S (rather than for the constant transport rate Q_s). In fact, according to Ref. [31], the regime state establishes itself when the (variable) transport rate Q_s ($= A_*^{-1}$) or, to be more exact, the sediment concentration ($\sim Q_s$), reaches its maximum. Broadly speaking, one can say that the regime channel R_2 is determined as the solution of the following system of equations:

---

[7] In the case of a (finite-length) laboratory channel, it must thus be ensured that the same (constant) rate of sediment, Q_s, is fed into the channel entrance for any t.

$$\left.\begin{array}{l}\text{(resistance equation)}\\ \text{(transport equation)}\\ \text{(minimum } Q_s^{-1}\text{)}\ .\end{array}\right\} \qquad (4.13)$$

All possible regime channels R_i of the extremal theories are determined by (at least) seven parameters whose numerical values must be known beforehand – a fact that can present some difficulties when using these theories in practice. For example, neither Q nor Q_s is constant in a natural river. Yet, although some methods are available to estimate a channel-forming Q (dominant discharge), no method is available to estimate a "channel-forming Q_s". Similarly, the case $S = const$ is also of a limited practical interest, for (as will be apparent in Chapter 5) it is exactly the *variation* of S which induces the river formation in general, and the development of meandering and braiding in particular. In the following, we will deal only with the (classical) regime channel R where Q_s establishes itself as an outcome of the two-phase motion.

It should be noted that the regime channels R_1 and R_2 are identical if R_1 is based on $A_* = S \rightarrow$ min. Indeed, consider Bagnold's form[8]

$$Q_s \sim Bv[\tau_0 - (\tau_0)_{cr}] = Bv[\gamma Sh - (\tau_0)_{cr}] = \gamma Q\,[S - f(h)] \qquad (4.14)$$

$$(\text{with } f(h) = (\tau_0)_{cr}/\gamma h)$$

which yields, for $Q = const$,

$$dQ_s \sim [dS - f'(h)dh]\,. \qquad (4.15)$$

In the case of R_1 we have $Q_s = const$, i.e. $dQ_s = 0$, and (4.15) gives

$$dS_1 = f'(h_1)dh_1\,. \qquad (4.16)$$

In the case of R_2 we have $S = const$, i.e. $dS = 0$, and (4.15) gives

$$d(Q_s)_2 \sim -f'(h_2)dh_2\,. \qquad (4.17)$$

According to Ref. [31], R_1 and R_2 must be compared for "the same values of breadth, velocity and depth". Hence $h_1 = h_2$, and from (4.16) and (4.17) one obtains

$$dS_1 \sim -\ d(Q_s)_2\,, \qquad (4.18)$$

which indicates that the decrement of S_1 and the increment of $(Q_s)_2$ are interrelated by a constant proportionality, and thus that the minimum of S_1 and the maximum of $(Q_s)_2$ must occur for the same B_R ($= B_{R_1} = B_{R_2}$). A more elaborate version of this demonstration is given in Ref. [31].

4.3 Dimensionless Formulation of the Regime Channel *R*

4.3.1 *Basic dimensionless forms*

i- From the critical stage relation $Y_{cr} = \Psi(\xi) = \Psi^*(\xi^3)$, i.e.

---

[8] It is assumed, for the sake of simplicity, that $u_b \sim v$ and $\lambda_c \approx 1$ (see (1.60)).

$$\frac{\rho v_{*cr}^2}{\gamma_s D} = \Psi^*\left(\frac{\gamma_s D^3}{\rho \nu^2}\right), \tag{4.19}$$

it is clear that γ_s is a function of ρ, ν, v_{*cr} and D only:

$$\gamma_s = f(\rho, \nu, v_{*cr}, D). \tag{4.20}$$

Substituting (4.20) in (4.2), one realizes that a property A_R of the regime channel R can be expressed equally well as

$$A_R = \bar{f}_{A_R}(Q, \rho, \nu, v_{*cr}, D, g). \tag{4.21}$$

From a physical standpoint, the relation (4.21) is more advantageous than its mathematical equivalent (4.2), for v_{*cr} reflects the resistance of alluvium to erosion in a more accomplished manner than γ_s.

Selecting Q, ρ and v_{*cr} as basic quantities and using the π-theorem, one determines the following dimensionless form of (4.21):

$$\Pi_{A_R} = Q^x \rho^y v_{*cr}^z A_R = \Phi^*_{A_R}(X_\nu, X_D, X_g), \tag{4.22}$$

where the dimensionless variables

$$X_\nu = \frac{v_{*cr}\lambda}{\nu}, \quad X_D = \frac{D}{\lambda} \quad \text{and} \quad X_g = \frac{v_{*cr}^2}{g\lambda} \tag{4.23}$$

reflect the influence of ν, D and g respectively; the "length"

$$\lambda = \sqrt{\frac{Q}{v_{*cr}}} \tag{4.24}$$

is the *typical length* of the phenomenon.

The material number ξ is completely determined by the dimensionless variables X_ν and X_D. Indeed,

$$\xi^3 = \frac{\gamma_s D^3}{\rho \nu^2} = \frac{(X_\nu X_D)^2}{Y_{cr}} = \frac{(X_\nu X_D)^2}{\Psi(\xi)}, \tag{4.25}$$

and therefore the consideration of X_D and ξ is equivalent to that of X_D and X_ν. Hence, (4.22) can be replaced by

$$\Pi_{A_R} = Q^x \rho^y v_{*cr}^z A_R = \Phi^{**}_{A_R}(\xi, X_D, X_g), \tag{4.26}$$

which yields for the dimensionless regime width, depth and slope

$$\left.\begin{aligned}\Pi_{B_R} &= \frac{B_R}{\lambda} = \Phi^{**}_{B_R}(\xi\,,X_D,X_g)\\ \bar{\Pi}_{h_R} &= \frac{h_R}{\lambda} = \bar{\Phi}^{**}_{h_R}(\xi\,,X_D,X_g)\quad{}^{9}\\ \Pi_{S_R} &= S_R = \Phi^{**}_{S_R}(\xi\,,X_D,X_g)\,.\end{aligned}\right\}\tag{4.27}$$

ii- Let $x_R = \Phi^{**}_{x_R}(\xi, X_D, X_g)$ be the regime value of a particular dimensionless property Π_A ($= x$). Eliminating X_D between $x_R = \Phi^{**}_{x_R}(\xi, X_D, X_g)$ and (4.26), one can express any Π_{A_R} ($\neq x_R$) as

$$\Pi_{A_R} = Q^x \rho^y v_{*cr}^z A_R = \Phi^{***}_{A_R}(\xi, x_R, X_g)\,. \tag{4.28}$$

In particular, one can express (4.27) as

$$\left.\begin{aligned}\Pi_{B_R} &= \frac{B_R}{\lambda} = \Phi^{***}_{B_R}(\xi, x_R, X_g)\\ \bar{\Pi}_{h_R} &= \frac{h_R}{\lambda} = \bar{\Phi}^{***}_{h_R}(\xi, x_R, X_g)\\ \Pi_{S_R} &= S_R = \Phi^{***}_{S_R}(\xi, x_R, X_g)\,.\end{aligned}\right\}\tag{4.29}$$

Clearly, out of all the mathematically possible x_R, it is only one which renders the relations (4.29) physically meaningful; and the next subsection concerns its determination.

iii- We go over to reveal how B_R and h_R are affected by $X_g = v_{*cr}^{5/2}/gQ^{1/2}$. And for this purpose we express the functions $\Phi^{***}_{B_R}$ and $\bar{\Phi}^{***}_{h_R}$ as

$$\Phi^{***}_{B_R}(\xi, x_R, X_g) = X_g^p\,\Phi_{B_R}(\xi, x_R) \quad \text{and} \quad \bar{\Phi}^{***}_{h_R}(\xi, x_R, X_g) = X_g^q\,\Phi_{h_R}(\xi, x_R)\,,$$

which makes it possible to express B_R and h_R in the following manner:

$$B_R = \sqrt{\frac{Q}{v_{*cr}}}\left(\frac{v_{*cr}^{5/2}}{gQ^{1/2}}\right)^p \Phi_{B_R}(\xi, x_R)\,, \tag{4.30}$$

$$h_R = \sqrt{\frac{Q}{v_{*cr}}}\left(\frac{v_{*cr}^{5/2}}{gQ^{1/2}}\right)^q \Phi_{h_R}(\xi, x_R)\,. \tag{4.31}$$

Experiment shows (see Table 4.1) that $B_R \sim Q^{1/2}$ in all cases, and (4.30) indicates that an explicit proportionality between B_R and $Q^{1/2}$ can be achieved

---

[9] The symbol Π_h, without "bar", is reserved for the final version of the dimensionless expression of h, which will be developed in the next section.

only if $p = 0$ (a possible (secondary) influence of Q, via x_R, is to be disregarded in the present context). Experiment also shows (Table 4.1) that, basically, $h_R \sim Q^{1/3}$ for sand channels,[10] and this means that $q = 1/3$. Hence,

$$\Pi_{B_R} = B_R \sqrt{\frac{v_{*cr}}{Q}} = \Phi_{B_R}(\xi, x_R) \tag{4.32}$$

and

$$\Pi_{h_R} = h_R \left(\frac{g}{Q v_{*cr}} \right)^{1/3} = \Phi_{h_R}(\xi, x_R), \quad ^{11} \tag{4.33}$$

where

$$\Pi_{h_R} = \overline{\Pi}_{h_R} X_g^{-1/3}. \tag{4.34}$$

4.3.2 *Variable, x_R; functions, Π_{B_R}, Π_{h_R} and S_R*

i- Consider now the flow resistance equation

$$\frac{Q}{B_R h_R} = c_R \sqrt{g S_R h_R} \tag{4.35}$$

which corresponds to "large" B_R/h_R. This equation can be brought identically into the dimensionless form

$$\left(\frac{Q/v_{*cr}}{B_R^2} \right) \left(\frac{Q v_{*cr}}{g h_R^3} \right) = c_R^2 S_R,$$

i.e.

$$\Pi_{B_R}^{-2} \Pi_{h_R}^{-3} = c_R^2 S_R, \tag{4.36}$$

where $c_R^2 S_R$ is the regime value of the Froude number

$$Fr = v^2/gh = c^2 S. \tag{4.37}$$

1- If the flow is rough turbulent, then $\xi \sim \nu^{-2/3}$ is no longer a variable, and the functions $\Phi_{B_R}(\xi, x_R)$ and $\Phi_{h_R}(\xi, x_R)$ become dependent on x_R alone:

$$\Pi_{B_R} = \phi_{B_R}(x_R) \quad ; \quad \Pi_{h_R} = \phi_{h_R}(x_R). \tag{4.38}$$

---

[10] In gravel bed channels, $K_s \sim D$ ($= const$) while $(\tau_0)_R \approx (\tau_0)_{cr}$. Consequently, knowing that the regime width of gravel channels is proportional to $Q^{1/2}$, one can determine the Q-exponent of their h_R (viz ≈ 0.43 [see (4.6)]) directly from the resistance equation $Q = B_R h_R c_R \sqrt{(\tau_0)_R/\rho}$. Indeed, using $B_R \sim Q^{1/2}$, $c_R \sim (h/D)^{1/6}$ and $(\tau_0)_R = (\tau_0)_{cr} = const$, one obtains $Q \sim Q^{1/2} h_R h_R^{1/6}$, and thus $h_R \sim [Q^{1/2}]^{6/7} = Q^{3/7} \approx Q^{0.43}$ (more details in 4.5.2). Hence, the present analysis of sand channels (whose K_s is not constant) covers automatically the gravel channels (having $K_s \approx const$).

[11] For a given Q, and the same remaining conditions, the width of rivers should thus be the same in different planets; their depth should decrease in inverse proportion to $g^{1/3}$.

Substituting (4.38) in (4.36), one determines

$$[\phi_{B_R}(x_R)]^{-2}[\phi_{h_R}(x_R)]^{-3} = (Fr)_R. \tag{4.39}$$

The left-hand side of this equation varies with x_R only; the right-hand side, with $(Fr)_R$ only. Hence x_R must be $(Fr)_R$:[12]

$$x_R = (Fr)_R. \tag{4.40}$$

Hereafter, Fr will be used instead of x.

Note from (4.39) (by considering (4.40)) that the determination of only one function is sufficient to furnish the expressions of both Π_{B_R} and Π_{h_R}. For example, if

$$\Pi_{B_R} = \phi_{B_R}((Fr)_R) \tag{4.41}$$

is determined, then

$$\Pi_{h_R} = \phi_{h_R}((Fr)_R) = \frac{1}{(Fr)_R^{1/3}}[\phi_{B_R}((Fr)_R)]^{-2/3}. \tag{4.42}$$

2- If the flow is not rough turbulent, then ξ is also a variable (in addition to $(Fr)_R$). The right-hand side of (4.36) consists only of $(Fr)_R$, and therefore ξ must appear in the expressions of Π_{B_R} and Π_{h_R} so that it is cancelled when $\Pi_{B_R}^2$ is multiplied with $\Pi_{h_R}^3$. In other words, ξ must affect Π_{B_R} and Π_{h_R} by means of the same multiplier-function $\phi_\xi(\xi)$ albeit in different powers:

$$\Pi_{B_R} = \Phi_{B_R}(\xi,\ (Fr)_R) = \phi_\xi(\xi)\ \phi_{B_R}((Fr)_R) \tag{4.43}$$

$$\Pi_{h_R} = \Phi_{h_R}(\xi,\ (Fr)_R) = \frac{[\phi_\xi(\xi)]^{-2/3}}{(Fr)_R^{1/3}}[\phi_{B_R}((Fr)_R)]^{-2/3}. \tag{4.44}$$

Clearly, $\phi_\xi(\xi)$ must approach unity with the increasing values of ξ.

ii- The regime slope S_R does not require the introduction of any "new" function(s): if Π_{B_R} and Π_{h_R} are known, then S_R is virtually known. Indeed, it has been shown in Chapter 3 that c can be computed from the transcendental function (3.88), which can be shown symbolically as $c = \phi_c(\xi, \eta, N)$. Since

$$\eta_R = \frac{gS_Rh_R}{v_{*cr}^2} = X_g^{-2/3}\Pi_{h_R}S_R \quad \text{while} \quad N_R = \frac{Q}{B_R D v_{*cr}} = \Pi_{B_R}^{-1}X_D^{-1}, \tag{4.45}$$

one can express $S_R = (Fr)_R c_R^{-2} = \Pi_{B_R}^{-2}\Pi_{h_R}^{-3}c_R^{-2}$ as

$$S_R = (\Pi_{B_R}^{-2}\Pi_{h_R}^{-3})\,[\phi_c(\xi,\ X_g^{-2/3}\Pi_{h_R}S_R,\ \Pi_{B_R}^{-1}X_D^{-1})]^{-2}. \tag{4.46}$$

---

12 It is pointless to insist that x_R should be identified with a function of $(Fr)_R$ rather than with $(Fr)_R$ itself, for the forms of $\phi_{B_R}(x_R)$ and $\phi_{h_R}(x_R)$ are not specified.

Knowing Π_{B_R} and Π_{h_R}, one can always compute S_R from this relation.

Hence, the determination of a (wide) regime channel (i.e. of B_R, h_R and S_R) corresponding to a given set of characteristic parameters (4.1) rests on the knowledge of two functions, viz ϕ_{B_R} (or ϕ_{h_R}) and ϕ_ξ; and of the regime Froude number $(Fr)_R$. If the flow is rough turbulent, then the knowledge of only ϕ_{B_R} (or ϕ_{h_R}) and $(Fr)_R$ is sufficient.

4.4 ($Fr \rightarrow$ min) – the Basis of Regime Development

4.4.1 *The quantity A_**

i- Consider now the channel which is still at the stage of its regime development ($t < T_R$; $\theta < 1$). The expressions of its dimensionless properties Π_A, including $Fr = c^2S$, can differ from those of Π_{A_R} only because of θ (see (4.2) and (4.3)). Thus, instead of (4.26), we have

$$\Pi_A = \Phi_A^{**}(\xi, X_D, X_g, \theta)\,, \tag{4.47}$$

and, in particular,

$$Fr = \Phi_{Fr}^{**}(\xi, X_D, X_g, \theta)\,. \tag{4.48}$$

The elimination of X_D from these two equations gives, formally,

$$\Pi_A = \Phi_A^{***}(\xi, Fr, X_g, \theta)\,, \tag{4.49}$$

which is but the extended version of (4.28) (with $x = Fr$). Now, the introduction of θ in the expressions (4.47) and (4.48) is necessary. Indeed, ξ, X_D and X_g remain constant in the course of an experiment, and therefore they cannot by themselves reflect the time-variation of Π_A and Fr during that experiment. This is not so in the case of (4.49), where we have Fr which varies with time and which can thus act as a "clock". The regime channel formation is a gradual (non-impulsive) process, and therefore it is unlikely that its properties Π_A should still be dependent on time ($t \sim \theta$) per se – if the time-varying Fr is already present in their expressions.

Out of all Π_A, we will be concerned here with Π_B and Π_h only; and it appears that their time variation is completely reflected by Fr alone. Indeed, it is generally accepted today that a developing regime flow can be treated as uniform, and thus that it satisfies the resistance equation

$$\Pi_B^{-2}\,\Pi_h^{-3} = Fr\,, \tag{4.50}$$

for *all* θ. Since this equation does not contain explicitly any time parameter, a possible presence of θ in the expressions of Π_B and Π_h must be such that it is cancelled in the product $\Pi_B^2\Pi_h^3$. This, however, would mean that one of Π_B and Π_h increases, whereas the other decreases with the passage of time. But such a scheme is unacceptable, for experiment shows that both $\Pi_B \sim B$

and $\Pi_h \sim h$ only increase with time in the (pertinent) interval $T_B < T < T_R$ (Fig. 4.12). Hence Π_B and Π_h cannot depend on θ (in addition to Fr): their expressions must be of the form

$$\Pi_B = \Phi_B^{***}(\xi, Fr, X_g) \quad \text{and} \quad \Pi_h = \Phi_h^{***}(\xi, Fr, X_g). \tag{4.51}$$

Since (4.50) and (4.51) are but (4.36) and (4.29) without the subscript R, the subscript-R-removed version of any relation derived from (4.36) and (4.29) in the preceding section must be valid also for the developing channels. Hence, we have (on the basis of (4.43), (4.44) and (4.46))

$$\Pi_B = B\sqrt{\frac{v_{*cr}}{Q}} = \Phi_B(\xi, Fr) = \phi_\xi(\xi)\,\phi_B(Fr) \tag{4.52}$$

$$\Pi_h = h\left(\frac{g}{Qv_{*cr}}\right)^{1/3} = \Phi_h(\xi, Fr) = \frac{[\phi_\xi(\xi)]^{-2/3}}{Fr^{1/3}}[\phi_B(Fr)]^{-2/3} \tag{4.53}$$

and

$$S = (\Pi_B^{-2}\Pi_h^{-3})\,[\phi_c(\xi,\ X_g^{-2/3}\Pi_h S,\ \Pi_B^{-1}X_D^{-1})]^{-2}. \tag{4.54}$$

Attention is drawn to the fact that the functions

$$\phi_B(Fr) \quad \text{and} \quad \phi_h(Fr) \tag{4.55}$$

are *essentially different* from the functions

$$\phi_{B_R}((Fr)_R) \quad \text{and} \quad \phi_{h_R}((Fr)_R). \tag{4.56}$$

The functions (4.56) indicate how the *developed* regime values of Π_B and Π_h corresponding to *different* experiments vary with $(Fr)_R$; whereas the functions (4.55) indicate how the *developing* values of Π_B and Π_h vary with Fr in the course of the *same* experiment. The forms of ϕ_{B_R} and ϕ_{h_R} are knowable (they can be discovered); the forms ϕ_B and ϕ_h are not knowable: they vary depending on the nature of experiment, i.e. depending on the nature of the initial channel (h_0, B_0, S_0).

ii- Among all the quantities involved in Eqs. (4.52), (4.53) and (4.54), only B, h, S and Fr vary with time during the regime channel formation. Hence, if the development of a regime channel − which manifests itself by the time variation of B, h and S − takes place because a (necessarily time-varying) quantity A_* tends to acquire its minimum, as suggested by the extremal theories, then this quantity A_* cannot be anything else but the Froude number Fr:

$$A_* = Fr. \tag{4.57}$$

The Froude number is a "measure" of the energy structure of flow. Indeed, if the kinetic energy of the unit fluid volume is characterized by $e_k = (1/2)\,\rho\, v^2$ and its cross-sectional potential energy by $e_p = \rho g h$, then the ratio e_k/e_p is (half) the Froude number:

$$\frac{e_k}{e_p} = \frac{1}{2} Fr. \tag{4.58}$$

One can say that the regime channel is forming because the flow tends to reduce its ratio e_k/e_p to a minimum:

$$(A_*)_{\min} = (Fr)_{\min} = (Fr)_R. \tag{4.59}$$

Note that the regime criterion $Fr = c^2S \to \min$ can be viewed as a "unification" of the previously proposed criteria, $S \to \min$ and $c^2 \to \min$ (see Section 4.2).[13]

4.4.2 *Fr-curves*

i- Using the flow resistance equation (4.35), in conjunction with

$$Z = \frac{h}{D}, \quad \eta = \frac{gSh}{v_{*cr}^2}, \quad Fr = c^2S, \quad N = \frac{Q}{BDv_{*cr}}, \tag{4.60}$$

and introducing

$$\alpha = \frac{\gamma_s}{\gamma} Y_{cr} = \frac{v_{*cr}^2}{gD} = \frac{X_g}{X_D}, \tag{4.61}$$

one determines

$$N = Z(c^2\eta)^{1/2} \quad \text{and} \quad Fr = \frac{\alpha}{Z}(c^2\eta) \tag{4.62}$$

and, consequently,

$$Fr = \frac{\alpha}{N}(c^2\eta)^{3/2}.^{14} \tag{4.63}$$

Eliminating (in principle) c between (4.63) and (3.89), one realizes that (for a specified α)

$$Fr = \phi_{Fr}(\xi, \eta, N). \tag{4.64}$$

Hence, one can always compute (in analogy to c) a family of *Fr-curves* for the $(Fr;\eta)$-plane corresponding to a given $\xi = (const)_i$, each individual curve implying

---

[13] The only work known to the author where it is explicitely stated that the regime formation is due to $Fr \to \min$ is due to Y. Jia [18] (1990). In this paper, the validity of $Fr \to \min$ is demonstrated by an independent approach: computer simulation in conjunction with the field data.

[14] Observe that although each of α, Z and N is dependent on D, the relations (4.62) and (4.63) are not dependent on it (D is cancelled from them). Note also that in the case of $\gamma_s/\gamma \approx 1.65$ (sand-or-gravel and water) the variation of α is insignificant, as it is due to Y_{cr} only.

$$Fr = \phi_{Fr}((const)_i\ , \eta\ , \ (const)_j) = \phi_{Fr,\ ij}(\eta) \tag{4.65}$$

(compare with (3.90)).

From the content of 3.6.2, it should be clear that in the case of a gravel bed (where bed forms are negligible) the c-curves are monotonous, as c_g in Fig. 4.1a, whereas in the case of a sand bed (where bed forms are prominent) they exhibit "dips", as c_s in Fig. 4.1a. The analogous is valid for the Fr-curves: in the case of gravel channels they are as $(Fr)_g$ in Fig. 4.1b; in the case of sand channels, as $(Fr)_s$. The main difference between the c-curves

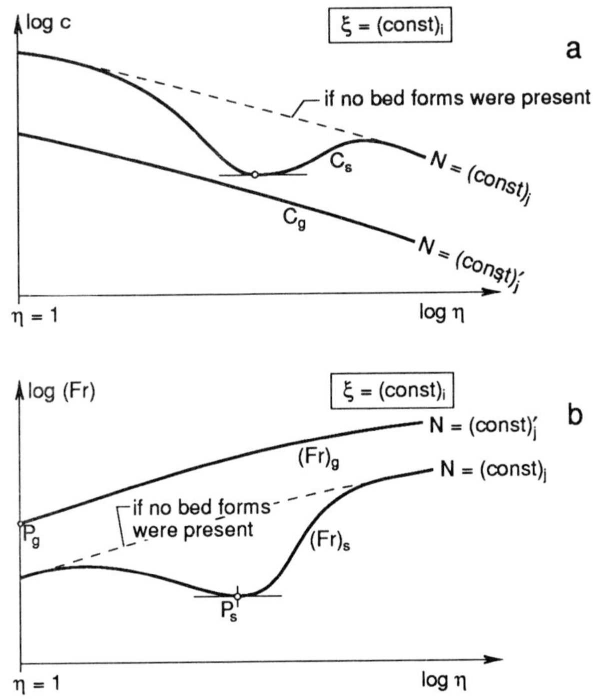

Fig. 4.1

and the Fr-curves is because the ordinates of the former curves (as a whole) increase, whereas those of the latter decrease with the decrement of η.

No transport of sediment is present, and thus no channel formation can take place if $\eta < 1$. Hence, in the present context, the Fr-curves are meaningful only in the region $1 \leq \eta < \infty$, which is treated in the following as the "existence region" of these curves.

Let the *regime-point* be that point P_R on the $(Fr;\eta)$-plane where $Fr = (Fr)_{\min}$ (Eq. (4.59)):

1) If a regime channel experiment is associated with a Fr-curve of the

type $(Fr)_s$ (sand),[15] then the regime point P_R is the lowest point P_s of the curve $(Fr)_s$ (Fig. 4.1b):[16]

$$P_R = P_s . \tag{4.66}$$

2) If, on the other hand, the experiment is associated with a Fr-curve of the type $(Fr)_g$ (gravel), then P_R is the lowest point P_g of the curve $(Fr)_g$, at $\eta = 1$:

$$P_R = P_g \quad (\text{at } \eta = 1). \tag{4.67}$$

4.4.3 *(Fr → min) and experimental data*

The sources of field and laboratory data which appear to be suitable to test the regime relations are given in "References B" (Refs. [1b] to [18b]). This data-set, which ranges from less than one-meter-wide laboratory streams to more than one-kilometer-wide rivers, corresponds to sand-or-gravel and water only ($\alpha \approx (1.65)(0.04) \approx 0.07$). The channel-forming Q of a natural river was identified with its bankfull flow rate. Only the data-points which correspond to supercritical flows ($(Fr)_R > 1$) and which remain outside the range $5 < B_R/h_R < 100$ were excluded.[17]

The values of $(Fr)_R$ supplied by the aforementioned data have been plotted on the $(Fr;\eta)$-planes corresponding to various $\xi \sim D$: Figs. 4.2, 4.3 and 4.4 are three examples of these plots.[18] The Fr-curves forming the background in these graphs were computed, for each $\xi = (const)_i$ and $N = (const)_j$, as implied by (4.64) and (4.65).

i- Fig. 4.2 contains the regime data of gravel channels. The scatter is gross; nonetheless, it can be observed that the data-points cluster around $\eta = 1$, so as to justify (4.67).

---

[15] This association will be clarified in the next section.

[16] Few exceptions to this general rule will be discussed in Section 4.7.

[17] The lower limit (5) is to ensure that the channel can be regarded as "wide", the upper limit (100) to eliminate a possible tendency for braiding, which may affect the regime-formation of the (single) channel under study.

[18] Figs. 4.2, 4.3 and 4.4 typify gravel, fine sand and coarse sand respectively: their Fr-curves were determined for $D = 50mm$, $0.17mm$ and $1mm$. In order to increase the number of points plotted, here too (in analogy to the plots presented in 3.6.2), the data of comparable D were also included. (The D-range plotted is particularly large in Fig. 4.2 where it is permissible. Indeed, if D is "large" and N is "moderate" ($D > 50mm$; $N < 10^2$, say) then the shape of the c_g- and $(Fr)_g$-curves remains invariant; only their level, which has no bearing on the determination of η_R, varies). Also in analogy to 3.6.2, the present data-points are represented by digits, each digit being that integer n in $N = 10^n$ which is the nearest to the N-value of the point plotted.

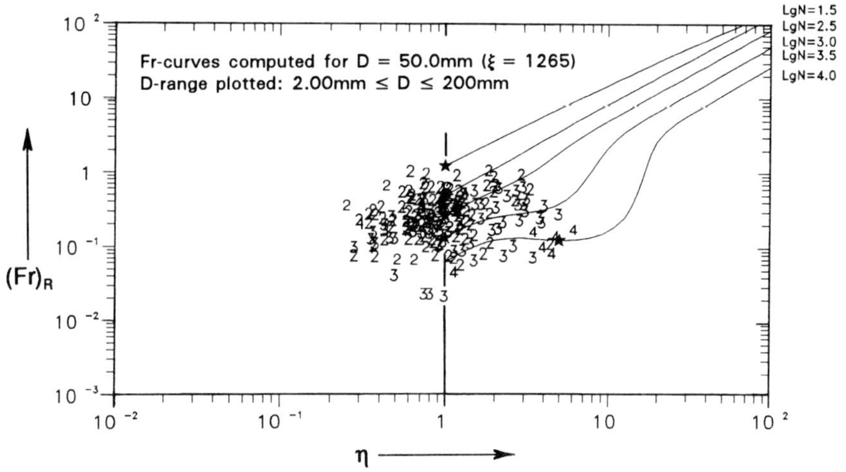

Fig. 4.2

ii- Fig. 4.3 contains the fine sand regime channel data. Thus the channel bed is covered by both ripples and dunes. Here too, the scatter is gross. Yet, it is noticeable that the data-points tend to locate themselves around the line ("valley") L_s formed by the points P_s (Eq. (4.66)). The line L_c in Fig. 4.3

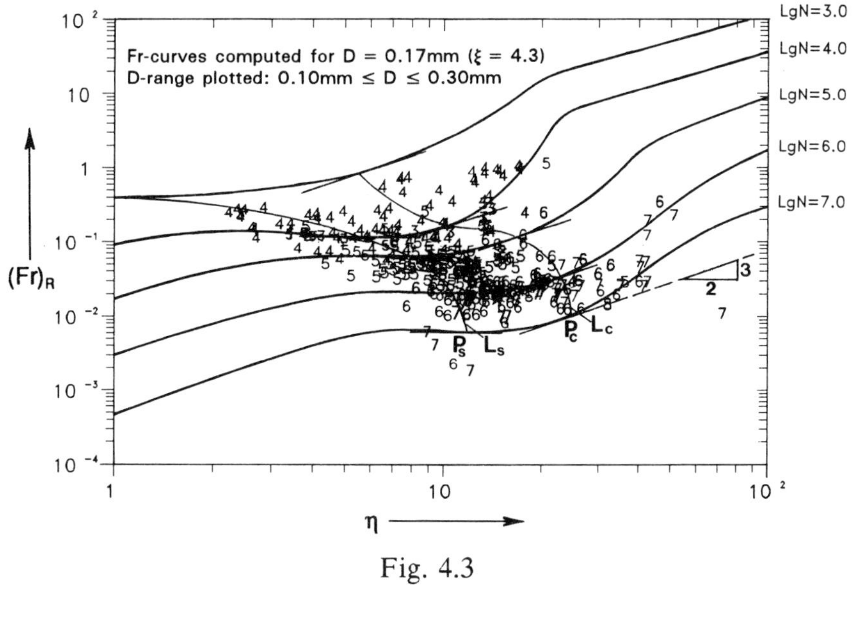

Fig. 4.3

connects those points P_c of the Fr-curves where $c = c_{\min}$. The difference between L_s and L_c is less than the degree of lateral scatter, and this may well be the reason for the emergence of the maximum friction factor ($\sim 1/c^2$) hypothesis (Refs. [9], [10]). The possibility that the regime channel formation is perhaps due to $c \rightarrow c_{\min}$, rather than to $Fr \rightarrow (Fr)_{\min}$, must be ruled out, at least because of the case (i) above. Indeed, the gravel channel data cluster around $\eta = 1$, where $Fr = (Fr)_{\min}$ but $c = c_{\max}$ (see the shape of the c_g-

curves): one can hardly accept a hypothesis which attributes the channel formation to $c \rightarrow c_{min}$ if D is "small", and to $c \rightarrow c_{max}$ if D is "large".[19]

iii- Fig. 4.4 contains the coarse sand channel data, and the regime channel bed is covered mainly by dunes. These dunes are not prominent even when

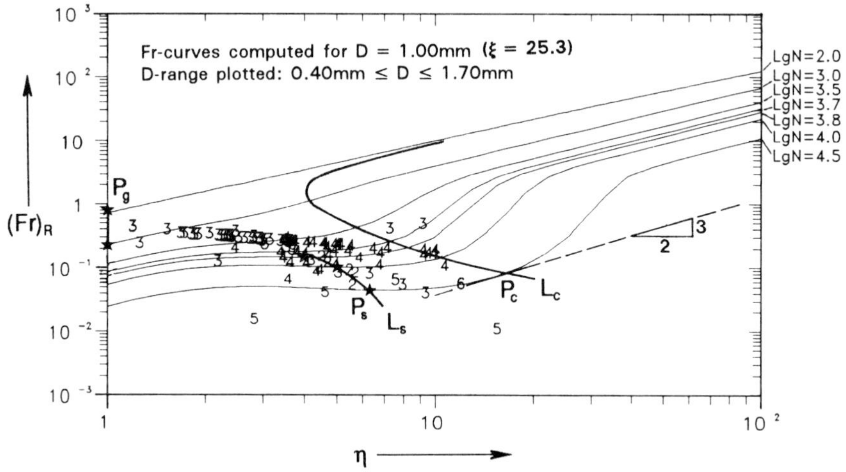

Fig. 4.4

N is large ($N = 10^{3.5}$ to $10^{4.5}$), and their steepness is certainly negligible when N is small ($N = 10^{2.0}$ to $10^{3.0}$). Thus, this (deliberately selected) case is transitional between (i) and (ii): the regime points ("stars") of the *Fr*-curves corresponding to $N = 10^{2.0}$ and $10^{3.0}$ are at $\eta = 1$, those corresponding to $N = 10^{3.5}$ to $10^{4.5}$ are on the (short) line L_s. Hence, the data must be expected to be scattered between L_s and $\eta = 1$: smaller N to the left, larger N to the right. In general, the points in Fig. 4.4 tend to justify this expectation.

[If the curvature of the (generally inclined) *Fr*-curve is only slight (faint dunes), then this *Fr*-curve may have a 3/2-inclined tangent but no horizontal tangent. The point P_c of such a *Fr*-curve may correspond to a substantial η although its regime point (P_g) is at $\eta = 1$ (*Fr*-curve corresponding to $N = 10^3$ in Fig. 4.4). Consequently, the curve L_c diverges from the regime locations corresponding to small N].

4.4.4 $(Fr;\eta)$*-plane; Sand-like and gravel-like behaviour*

i- The *Fr*-curves on the $(Fr;\eta)$-plane corresponding to a constant ξ are shown schematically in Fig. 4.5. The transition from the $(Fr)_s$-type curves to the extreme $(Fr)_g$-type curves (which do not have any inflection points) with the

---

[19] The points P_c, forming the line L_c, are determined as follows. Put $N = (const)_j$ in (4.63). One obtains $Fr \sim [c^3]\eta^{3/2}$ which, for any given c, is the equation of a 3/2-inclined straight line (in the log-log $(Fr;\eta)$-plane). Hence $c = c_{min}$, if it exists, is at that point P_c of the *Fr*-curve where its tangent has the 3/2-inclination.

decrement of N is continuous. In addition to their local minimum points P_s, the $(Fr)_s$-type curves also have their local maximum points P_s'. When N decreases, then the curvature of the Fr-curves decreases as well, and the points P_s and P_s' come closer and closer to each other. At a certain $N = N_*$ they merge into a single point P_*, which is the inflection point (with a horizontal tangent) of the Fr_*-curve corresponding to N_*. From Fig. 4.5, one infers that

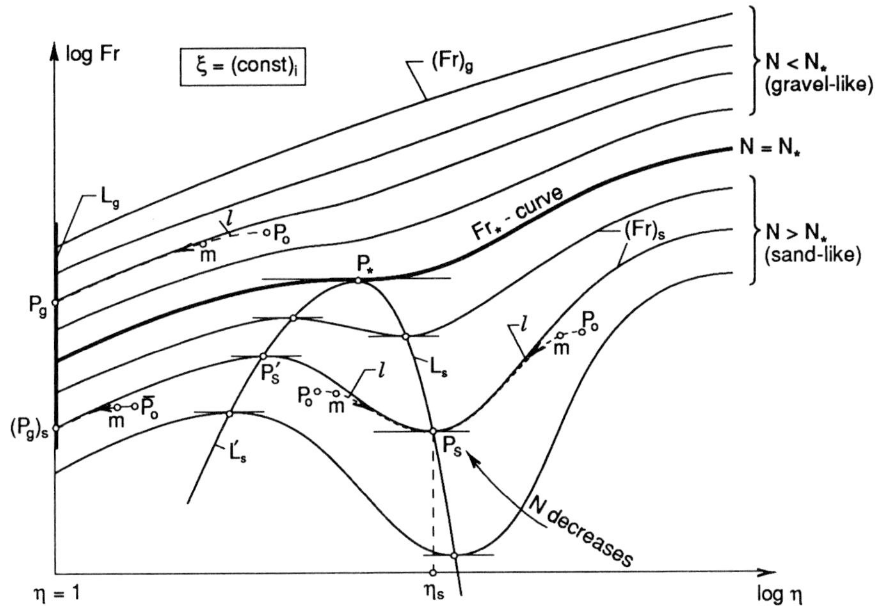

Fig. 4.5

the curves L_s and L_s' (formed by the points P_s and P_s') are, in fact, but two branches of a single curve, which is tangent to the Fr_*-curve at P_*. Clearly, the Fr-curves situated above the Fr_*-curve do not have horizontal tangents, i.e. they do not have the proper minimums in the sense $\partial Fr/\partial\eta = 0$ – they merely have the smallest values of Fr at $\eta = 1$. The Fr-curves situated below the "boundary-curve" Fr_* will be called *sand-like curves*; those situated above Fr_*, *gravel-like curves*.

ii- Consider an "experiment" which is determined by some specified values of the characteristic parameters (4.1). It starts, at $t = 0$, in an initial channel which is, to some extent, arbitrary.[20] The development of a regime channel can be pictured by the motion of a point m in the $(Fr;\eta)$-plane. Let P_0 be the point representing the values $Fr = (Fr)_0$ and $\eta = \eta_0$ of the initial channel:

---

[20] The restrictions imposed on the initial channel will be discussed in Section 4.7.

at $t = 0$, the point m is at P_0. With the passage of time, the point m moves, along a path l, towards its "target" regime location P_R: at $t = T_R$, it reaches it.

Each experiment generates "its own" regime width B_R. Hence, it is associated with its own value of $N_R = Q/(B_R D v_{*cr})$ and consequently with its own Fr-curve, which will be referred to in the following as the *associated Fr-curve* or the *regime Fr-curve* ($(Fr)_R$-curve). Each regime point P_R, to which all possible paths l must merge when t approaches T_R, is, of course, on the associated Fr-curve. If the associated curve is above the curve Fr_*, then P_R is the point P_g at $\eta = 1$ and m will move toward it: it will exhibit the *gravel-like behaviour.* If, on the other hand, the associated curve is below the curve Fr_*, then m will move toward the point P_s ($= P_R$): it will exhibit the *sand-like behaviour* (Fig. 4.5).

iii- There is no doubt that if e.g. $D = 2cm$, then we are dealing with gravel (and $P_R = P_g$ is on $\eta = 1$) and if $D = 0.2mm$, then we are dealing with sand (and $P_R = P_s$ is on L_s). The situation is not so obvious, however, if D is $1mm$ or $3mm$, say. From the explanations above, it should be clear that the gravel- or sand-like behaviour does not depend on $D \sim \xi$ alone; it depends also on N_R:

$$\text{if } N_R > N_*, \text{ then sand-like} \;;\; \text{ if } N_R < N_*, \text{ then gravel-like.} \quad (4.68)$$

The criterion for the prediction of sand- or gravel-like behaviour is thus given solely on the basis of the "geometry" of the $(Fr)_R$-curves. In some (very rare) cases, the initial point of a sand-like associated Fr-curve may be so near to $\eta = 1$, that the point m may "choose" to minimize Fr by moving to the (nearby) point $(P_g)_s$ (see $\overline{P}_0$ in Fig. 4.5) rather than to P_s, and thus it may exhibit the gravel-like behaviour. However, we will not encumber the present explanations by considering such (rare) cases, and we will continue to identify the sand-like behaviour with the sand-like $(Fr)_R$-curves. (The conditions presented by various P_0 will be discussed in 4.7.2).

Let η_s be the abcissa of a sand-like regime point P_s. As can be inferred from Figs. 4.3 and 4.4, the values of η_s are usually larger than ≈ 3, say, and thus they are much larger than $\eta = 1$. In the regime data plots presented in the next section, the sand- and gravel-like points were distinguished by using the regime values of η: if $\eta_R > 3$, then sand-like; otherwise, gravel-like. In the regime channel computation method presented in Section 4.6, the gravel-like behaviour is revealed by the fact that the computer does not yield $\partial Fr/\partial\eta = 0$ for $\eta_R > 1$ (and if it does, then the behaviour is sand-like).

iv- The conversion of the sand-like behaviour to the fundamentally different gravel-like behaviour (or vice-versa) does not occur gradually along a transitional interval of N; but rather at a certain value N_* (which varies depending on ξ [26]). This is analogous to the (sudden) conversion of a laminar flow to a turbulent one at a certain Re, or to the conversion of a supercritical flow to a subcritical one at $Fr = 1$.

v- The curve L_s, which connects the regime points P_s of various experiments, is predictable (it can be computed from (4.63) and (4.64)): the paths l are not. Indeed, each l connects a certain regime point P_s to a non-certain point P_0 (whose location on the $(Fr;\eta)$-plane is "to some extent arbitrary").
It has been stated in 4.4.1 (i) that the functions (4.55) show how the *developing* values of Π_B and Π_h vary with Fr in the course of a given experiment, whereas the functions (4.56) indicate how the *developed* regime values Π_{B_R} and Π_{h_R} vary with $(Fr)_R$ from one experiment to another. From the aforementioned, it should be clear that the functions (4.55) represent the variations of Π_B and Π_h along a path l: when the point m moves along l, then its properties Π_B and Π_h vary (develop) as functions of its continually decreasing ordinates Fr. Since P_0, and thus l, is unpredictable, the functions (4.55) are also unpredictable. In contrast to this, the functions (4.56) indicate how Π_{B_R} and Π_{h_R} vary with the ordinates $(Fr)_R$ of the (predictable) curve L_s. The explanations above are given in terms of the sand-like curve L_s. Clearly, the analogous is valid also for the vertical line, L_g say, which connects the points P_g at $\eta = 1$.

4.5 Regime Equations

4.5.1 *Formulation of* Π_{B_R} *and* Π_{h_R}

It is intended now to reveal, with the aid of experimental data, the nature of the functions (4.43) and (4.44), viz

$$\Pi_{B_R} = \phi_\xi(\xi)\,\phi_{B_R}((Fr)_R) \quad \text{and} \quad \Pi_{h_R} = [\phi_\xi(\xi)]^{-2/3}\phi_{h_R}((Fr)_R)\,,$$

where

$$\phi_{h_R}((Fr)_R) = [\phi_{B_R}((Fr)_R)]^{-2/3}(Fr)_R^{-1/3}. \qquad (4.69)$$

The combinations $\Pi_{B_R}/\phi_\xi(\xi)$ and $\Pi_{h_R}[\phi_\xi(\xi)]^{2/3}$ are functions of $(Fr)_R$ alone. In the case of gravel channels, the turbulent flow is rough; hence, $\phi_\xi(\xi) \equiv 1$ (see 4.3.2), and the combinations mentioned reduce into Π_{B_R} and Π_{h_R} themselves.

The sand-channel combinations $\Pi_{B_R}/\phi_\xi(\xi)$ and $\Pi_{h_R}[\phi_\xi(\xi)]^{2/3}$ and their gravel-channel counterparts Π_{B_R} and Π_{h_R} computed from the available regime-data (Refs. [1b] to [18b]) are plotted versus $(Fr)_R$ in Figs. 4.6 to 4.9. The symbols, sources and the D-ranges of the data-points are given in Tables 4.2 and 4.3.

The scatter of experimental points in all four dimensionless plots is gross.[21] Yet it is entirely experimental:[22] the points do no tend to sort themselves out depending on $D \sim \xi$.

Table 4.2
(Sand-like)

Symbol	Range of D (mm)	No. of points	Sources
■	0.02 ≤ D ≤ 0.10	71	[4b], [5b], [6b], [7b], [10b], [14b], [16b], [17b]
+	0.11 ≤ D ≤ 0.20	84	[1b], [2b], [4b], [5b], [6b], [7b], [10b], [14b], [15b], [18b]
✶	0.21 ≤ D ≤ 0.60	180	[1b], [3b], [4b], [5b], [6b], [7b], [8b], [9b], [10b], [11b], [12b], [14b], [15b], [16b], [17b], [18b]
□	0.61 ≤ D ≤ 1.00	57	[1b], [3b], [4b], [5b], [6b], [7b], [8b], [9b], [10b], [14b], [16b], [17b], [18b]
×	1.20 ≤ D ≤ 2.05	28	[1b], [3b], [4b], [5b], [7b], [8b], [10b], [11b], [14b], [16b], [18b]

Table 4.3
(Gravel-like)

Symbol	Range of D (mm)	No. of points	Sources
✶	3.80 ≤ D ≤ 7.60	15	[2b], [6b], [7b], [10b], [17b]
+	12.0 ≤ D ≤ 400	169	[3b], [9b], [10b], [13b]

i- *Sand-like behaviour*

Being a substitute of the Reynolds number $X \sim \nu^{-1}$, the variable ξ must be expected to affect the regime channel formation (via earlier occurring ripples) when its values are small. On the basis of trial-plots (see Ref. [26]), it has been found that the least scatter of Π_{B_R} and Π_{h_R} occurs when $\phi_\xi(\xi)$ is of the form

---

21 The scatter in dimensionless plots is, as a rule, larger by far than that in corresponding dimensional plots.

22 Actually, the experimental measurements as such form only a minor reason for the gross scatter in regime plots. The major reason is usually due to the subjective assessements (*which* depth and width of an irregular stream is *the* depth and width of that stream; *which* value of its time varying Q is the channel-forming Q, etc.).

$$\phi_\xi(\xi) = 0.45\xi^{0.3} \text{ if } \xi \le 15 \text{ and } \phi_\xi(\xi) \equiv 1 \text{ if } \xi > 15. \qquad (4.70)$$

The function $\phi_\xi(\xi)$ in the ordinates of Figs. 4.6 and 4.7 was evaluated according to (4.70).

1. *Dimensionless flow width*

Consider Fig. 4.6. The first impression is that the (very scattered) point-pattern is simply parallel to the $(Fr)_R$-axis; i.e. that $\phi_{B_R} = \Pi_{B_R}/\phi_\xi(\xi)$ is independent of $(Fr)_R$ and thus of Q (for $(Fr)_R = Q^2/gh_R^3 B_R^2$) – which would be precisely what is meant by the trend $B_R \sim Q^{1/2}$ (4.1.2). However, a more careful examination of the pattern reveals that it is not exactly straight and

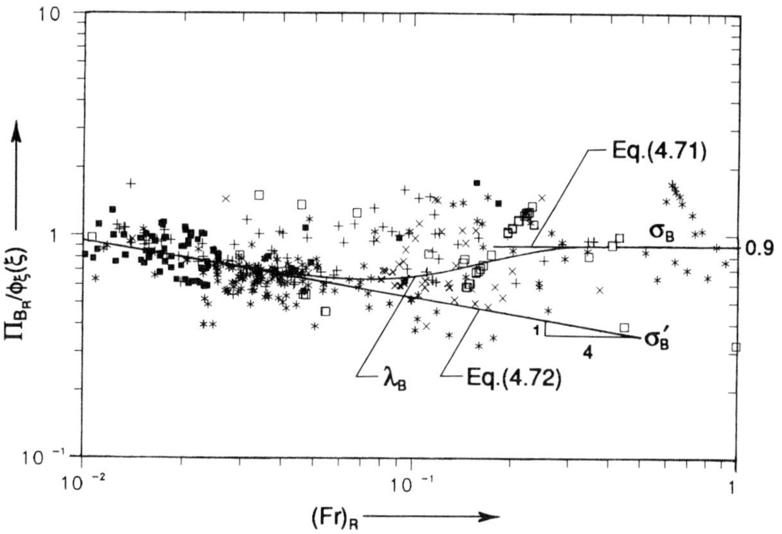

Fig. 4.6

horizontal: it can be better reflected by a faintly curved line, λ_B say, which exhibits a "dip" and which can be taken as parallel to the $(Fr)_R$-axis (as implied by the line σ_B) only for sufficiently large values of $(Fr)_R$. For small $(Fr)_R$, the line λ_B tends to become indistinguishable from the $\approx$1/4-declining straight line σ_B'.

a) The upper limit σ_B is of no particular interest: it merely conveys that if $(Fr)_R > \approx 0.2$, then the conventional proportionality $B_R \sim Q^{1/2}$ holds, and

$$\phi_{B_R}((Fr)_R) \approx 0.9 \qquad (4.71)$$

which is the equation of σ_B.

b) The lower limit σ_B' indicates that if $(Fr)_R < \approx 0.05$ (large lowland rivers; very fine sand), then

$$\phi_{B_R}((Fr)_R) \approx \frac{0.3}{(Fr)_R^{1/4}} , \tag{4.72}$$

which is the equation of σ_B'.

Extensive series of measurements were carried out over a long period of time in large lowland rivers of western Russia (including Volga, Oka, Kama, Dnepr, etc. [21], [22], [29], [30], [42]). Using this voluminous data, M.A. Velikanov [30] produced the following dimensionally homogeneous relation:

$$B_R = a \frac{Q^{1/2}}{(DgS_R)^{1/4}} , \tag{4.73}$$

where the dimensionless factor a is treated as a constant (≈ 2.5). The validity of the form (4.73) was subsequently verified by R.J. Garde [12], [13], who has compared it with additional data from various American rivers. Similarly, using the data obtained from equally extensive measurements carried out in rivers of Central Asia (including Amu-Darya, Syr-Darya, Kara-Darya, etc.), S.T. Altunin [1] suggested the form (see [42])

$$B_R \sim \frac{Q^p}{S_R^{p/2}} , \tag{4.74}$$

which is the same as (4.73) as far as the interrelation between B_R, S_R and Q is concerned, if p is identified with 1/2. (The largest channel-forming Q in the data-set used to determine the relations above was $\approx 8000 m^3/s$, the smallest S was ≈ 0.00002).[23] The relations of Artamanov and Annayev (reported in [42]) have also similar structure.

Let us now return to the equation (4.72), which can be expressed as

$$\frac{\Pi_{B_R}}{\phi_\xi(\xi)} = \frac{B_R}{\phi_\xi(\xi)} \left(\frac{v_{*cr}}{Q} \right)^{1/2} = \frac{0.3}{(Fr)_R^{1/4}} . \tag{4.75}$$

Considering that for $\gamma_s/\gamma = 1.65$ (sand-or-gravel, and water), the value of v_{*cr} is given by

$$v_{*cr} = \sqrt{(\gamma_s/\gamma) gDY_{cr}} = \sqrt{(1.65) gDY_{cr}} = 1.28\sqrt{gDY_{cr}} , \tag{4.76}$$

and taking into account that $(Fr)_R = c_R^2 S_R$, one determines from (4.75)

[23] The word "regime" is seldom used in Russian literature. Thus the original "river-bed-forming" flow rate is used here as the regime-channel-forming Q, while the width and slope of the "settled" river are identified with B_R and S_R.

$$\frac{1.13}{\phi_\xi(\xi)} \frac{B_R (gDY_{cr})^{1/4}}{Q^{1/2}} = \frac{0.3}{c_R^{1/2} S_R^{1/4}} , \tag{4.77}$$

and thus

$$B_R = \beta \frac{Q^{1/2}}{(DgS_R)^{1/4}} , \tag{4.78}$$

which is the same as (4.73). Here,

$$\beta = \left[\frac{0.26}{c_R^{1/2}} \frac{\phi_\xi(\xi)}{Y_{cr}^{1/4}} \right] . \tag{4.79}$$

It follows that the dimensionless factor a in (4.73) is not really a constant: it varies, however insignificantly, with c_R and ξ.[24]

Hence, the 1/4-declining straight line σ_B' in Fig. 4.6 can be viewed as the graph of the form (4.73). The total curve λ_B in Fig. 4.6 is the graph of the comp-eq.

$$\phi_{B_R}((Fr)_R) = \frac{0.3}{(Fr)_R^{1/4}} e^{-40(Fr)_R^2} + 0.9(1 - e^{-40(Fr)_R^2}) . \tag{4.80}$$

2. *Dimensionless flow depth*

The point pattern in Fig. 4.7 as a whole exhibits roughly a 1/3-declination (which would justify the trend $h_R \sim Q^{1/3}$). However, here too, the deviations from this overall trend are detectable. Substituting (4.71) and (4.72) into the second equation of (4.69), one obtains the following expressions for the upper and lower limits σ_h and σ_h' of the curve λ_h (which are the h-counterparts of σ_B, σ_B' and λ_B respectively):

$$\sigma_h: \ \phi_{h_R}((Fr)_R) = \frac{1.07}{(Fr)_R^{1/3}} . \tag{4.81}$$

$$\sigma_h': \ \phi_{h_R}((Fr)_R) = \frac{2.23}{(Fr)_R^{1/6}} . \tag{4.82}$$

The substitution of (4.80) in (4.69) yields the comp-eq. of the (total) curve λ_h itself:

---

[24] The (fine-sand) bed of a lowland river having small Fr is invariably covered by dunes and ripples; and their steepness in the regime state is nearly the largest. Hence, $\approx 8 < c < \approx 15$ (see the "dips" in Figs. 3.34 and 3.35) and thus $\approx 3 < c_R^{1/2} < \approx 4$, which does not reflect a significant variation. The same can be said for $\phi_\xi(\xi)/Y_{cr}^{1/4} \sim \xi^{0.3}/[\Psi(\xi)]^{1/4}$. (It is thus not surprising that a was treated as a constant).

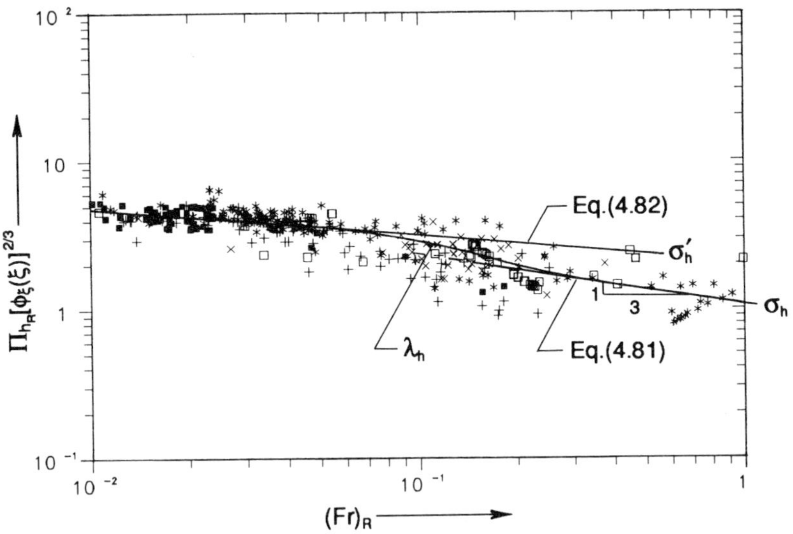

Fig. 4.7

$$\phi_{h_R}((Fr)_R) = \left[\frac{0.3}{(Fr)_R^{1/4}}\, e^{-40(Fr)_R^2} + 0.9(1 - e^{-40(Fr)_R^2})\right]^{-2/3} (Fr)_R^{-1/3}. \quad (4.83)$$

The lines σ_h, σ_h' and λ_h representing the above derived relations are in agreement with the pattern of data-points (Fig. 4.7).

From the content of this paragraph it should be clear that, in the case of sand channels, the proportionalities $B_R \sim Q^{1/2}$ and $h_R \sim Q^{1/3}$, strictly speaking, are valid when $(Fr)_R$ is sufficiently large.

ii- *Gravel-like behaviour*

In this case, $\phi_\xi(\xi) \equiv 1$ and the ordinates of the plots in Figs. 4.8 and 4.9 are Π_{B_R} and Π_{h_R} themselves.

1. *Dimensionless flow width*

The point-pattern in Fig. 4.8 is practically horizontal at the level $\approx$1.42. Hence

$$\Pi_{B_R} \approx 1.42\,, \quad (4.84)$$

which yields

$$B_R \approx 1.42\sqrt{\frac{Q}{\upsilon_{*cr}}}\,. \quad (4.85)$$

2. *Dimensionless flow depth*

Using (4.84) in (4.69), one obtains

$$\Pi_{h_R} = (1.42)^{-2/3}(Fr)_R^{-1/3} = 0.79(Fr)_R^{-1/3}. \quad (4.86)$$

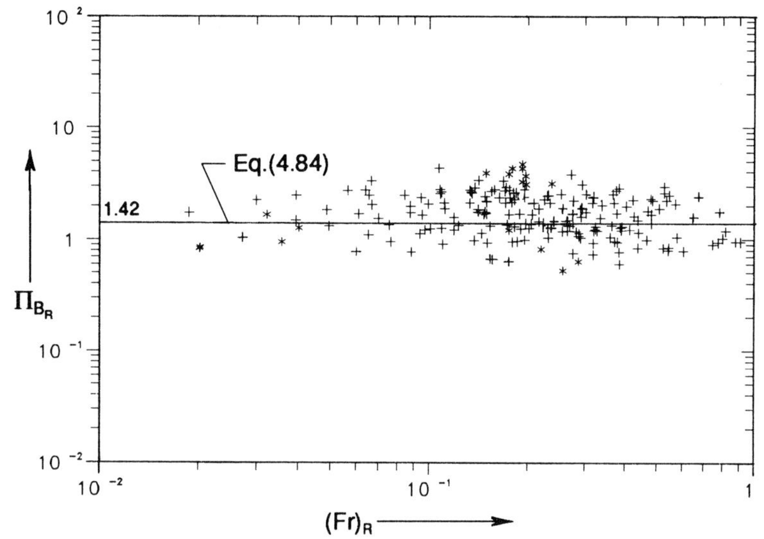

Fig. 4.8

The 1/3-declining straight line representing (4.86) and the corresponding pattern of experimental points are shown in Fig. 4.9.

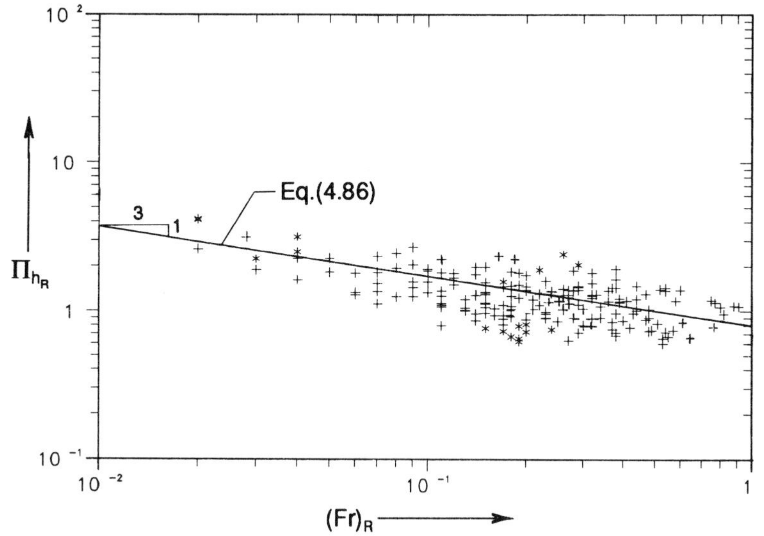

Fig. 4.9

It follows that the determination of Π_{B_R} and Π_{h_R}, for both sand- and gravel-like behaviours, rests entirely on the knowledge of $(Fr)_R$. If $(Fr)_R$ is known and Π_{B_R} and Π_{h_R} are determined (with the aid of the expressions or graphs presented in this subsection), then the regime slope S_R can be computed from (4.46). $(Fr)_R$ is thus the "key" to all three regime-channel characteristics, and its determination is explained in Section 4.6.

4.5.2 *Gravel channels with no bed forms*

In this (upper) limiting case of gravel-like behaviour, which is symbolized in Fig. 4.5 by the curve $(Fr)_g$ having no points of inflection, the energy losses are due to the granular skin roughness $k_s \approx 2D$ only, and the friction factor is given by

$$c_R = 7.66\left(\frac{h_R}{k_s}\right)^{1/6} \approx 6.82\left(\frac{h_R}{D}\right)^{1/6}. \tag{4.87}$$

Furthermore,

$$\eta_R = 1 \quad \text{i.e.} \quad gS_R h_R = v_{*cr}^2 . \tag{4.88}$$

The relations (4.87) and (4.88) are sufficient to furnish the expressions of h_R and S_R – the expression of B_R is already known (Eq. (4.85)). Indeed, substituting (4.85), (4.87) and (4.88) in the resistance equation

$$Q = B_R h_R c_R \sqrt{gS_R h_R} ,$$

one obtains

$$h_R = (0.143 D^{1/7})\left(\frac{Q}{v_{*cr}}\right)^{3/7} \tag{4.89}$$

which, with the aid of (4.88), yields

$$S_R = \left(6.993 \frac{v_{*cr}^2}{gD^{1/7}}\right)\left(\frac{Q}{v_{*cr}}\right)^{-3/7}. \tag{4.90}$$

Adopting for gravel channels $Y_{cr} = 0.05$ and taking $g = 9.81 m/s^2$, we obtain from (4.76)

$$v_{*cr} = 0.90 D^{1/2} \quad (\text{with } [v_{*cr}] = m/s). \tag{4.91}$$

Using this value of v_{*cr} in (4.85), (4.89) and (4.90), one arrives at the following (dimensionally non-homogeneous) relations:

$$B_R = 1.50 D^{-0.25} Q^{0.50} \tag{4.92}$$

$$h_R = 0.15 D^{-0.07} Q^{0.43} \tag{4.93}$$

$$S_R = 0.55 D^{1.07} Q^{-0.43}, \tag{4.94}$$

which are valid for metric units only. Note from Table 4.1 that these relations can be regarded as "averages" of the regime equations produced for width, depth and slope of gravel channels (Refs. [13a] to [20a]).

4.6 Computation of B_R, h_R and S_R

Since the friction factor $c = \phi_c(\xi, \eta, N)$ is determined by a transcendental equation (viz Eq. (3.88)), neither the regime value of c nor that of $Fr = \alpha(c^2\eta)^{3/2}/N$ can be computed readily. Considering this, the following computational procedure is suggested for the determination of B_R, h_R and S_R.

It is assumed that the numerical values of the characteristic parameters (4.1) are given, and thus that the values of the dimensionless variables X_g, X_D and ξ (and α) are known. Observe that N_R can be expressed as $N_R^{-1} = X_D \Pi_{B_R}$, indicating that the values of N_R and Π_{B_R} are unambiguously interrelated.

i- *Sand-like behaviour*:

1- Adopt $(B_R)_i$ and thus $(\Pi_{B_R})_i$ and compute $N_{Ri}^{-1} = X_D(\Pi_{B_R})_i$.

2- Knowing N_{Ri}, compute from

$$c = \phi_c(\xi, \eta, N_{Ri}) \text{ and } Fr = \alpha(c^2\eta)^{3/2}/N_{Ri}$$

(by eliminating c) such an $\eta = \eta_i$ which yields $Fr = (Fr_i)_{\min}$.

3- Using this $(Fr_i)_{\min}$ (as $(Fr)_R$) in (4.80), compute $\phi_{B_R} = (\Pi_{B_R})_{i+1}/\phi_\xi(\xi)$ and thus $(\Pi_{B_R})_{i+1}$:
If $(\Pi_{B_R})_{i+1} = (\Pi_{B_R})_i$, then the problem is solved: $\Pi_{B_R} = (\Pi_{B_R})_i$.
If $(\Pi_{B_R})_{i+1} \neq (\Pi_{B_R})_i$, then repeat the procedure by using $(\Pi_{B_R})_{i+1}$ until such an $i = j$ is reached which yields $(\Pi_{B_R})_{j+1} = (\Pi_{B_R})_j$ ($= \Pi_{B_R}$).

4- Knowing Π_{B_R} and consequently B_R, as well as $(Fr)_R$, compute h_R from $(Fr)_R = Q^2/gB_R^2h_R^3$. And knowing η_R, compute S_R from $\eta_R = gS_Rh_R/v_{*cr}^2$.

ii- *Gravel-like behaviour*:

1- Compute Π_{B_R} (from (4.84)), and $N_R^{-1} = X_D\Pi_{B_R}$.

2- Knowing N_R (together with ξ and α), compute c_R and thus $(Fr)_R$ from

$$c_R = \phi_c(\xi, 1, N_R) \quad \text{and} \quad (Fr)_R = \alpha(c^2\, 1)^{3/2}/N_R.$$

3- Knowing Π_{B_R} and consequently B_R, as well as $(Fr)_R$, compute h_R from $(Fr)_R = Q^2/gB_R^2h_R^3$ and S_R from $1 = gS_Rh_R/v_{*cr}^2$.

In case of doubt, it is advisable to start from (i), i.e. to start assuming that the behaviour is sand-like. If the relations $c = \phi_c(\xi, \eta, N)$ and $Fr = \alpha(c^2\eta)^{3/2}/N$ do not yield a minimum for Fr in the sense of $dFr = 0$ for $\eta \gg 1$ (as required by item (2)), then the regime channel formation is gravel-like. In this case the program should switch over to the gravel-like version (ii) and solve the problem accordingly.

Some examples of regime characteristics computed with the aid of the aforementioned method are shown below ($\gamma_s/\gamma = 1.65$, $\nu = 10^{-6} m^2/s$).

1) $Q = 1669.7 m^3/s$, $D = 0.18 mm$ (Bhagirathi River; Ref. [7b]):

Computed: $B_R = 230.0 m$ $h_R = 8.00 m$ $S_R = 0.000025$
$[N_R \approx 10^{6.5};\ (Fr)_R = 0.010;\ \eta_R = 11.0;\ c_R = 20.0]$

Reported in [7b]:
$B = 218.1 m$ $h = 5.95 m$ $S = 0.000058$
$[N \approx 10^{6.5};\ (Fr) = 0.028;\ \eta = 18.6]$

2) $Q = 1750 m^3/s$, $D = 0.40 mm$ (Solo River; V. Galay, personal communication):

Computed: $B_R = 300.0 m$ $h_R = 7.10 m$ $S_R = 0.000027$
$[N_R \approx 10^{6.0};\ (Fr)_R = 0.0097;\ \eta_R = 9.1;\ c_R = 19.0]$

Reported:
$B = 160.0 m$ $h = 5.20 m$ $S = 0.000089$
$[N \approx 10^{6.3};\ (Fr) = 0.0870;\ \eta = 22]$

3) $Q = 40 m^3/s$, $D = 0.70 mm$ (Solomon River; Ref. [15b]):

Computed: $B_R = 38.7 m$ $h_R = 1.96 m$ $S_R = 0.00015$
$[N_R \approx 10^{4.9};\ (Fr)_R = 0.0144;\ \eta_R = 7.3;\ c_R = 9.8]$

Reported in [15b]:
$B = 38.0 m$ $h = 2.30 m$ $S = 0.00026$
$[N \approx 10^{4.9};\ (Fr) = 0.0093;\ \eta = 14.8]$

4) $Q = 4386.0 m^3/s$, $D = 3.10 cm$ (North Saskatchewan River; Ref. [11b])

Computed: $B_R = 232.7 m$ $h_R = 6.60 m$ $S_R = 0.000385$
$[N_R \approx 10^{3.6};\ (Fr)_R = 0.1206;\ \eta_R = 1;\ c_R = 17.7]$

Reported in [11b]:
$B = 244.0 m$ $h = 7.62 m$ $S = 0.00035$
$[N \approx 10^{3.6};\ (Fr) = 0.0740;\ \eta = 1.12]$

Special case: If the flow width is constant (protected river banks, laboratory flume experiments), then it is only S and h which are subjected to the regime development. In this case the points P_0 and P_R are on the same (associated) Fr-curve: the computations are simplified because Π_B ($= \Pi_{B_R}$) is known beforehand.

Since the regime Froude number $(Fr)_R$ is a property (Π_{A_R}) of the phenomenon determined by the parameters (4.1), it must be a function (to the most) of ξ, X_D and X_g (see (4.26)). It is unlikely that $Fr \sim g^{-1}$ should be dependent on $X_g \sim g^{-1}$, and one would expect that

$$(Fr)_R = \Phi^{**}_{Fr}(\xi\,,\, X_D)\,. \tag{4.95}$$

The values of $(Fr)_R$ given by the available data (for both sand and gravel) are plotted versus X_D in Fig. 4.10. This graph may be helpful for estimation of the expected $(Fr)_R$ (shaded ribbon).

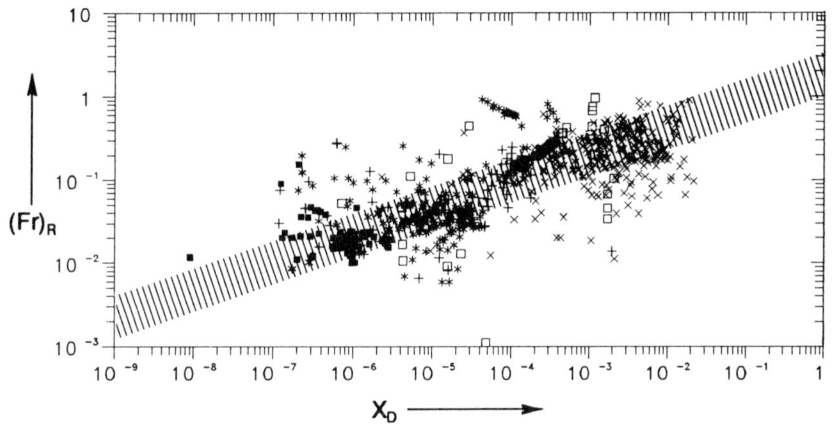

Fig. 4.10

4.7 Regime Channel Development

4.7.1 *Development of the regime width and slope*

i- The formation of a wide alluvial channel (itself) is due to the time-development of its width B and its slope S. The sediment transporting flow actively alters B and S so as to achieve the minimum value of c^2S ($= Fr$). In the process, h and K_s merely "adjust" themselves by satisfying the resistance formula $Q = Bhc\sqrt{gSh}$ and the friction factor expression $c = = \Phi_c(h/K_s) = \phi_c(\xi, \eta, N)$.[25]

Experiment shows (see e.g. [2], [20]) that B can only increase during the formation of the regime channel R. (The decrement of B can take place only if (the imposed) Q_s at the entrance of the channel is larger than Q_s carried (naturally) by the flow in that channel; this, however, is not consistent with the definition of R).

Since N implies Q/BDv_{*cr}, the increment of B in the course of a given experiment (i.e. for given Q, D and v_{*cr}) means the decrement of N. Each Fr-curve (on the $(Fr;\eta)$-plane corresponding to a given ξ) is associated with a certain constant value of N, and therefore the Fr-curves are but the $N = const$ curves determining the N-field in a $(Fr;\eta)$-plane: the larger is $N = const$, the lower is the Fr-curve (Figs. 4.2 to 4.4). Hence a moving point m (in Fig. 4.5) must approach its "target" P_R (P_s or P_g on the regime Fr-curve) in such a way that both Fr and $N \sim B^{-1}$ continually decrease in the process (the paths l of the points m in Fig. 4.5 are drawn accordingly).

ii- From a series of special (recirculating-flume) experiments [26], it follows that the development of S must not necessarily be by means of its decrement.

---

[25] It is assumed that the depth of the channel itself is only "slightly larger" than the flow depth h, at all stages.

If P_0 is between P_s and P_s' (see Fig. 4.5), then the decrement of $Fr = c^2 S$, during the regime development of flow bed, is achieved by a (strong) decrement of c^2, and a (not so strong) increment of S.[26]

Eliminating c between $c = \phi_c(\xi, \eta, N)$ and $c^2 S = (\alpha/N)(c^2\eta)^{3/2}$, one can compute the relation between S and η for various constant values of ξ and N. By doing so, it has been found that η varies with S as its increasing function – for all practically possible ξ and N (Fig. 4.11). Hence, the increment and decrement of S can always be interpreted as the increment and decrement of η (in a $(Fr;\eta)$-plane).[27]

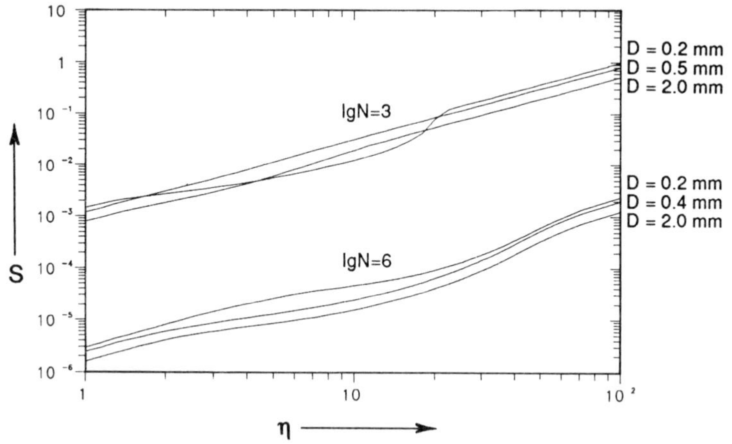

Fig. 4.11

It follows that although the point m must always move in the direction of decreasing Fr and N, no restriction is imposed on the direction of its motion along $\eta \sim S$.

iii- From laboratory research, it follows that the regime development of B takes place much faster than that of S. In Ref. [20] it is reported that in more than 20 runs (with $B_0 < B_R$) "the adjustement of the channel width by bank erosion and the consequent change of mean depth took place, on average, in less than 5 minutes ($= T_B$)"; although "the runs lasted not less than 5 hours and often as long as 30 hours. After the adjustement in channel width within the first 5 minutes, no subsequent change in width took place during the rest of the run" (p. 64, Ref. [20]). The conditions described are shown schematically in Fig. 4.12a, where the duration of development of regime channel (T_R) is identified with the duration of transition $S_0 \rightarrow S_R$. The

---

26 In such cases, $c^2 \sim (h/K_s)^{1/3}$ decreases because of both the decrement of h and the increment of $K_s \sim \Delta$.

27 Although a simultaneous increment of S and η may be intuitively expected, it is by no means obvious; for η is proportional to the product hS, where h usually decreases when S increases, and vice versa.

characteristics h and c simply "follow" the development of B and S by satisfying the relations mentioned in (i). (The time T_Δ in Fig. 4.12 is the duration of development of bed forms). It should be noted (for future reference) that, since the transition $B_0 \to B_R$ is virtually accomplished (at T_B) when the transition $S_0 \to S_R$ has only just started, *the regime development of S takes place for a nearly constant $B \approx B_R$.*

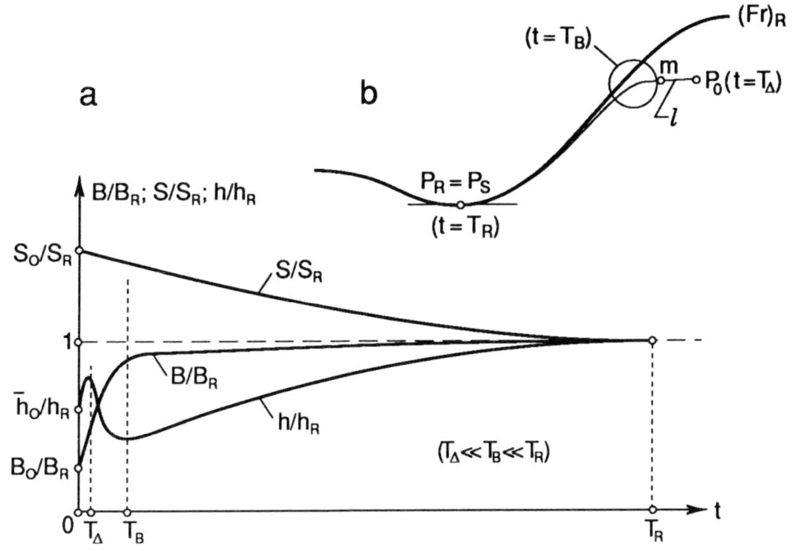

Fig. 4.12

Fig. 4.12 shows the path l of the moving point m in relation to typical times.

4.7.2 *Initial channel*

We assume that the "experiment" (determined by the parameters (4.1)) begins at $t = 0$ in the initial channel (B_0, h_0, S_0) having a flat (scraped) bed.

i- *Sand-like behaviour*

Let the $(\overline{Fr})_0$-curve and the $(Fr)_0$-curve be the Fr-curves of the initial channel: $(\overline{Fr})_0$ – when its bed is flat; $(Fr)_0$ – after it became undulated (Fig. 4.13a). At $t = 0$, the bed is flat and the point m is either at $\overline{P}_{01}$, or at $\overline{P}_{02}$, or ... etc., (in short, at $\overline{P}_{0i}$) on the $(\overline{Fr})_0$-curve. By the time $t = T_\Delta$, the bed becomes covered by sand waves and m is displaced to P_{0i} on the $(Fr)_0$-curve.

No detectable change in B_0 or S_0 can occur during T_Δ:[28] the occurrence of bed forms leads only to the increment of the initial flow depth from $\bar{h}_0$ to h_0 and consequently to the increment of $\bar{\eta}_{0i}$ to η_{0i} (which is the reason for P_{0i} to be on the right of $\bar{P}_{0i}$ in Fig. 4.13a). Clearly, it is the points P_{0i} (rather than $\bar{P}_{0i}$) which signify the locations of the initial channels on the $(Fr;\eta)$-plane.

Consider the initial $(Fr)_0$-curve (containing a point P_{0i}) and the associated $(Fr)_R$-curve (containing the points P_s and $(P_g)_s$). The $(Fr)_R$-curve is above the $(Fr)_0$-curve, for its N-value is smaller (B_R is larger than B_0). The horizontal lines λ_s and λ_g pass through the points P_s and $(P_g)_s$: the vertical line λ_0 passes through the local maximum point M_0 of the initial $(Fr)_0$-curve.

From the graph in Fig. 4.13a it is easily realized that the (moving) point m can then only reach P_s by a continuous decrement of both Fr and N, if its motion starts from such a P_{0i} which is in the region between the line λ_s and the $(Fr)_R$-curve – to the right of λ_0 (paths l_1 and l_2 starting from P_{01} and P_{02}). If m starts from a point (P_{03}) in the region between λ_g and the $(Fr)_R$-curve – to the left of λ_0 – then it cannot possibly reach P_s by a continual decrement of Fr and N. Indeed, if its path is l_3', then Fr will increase in the process; and if it is l_3'', then N will increase. On the other hand, starting from P_{03}, the point m can reach $(P_g)_s$, by a continuous minimization of both Fr and N, if it moves along the path l_3. Hence, if the initial point (P_{03}) happens to be between λ_g and the $(Fr)_R$-curve – to the left of λ_0 – then the channel formation will be gravel-like – in spite of the fact that the $(Fr)_R$-curve is sand-like (i.e. having a minimum point P_s). It follows that the nature of regime channel formation (sand- or gravel-like) does not depend *only* on the geometry of the $(Fr)_R$-curves; it depends also on the nature of the initial channel (on the location of the initial point P_{0i} on the $(Fr;\eta)$-plane). [e.g. the point P_{03} "cannot see the valley P_s beyond the $(Fr)_R$-hill on its right; hence it slides down to the point P_g on its left"].

Clearly, neither P_s nor P_g can be reached (by minimizing both Fr and N) from a point P_0 that is above the $(Fr)_R$-curve or below the lines λ_g or λ_s: the regime development simply will not take place for such an initial channel.

ii- *Gravel-like behaviour*

In this case the point P_s on the $(Fr)_R$-curve is not present: nor the point M_0 on the $(Fr)_0$-curve (Fig. 4.13b). Consequently, there is no line λ_s, and the region between λ_g and the $(Fr)_R$-curve extends to the right indefinitely. The point m can always reach P_g by a continuous decrement of both Fr and N, as long as P_0 is in that region (path l_4).

---

[28] It is assumed, in accordance with reality, that the duration T_Δ required for the development of bed forms is "small" in comparison to T_B (Figs. 4.12) which, in turn, is "small" in comparison to T_R $(T_\Delta \ll T_B \ll T_R)$.

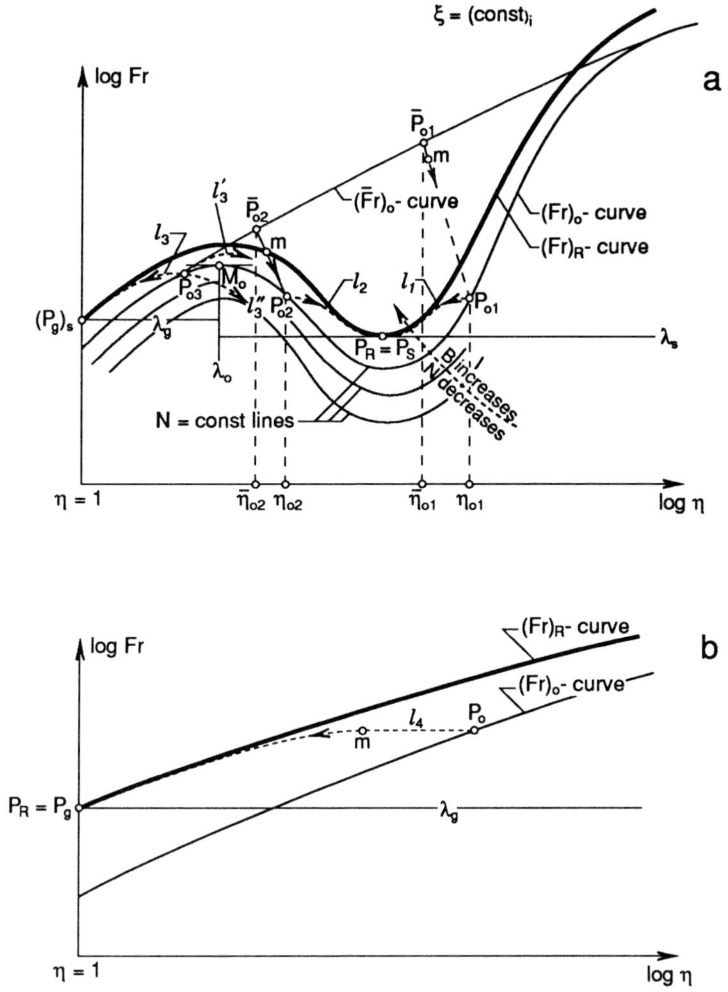

Fig. 4.13

It follows that the "arbitrariness" of an initial channel is thus rather restricted – as is also well known from experiment (see e.g. [2]).

PART II

4.8 Regime Cross-Section

Here we will be dealing exclusively with the regime cross-sections, and therefore the regime-subscript R will be omitted. We assume that the cross-section is symmetrical with respect to the center line: the bed is horizontal, the banks curvilinear. The cross-sectional coordinates y and z are shown in Fig. 4.14.

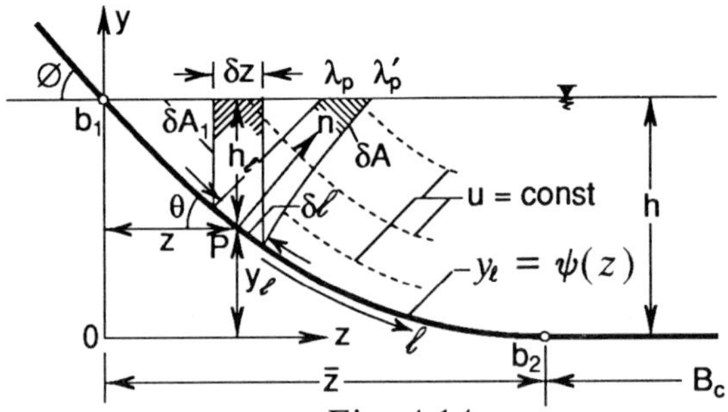

Fig. 4.14

Let P be a point on the curve ℓ forming the bank. The coordinates of P, viz y_ℓ and z, are interrelated by the bank equation $y_\ell = \psi(z)$: the bank inclination angle θ, at P, is given by $\tan\theta = -\,dy_\ell/dz = -\,\psi'(z)$. Clearly, the function $y_\ell = \psi(z)$ satisfies

$$\psi(0) = h \;\; ; \;\; \psi(\bar{z}) = 0 \;\; ; \;\; \psi'(0) = -\tan\phi \;\; ; \;\; \psi'(\bar{z}) = 0. \qquad (4.96)$$

As is clear from Fig. 4.14, the shear stress τ_ℓ acting on the bank surface at P is given by

$$\tau_\ell = \gamma S\left(\lim_{\delta\ell\to 0} \frac{\delta A}{\delta\ell}\right) = \gamma S\left(\lim_{\delta z\to 0} \frac{\delta A}{\delta z}\right)\cos\theta = \gamma S\frac{dA}{dz}\cos\theta \qquad (4.97)$$

where δA is the area between the (near to each other) lines λ_P and λ_P' which are orthogonal to the family of isovels ($u = const$). Usually, the elementary area δA is approximated by the area $\delta A_1 = h_\ell\,\delta z$ (between the vertical lines). Thus, one determines

$$\tau_\ell \approx \gamma S\, h_\ell \cos\theta \qquad (4.98)$$

which is commonly used today.

i- *Gravel channel*

In this case, the bed of the regime channel is in the critical stage ($\eta = 1$), and so are the banks. Dividing (4.98) with $\tau_0 = \gamma S h$, one obtains

$$\frac{\tau_\ell}{\tau_0} \approx \frac{(\tau_\ell)_{cr}}{(\tau_0)_{cr}} \approx \frac{h_\ell}{h}\cos\theta. \qquad (4.99)$$

On the other hand, the equilibrium of forces acting (at the critical stage) on a bank grain gives

$$\frac{(\tau_\ell)_{cr}}{(\tau_0)_{cr}} = \cos\theta\left[1 - \left(\frac{\tan\theta}{\tan\phi}\right)^2\right]^{1/2} \qquad (4.100)$$

(see e.g. [15], [16], [28]).[29] Equating (4.99) and (4.100), and bearing in mind that $h_\ell = h - y_\ell = h - \psi(z)$ while $\tan\theta = -\psi'(z)$, one determines

$$1 - \frac{\psi(z)}{h} = \left[1 - \left(\frac{\psi'(z)}{\tan\phi}\right)^2\right]^{1/2}, \qquad (4.101)$$

which is the differential equation of the regime bank curve $y_\ell = \psi(z)$ of a gravel channel. The integration of (4.101) yields the following (well known) trigonometric form:

$$\frac{y_\ell}{h} = 1 - \sin\left[\tan\left(\phi\frac{z}{h}\right)\right]. \qquad (4.102)$$

ii- *Sand channel*

In this case, the regime values of $\eta = \tau_0/(\tau_0)_{cr}$ are comparable with ≈ 10 (Fig. 4.3); and the regime channel transports the sediment – it is a "live-bed" channel. The realization of this fact led to much discussion in the past; it gave rise to the "open-channel paradox", which can be summarized as follows.

If $\eta \approx 10$, say, and the conditions along the flow bed B_c (Fig. 4.14) are two-dimensional, then at the bed-bank interface b_2 we have $\tau_0 \approx 10(\tau_0)_{cr} \gg (\tau_0)_{cr}$. But if so, then owing to the continuity of the τ_ℓ-diagram, there must exist such a (lower) part of the banks where $\tau_\ell \gg (\tau_0)_{cr}$, i.e. where sediment is transported (as a bank-load).[30] However, since anywhere on the banks we have a non-zero inclination angle θ, the direction of the bank-load transport cannot be parallel to x: owing to the action of gravity, the bank-load rate vector $\mathbf{q}_{sb}'$ must necessarily have a downward-to-the-right directed component $(\mathbf{q}_{sb}')_\ell$. But this means that the bank material must be continually removed downhill – with no replenishment from the top (from b_1). How, then, does the width of the channel not continue to widen with the passage of time?

The reason for the puzzle is the assumption of "no replenishment". Indeed, it is perfectly possible that while some material is being removed from the bank, other is being deposited on it. This possibility was first recognized by H.A. Einstein [11]; the first mathematical treatment of the topic is due to G. Parker [24]. The principal pattern of the derivation in Ref. [24] can be outlined as follows.

Since the steady state regime flow is assumed to be uniform, the time average volumetric concentration C of solids transported in suspension does not vary in the flow direction x. It varies as a function of y and z only:

$$C = \phi_C(y, z). \qquad (4.103)$$

---

[29] A more detailed analysis of the topic can be found in [23].

[30] In analogy to "bed-load", "bank-load" also consists of grains transported at the flow boundaries.

In Ref. [24] it is demonstrated (on the basis of river data) that, in the bank region $0 < z < \bar{z}$, the value of C progressively decreases when the banks are approached (at any level y), and at the banks themselves C is zero. In other words, (4.103) is an increasing function of $z \in [0\,;\bar{z}]$, which has the property $\phi_C(y_\ell, z) = \phi_C(\psi(z), z) = 0$. The positive concentration gradient $\partial C/\partial z$ inevitably induces the turbulent diffusion of suspended particles toward the banks ($(Q_{ss})_z$ in Fig. 4.15). When these particles reach the bank, they deposit themselves on it. According to Ref. [24], it is this continual deposition which compensates the eroding action of the bank-load component $(\mathbf{q}_{sb}')_\ell$ and keeps the regime banks unchanging.[31]

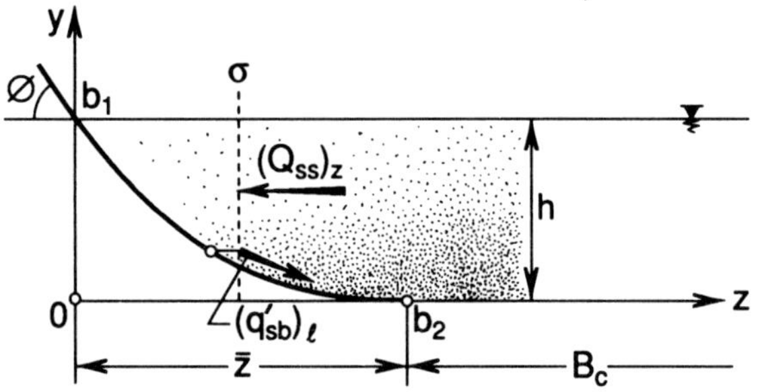

Fig. 4.15

Let $(q_{ss})_z$ be the volume of suspended solids passing per unit time (due to turbulent diffusion) through a unit area perpendicular to z. According to Fick's law of diffusion,

$$(q_{ss})_z = -\,\epsilon_z \frac{\partial C}{\partial z}, \tag{4.104}$$

where ϵ_z is the turbulent diffusion coefficient along z. Consider now a section, σ say, perpendicular to the z-axis (Fig. 4.15). The volume $(Q_{ss})_z$ of suspended solids passing per unit time through the area $h_\ell \times 1$ of this section is given by

$$(Q_{ss})_z = \int_{y_\ell}^{h} (q_{ss})_z\, dy. \tag{4.105}$$

Since under the (steady state) regime conditions, the volume of sediment in fluid and the shape of banks remain unchanged, the amount of sediment entering the fluid volume to the left of σ (per unit time), viz $(Q_{ss})_z$, must be equal to the amount $(q_{sb}')_\ell$ leaving it (per unit time). i.e. we must have

$$(Q_{ss})_z + (q_{sb}')_\ell = 0 \quad \text{for any} \quad z \in [0\,;\bar{z}], \tag{4.106}$$

where $(q_{sb}')_\ell$ is the magnitude of $(\mathbf{q}_{sb}')_\ell$.

---

31 It should thus be clear that the time-invariance of the cross-section of a gravel-bed regime channel is due to the static equilibrium; that of a sand-bed regime channel is due to the dynamic equilibrium.

The only drawback of the derivations above lies in their restrictive interpretation. The impression conveyed is that the widening of a sediment transporting stream cannot stop unless it carries a substantial suspended-load (which migrates toward the banks from the central region B_c). This is not so: there are many regime streams with a moderate or even insignificant suspended-load. Take, for example, the sand channels themselves. Their regime value of η is comparable with ≈ 10 and their sand waves have nearly maximum steepness. Hence, it is hardly likely that the suspended-load carried by these streams can always be "substantial".[32]

There is no need to suppose that the bank transport is necessarily a bed-load: it can be a total-load, formed by an (insignificant) suspended-load and a bed-load (Fig. 4.16). We assume that the suspended-load disperses

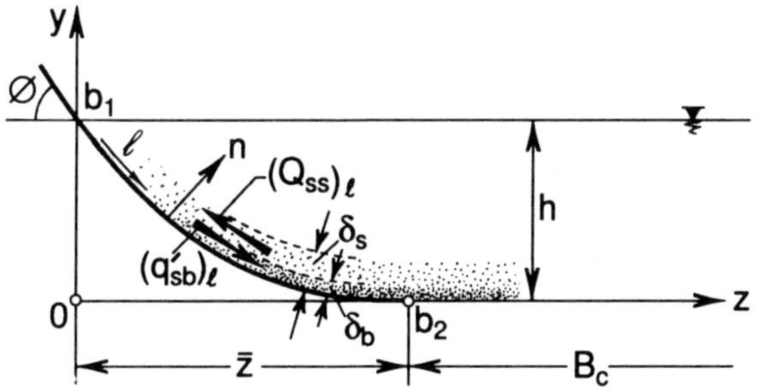

Fig. 4.16

along n only up to a limited "ceiling" δ_s, while the bed-load occupies a thickness δ_b ($< \delta_s$). The concentration of solids carried in suspension increases along ℓ. Thus, we have a positive gradient $\partial C/\partial \ell$, which must induce (in analogy to Ref. [24]) the turbulent diffusion of suspended particles in the negative ℓ-direction with the rate

$$(Q_{ss})_\ell = - \int_0^{\delta_s} \epsilon_\ell \frac{\partial C}{\partial \ell}\, dn. \tag{4.107}$$

Clearly, the bank will remain invariant in time if

$$(Q_{ss})_\ell + (q_{sb}')_\ell = 0 \tag{4.108}$$

which is essentially the same as (4.106). The difference between the two expressions lies in their interpretation. In the case of (4.106), $(Q_{ss})_z$ and $(q_{sb}')_\ell$ are from *different* sources (one from the central region, the other from the bank); in the case of (4.108), $(Q_{ss})_\ell$ and $(q_{sb}')_\ell$ are from the *same* source (they both are from the bank).

---

32 As can be inferred from 3.3.3, the emergence of suspended-load is correlated with the disappearance of sand waves (both are intensified with the increment of η; after it has well exceeded ≈ 15 to 20, say). The turbidity of a tranquil river is often due to the non-settling "washload" – not necessarily to the suspended-load.

The expressions such as (4.106) and (4.108), if appropriately developed and evaluated, can be used to determine the shape of mobile banks. However, for the present, experiment is more reliable for this purpose. On the basis of an extensive series of laboratory measurements, S. Ikeda [17] proposes the following exponential form for $y_\ell = \psi(z)$:

$$y_\ell = he^{-z/\Delta}, \tag{4.109}$$

which can be considered to be valid "regardless of the (magnitude) of discharge, scale of flume and kind of sand" [17]. The value of the length Δ can be determined[33] by differentiating (4.109) and by taking into account that when $z = 0$, then $dy_\ell/dz = -\tan\phi$. By doing so, one obtains $\Delta = h/\tan\phi$, which yields for (4.109)

$$\frac{y_\ell}{h} = e^{-(\tan\phi)z/h}. \tag{4.110}$$

References

1. Altunin, S.T.: *Formation of alluvial streams.* Proc. Academy of Sciences Uzb.S.S.R., Vol. 7, 1955.
2. Ackers, P.: *Experiments on small streams in alluvium.* J. Hydr. Div., ASCE, Vol. 90, No. HY4, July 1964.
3. Bettess, R., White, W.R.: *Extremal hypothesis applied to river regime.* in *Sediment Transport in Gravel-Bed Rivers,* C.R. Thorne, J.C. Bathurst and R.D. Hey eds., John Wiley and Sons, 1987.
4. Chang, H.H.: *Fluvial processes in river engineering.* John Wiley and Sons, 1988.
5. Chang, H.H.: *Stable alluvial canal design.* J. Hydr. Div., ASCE, Vol. 106, No. HY5, May 1980.
6. Chang, H.H.: *Geometry of gravel streams.* J. Hydr. Div., ASCE, Vol. 106, No. HY9, Sept. 1980.
7. Chang, H.H.: *Minimum stream power and river channel patterns.* J. Hydrology, Vol. 41, 1979.
8. Chang, H.H., Hill, J.C.: *Minimum stream power for rivers and deltas.* J. Hydr. Div., ASCE, Vol. 103, No. HY12, Dec. 1977.
9. Davies, T.H.R., Sutherland, A.J.: *Extremal hypothesis for river behaviour.* Water Resour. Res., Vol. 19, No. 1, 1983.
10. Davies, T.H.R., Sutherland, A.J.: *Resistence to flow past deformable boundaries.* Earth Surf. Proc., Vol. 5, 1980.
11. Einstein, H.A.: *Sedimentation.* in *River Ecology and Man,* R. Oglesby ed., Academic Press, 1972.
12. Garde, R.J., Raju, K.G.R.: *Mechanics of sediment transportation and alluvial stream problems.* Wiley Eastern, New Dehli, 1977.
13. Garde, R.J.: *Analysis of distorted river models with movable beds.* Irrigation and Power, No. 4, 1958.
14. Gupta, R.D.: *Total sediment load as a parameter in the design of stable channels.* M.E. Thesis, UOR, 1967.

---

[33] Eq. (5) in Ref. [17] is an identity: it does not supply any value for Δ.

15. Henderson, F.M.: *Open channel flows.* MacMillan, New York, 1966.
16. Ikeda, S.: *Lateral bed load transport on side slopes.* J. Hydr. Div., ASCE, Vol. 108, No. HY11, Nov. 1982.
17. Ikeda, S.: *Self-formed straight channels in sandy beds.* J. Hydr. Div., ASCE, Vol. 107, No. HY4, April 1981.
18. Jia, Y.: *Minimum Froude number and the equilibrium of alluvial sand rivers.* Earth Surf. Processes and Landforms, Vol. 15, 1990.
19. Kondap, D.M.: *Some aspects of flow in stable alluvial channels.* Ph.D. Thesis, UOR, 1977.
20. Leopold, L.B., Wolman, M.G.: *River channel patterns: braided, meandering and straight.* U.S. Geol. Survey Prof. Paper 282-B, 1957.
21. Levi, I.I.: *Dynamics of alluvial streams.* (2nd edition) Gosenergoizdat, Moscow 1957.
22. Makaveyev, N.I.: *River bed and erosion in its basin.* Press of the Academy of Sciences of the USSR, Moscow 1955.
23. Nakagawa, H., Tsujimoto, T., Murakami, S.: *Non-equilibrium bed load transport along side slope of an alluvial stream.* in *Study on interaction between flowing water and sediment transport in alluvial streams,* Kyoto Univ., March 1986.
24. Parker, G.: *Self-formed straight rivers with equilibrium banks and mobile bed. Part 1. The sand-silt river.* J. Fluid Mech., Vol. 89, 1978.
25. Ranga Raju, K.G., Dhandapani, K.R., Kondap, D.M.: *Effect of sediment load on stable sand canal dimensions.* J. Waterways and Harbours Div., ASCE, Vol. 103, WW2, May 1977.
26. Shizong, L.: *A regime theory based on the minimization of the Froude number.* Ph.D. Thesis, Dept. of Civil Engrg., Queen's Univ., Kingston, Canada (in preparation).
27. Song, C.C.S., Yang, C.T.: *Minimum stream power: theory.* J. Hydr. Div., ASCE, Vol. 106, No. HY9, July 1982.
28. Stephenson, D.: *Rockfill in hydraulic engineering.* Elsevier, Amsterdam, Oxford, 1977.
29. Velikanov, M.A.: *Alluvial processes: fundamental principles.* State Publishing House for Physical and Mathematical Literature, Moscow 1958.
30. Velikanov, M.A.: *Dynamics of alluvial streams, Vol. II, Sediment and bed flow.* State Publishing House for Theoretical and Technical Literature, Moscow 1955.
31. White, W.R., Bettess, R., Paris, E.: *Analytical approach to river regime.* J. Hydr. Div., ASCE, Vol. 108, No. HY10, Oct. 1982.
32. White, W.R., Paris, E., Bettess, R.: *River regime based on sediment transport concepts.* Rep. IT 201, Hydraulic. Res. Stn., Wallingford, U.K., 1981.
33. Yalin, M.S.: *Mechanics of sediment transport.* Pergamon Press, Oxford, 1977.
34. Yalin, M.S.: *Theory of hydraulic models.* MacMillan, London, 1971.
35. Yang, C.T.: *Energy dissipation rate approach in river mechanics.* in *Sediment Transport in Gravel Bed Rivers,* C.R.Thorne, J.C.Bathurst and R.D.Hey eds., John Wiley and Sons, 1987.
36. Yang, C.T.: *Unit stream power equation for gravel.* J. Hydr. Engrg., ASCE, Vol. 110, No. 12, Dec. 1984.
37. Yang, C.T., Molinas, A.: *Sediment transport and unit stream power function.* J. Hydr. Div., ASCE, Vol. 108, No. HY6, June 1982.
38. Yang, C.T., Song, C.C.S., Woldenberg, M.J.: *Hydraulic geometry and minimum rate of energy dissipation.* Water Resour. Res., 17, 1981.
39. Yang, C.T., Song, C.C.S.: *Theory of minimum rate of energy dissipation.* J. Hydr. Div., ASCE, Vol. 105, No. HY7, July 1979.
40. Yang, C.T.: *Minimum unit stream power and fluvial hydraulics.* J. Hydr. Div., ASCE, Vol. 102, No. HY7, July 1976.
41. Yang, C.T.: *Unit stream power equation and sediment transport.* J. Hydr. Div., ASCE, Vol. 98, No. HY10, Oct. 1972.
42. Znamenskaya, N.S.: *Sediment transport and alluvial processes.* Hydrometeoizdat, Leningrad, 1976.

References A
Table 4.1

1a. Leopold, L.B., Maddock, T.: *The hydraulic geometry of stream channels and some physiographic implications*. U.S. Geol. Survey Prof. Paper 252, 1953.

2a. Leopold, L.B., Miller, J.: *Ephemeral streams - hydraulic factors and their relation to the drainage net*. U.S. Geol. Survey Prof. Paper 282-A, 1956.

3a. Nixon, M.: *A study of the bank-full discharges of rivers in England and Wales*. Proc. Instn. Civ. Engrs., Vol. 12, 1959.

4a. Nash, E.A.: discussion on *A study of the bank-full discharges of rivers in England and Wales.* Proc. Instn. Civ. Engrs., Vol. 14, 1959.

5a. Lacey, G.: *Stable channels in alluvium.* Min. Proc. Instn. Civ. Engrs., Vol. 229, 1929.

6a. Lapturev, N.V.: *Computation of stable channel beds in weak, fine-graided ground.* Soviet Hydrology, Selected Papers, No. 3, 1969.

7a. Ackers, P.: *Experiments on small streams in alluvium.* J. Hydr. Div., ASCE, Vol. 90, No. HY4, 1964.

8a. Blench, T.: *Regime behaviour of canals and rivers.* Butterworth, London, England, 1957.

9a. Blench, T., Erb, R. B.: *Regime analysis of laboratory data on bed load transport.* La Houille Blanche, Vol. 2, 1957.

10a. Simons, D.B., Albertson, M.L.: *Uniform water convergence channels in alluvial materials.* J. Hydr. Div., ASCE, Vol. 86, No. HY5, 1960.

11a. Bose (from Ref. [7a]), 1936.

12a. Inglis, C.C., Allen, F.H.: *The regimen of the Thames Estuary as affected by currents, salinities and river flows.* Proc. Instn. Civ. Engrs., Vol. 7, 1957. See also
Inglis, C.C.: *The effects of variation in charge and grade on the slopes and shapes of channels.* Proc. Third IAHR Congress, Grenoble, 1949.

13a. Hey, R.D.: *Design equations for mobile gravel-bed rivers.* in *Gravel-Bed Rivers*, R.D. Hey, J.C. Bathurst and C.R. Thorne eds., John Wiley and Sons, 1982.

14a. (to 17a) Bray, D.I.: *Regime equations for gravel-bed rivers.* in *Gravel-Bed Rivers*, R.D. Hey, J.C. Bathurst and C.R. Thorne eds., John Wiley and Sons, 1982.

18a. Glover, R.E., Florey, Q.L.: *Stable channel profiles.* U.S. Bureau of Reclamation Hyd. Lab., Report Hyd-325, 1951.

19a. Ghosh, S.K.: *A study of regime theories for an alluvial meandering channel.* Proc. Second Int. Symp. on River Sedimentation, Nanjing, China, Vol. 1, paper C11, 1983.

20a. Hey, R.D., Thorne, C.R.: *Hydraulic geometry of mobile gravel-bed rivers.* Proc. Second Int. Symp. on River Sedimentation, Nanjing, China, Vol. 1, paper C11, 1983.

References B
Sources of Data Used for Regime Plots

1b. Ackers, P. et al.: *The geometry of small meandering streams.* Proc. Instn. Civ. Engrs. Paper 7328S, London, 1970.

2b. Ackers, P.: *Experiments on small streams in alluvium.* J. Hydr. Div., ASCE, Vol. 90, No. HY4, July 1964.

3b. Bray, D.: *Estimating average velocity in gravel-bed rivers.* J. Hydr. Div., ASCE, Vol. 105, No. HY9, Sept. 1979.

4b. Center Board of Irrigation and Power: *Library of canal data (Punjab and Sind).* Technical Report No. 15, June 1976.

5b. Center Board of Irrigation and Power: *Library of canal data (Upper Ganga).* Technical Report No. 15, June 1976.

6b. Center Board of Irrigation and Power: *Library of canal data (U.S. canals).* Technical Report No. 15, June 1976.

7b. Chitale, S.V.: *River channel patterns.* J. Hydr. Div., ASCE, Vol. 96, No. HY1, 1970.
8b. Einstein, H.A.: *Bed load transport in Mountain Creek.* U.S. Soil Conservation Service, SCS-TP-55, 1944.
9b. Hey, R.R., Thorne, C.R.: *Stable channels with mobile beds.* J. Hydr. Engrg., ASCE, Vol. 112, No. 8, Aug. 1986.
10b. Higginson, N.N.J., Johnston, H.T.: *Estimation of friction factor in natural streams.* Int. Conf. on River Regime, W.R. White ed., Wallingford, 1986.
11b. Neill, G.R.: *Hydraulic geometry of sand rivers in Alberta.* Proc. Hydrology Sym., Alberta, May 1973.
12b. Odgaard, A.J.: *Stream bank erosion along two rivers in Iowa.* Water Resour. Res., Vol. 23, No. 7, 1987.
13b. Pizzuto, J.E. et al.: *Evaluation of a linear bank erosion equation.* Water Resour. Res., Vol 25, No. 5, May 1989.
14b. Schumm, S.A., Khan, H.R.: *Experimental study of channel patterns.* Geol. Soc. Am. Bull., 83, 1972.
15b. Schumm, S.A.: *River adjustements to altered hydrologic regime – Murrumbridge River and Paleochannels, Australia.* U.S. Geol. Survey Prof. Paper 598, 1968.
16b. Shinohara, K., Tsubaki, T.: *On the characteristics of sand waves formed upon the beds of the open channels and rivers.* Reprinted from Reports of Research Inst. for Applied Mech., Kyushu Univ., Vol. 7, No. 5, 1959.
17b. Simons, D.B., Albertson, L.: *Uniform water conveyance channels in alluvial materials.* J. Hydr. Div., ASCE, Vol. 86, No. HY5, 1960.
18b. Struiksma, N., Olesen, K.W.: *Bed deformation in curved alluvial channels.* J. Hydr. Res., Vol. 23, No. 1, 1985.

CHAPTER 5

MEANDERING AND BRAIDING

PART I: Meandering

5.1 Meander Geometry

5.1.1 *General*

Before entering this chapter it might be useful to clarify first what the word "meandering" means. The usual classification of rivers as straight, meandering and braiding is not of much help for this purpose; for, according to this classification, every single-channel stream which is not straight must be meandering. In fluvial hydraulics we are interested not only in the geometry, but also in the physics of the processes involved. Consequently, meandering is not identified with *any* plan deformation of a stream – however induced. It is expected that the deformation of a meandering stream exhibits a traceable periodicity along the general flow direction x, and that this deformation is induced by the stream itself: it should not be "forced" upon the stream by its environment. Considering this, it might be appropriate to define meandering as a *self-induced* plan deformation of a stream that is (ideally) periodic and anti-symmetrical with respect to an axis, x say, which may or may not be exactly straight. According to this definition, which will be used in the following, an alluvial stream which deforms its initially straight channel into one of the periodic and anti-symmetrical plan forms such as those shown in Figs. 5.1, is meandering; whereas a stream flowing in a tortuous rocky terrain or in a rigid sinuous flume, whose curvilinear plan pattern has not been created by that stream itself, is not meandering.[1]

---

[1] The present definition of meandering does not involve any concept peculiar to alluvial streams only, and therefore it can be used for non-alluvial flows as well. Thus, according to the present definition, the flow between parallel walls in Fig. 5.17 is meandering (internally), for its deformation is induced by its own sequence of bursts. Yet, the water-line dragged by a drop rolling down an inclined dusty glass plate (though often used as an example to illustrate that "meanders can occur everywhere") is not meandering, for the arbitrary sideways wandering of the water-line is forced upon it by the environment (by the irregularities of dust distribution on the plate).

If the meander loops, or half waves, are symmetrical with respect to the z-axis, passing through the middle of OO', as in Fig. 5.1a, then it is said that the meandering is regular; otherwise, it is irregular or skewed (Fig. 5.1b).

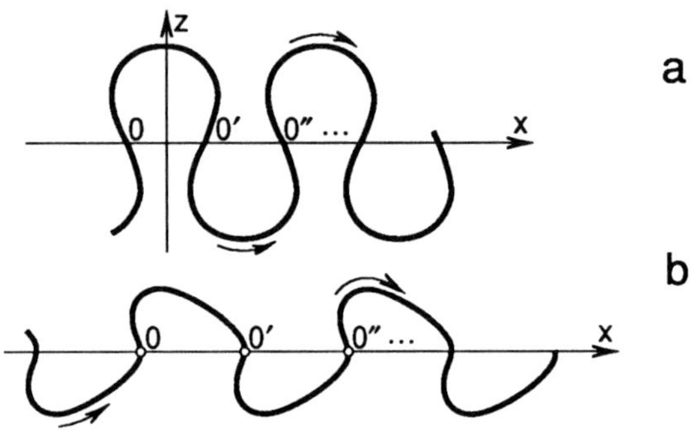

Fig. 5.1

The present chapter concerns the regular meandering of an ideal river in a cohesionless alluvium:[2] the pertinent characteristics and notation are summarized in Fig. 5.2.

5.1.2 *Sine-generated curve*

i- It appears to be generally accepted that the plan shape of a regular ideal river can be reflected best by the sine-generated curve

$$\theta = \theta_0 \cos\left(2\pi \frac{l}{L_m}\right), \tag{5.1}$$

which "closely approximates the shape of real river meanders" (p. 62 in [27]). The relation (5.1) can be taken as the equation of the center line;[3] and its origin can be explained as follows.

Consider a geometric plane containing two fixed points a and b, and also a moving point m. The point m, starting from a, must arrive at b after a given number N of equally long steps. At the end of each step, m changes its direction by an angle ϕ whose probability of occurrence is assumed to be distributed according to the normal law. Suppose now that each of the

---

[2] The reader interested in irregular meanders is referred to the recent works [37], [38], [39], [25].

[3] If B/R is finite, then the center line, the outer and inner banks and the thalweg are different curves in plan. In this case it becomes uncertain whether (5.1) is indeed the equation of the center line; and not of the thalweg, say. The present choice of the center line (which is easiest to define and detect) is thus only because of convenience.

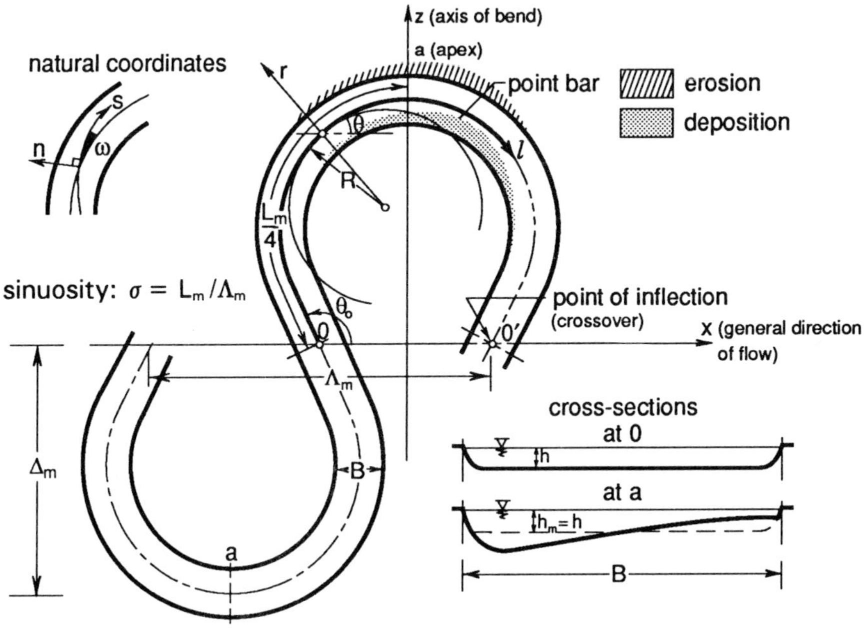

Fig. 5.2

possible (polygonal) paths of m is smoothed, i.e. it is replaced by an appropriate continuous curve. It has been shown by H. Von Schelling [47], [48] that among these curves the likeliest to occur is the one which has the minimum overall curvature, i.e. which satisfies

$$\int_a^b \frac{dl}{R^2} \rightarrow \min . \tag{5.2}$$

L. Leopold and W. Langbein [27] have found that the most probable path implied by (5.2) can be well approximated by the sine-generated curve (5.1), which was adopted by them for the formulation of the plan shape of meanders.

Although the realization of a river pattern in nature is associated with a very strong "random element", the meandering as such is not an inherently random process. Nor is it a process which occurs in the form of discrete steps. But if so, then the sine-generated curve should be derivable also on a non-probabilistic and continuous basis; such a derivation is presented below.

ii- Since the sign of the curvature $1/R$ and of its derivative $d(1/R)/dl = (1/R)'$ vary along l, the integrals of these quantities along an l-region can acquire positive, negative or zero values − without throwing any light on the actual magnitude of $1/R$ and $(1/R)'$ in that l-region. Consider, for example, the lines l and $\bar{l}$ in Fig. 5.3. The line $\bar{l}$ is, on average, more curved than l, and yet the integrals of $1/R$ and $(1/R)'$ along $\bar{l}_{ab}$ may turn out to be larger, or smaller, or equal to their counterparts along l_{ab}. As is well known, such situations are remedied by operating with the squares of the quantities involved, rather than with those quantities themselves. Hence, $[1/R]^2$ and $[(1/R)']^2$ will be used below in lieu of $1/R$ and $(1/R)'$ (as e.g. in (5.2)).

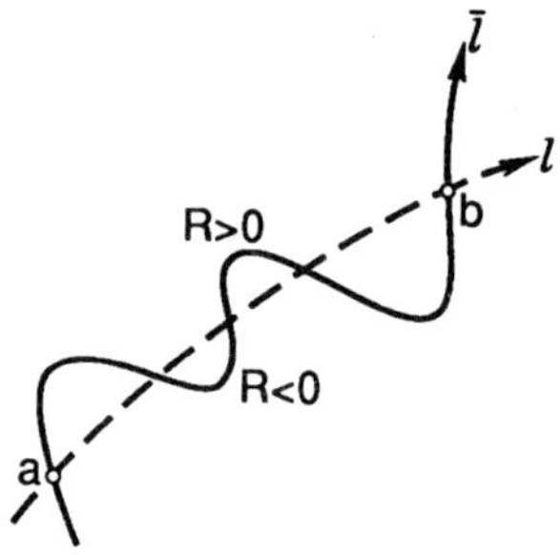

Fig. 5.3

Following Ref. [27], we assume that "... meanders are not mere accidents of nature, but the form in which a river does the least work in turning ..." (p. 60 in [27]). Consequently, we postulate that if a river is to turn in a meander loop having a given average curvature (square)

$$\frac{1}{R_{av}^2} = \frac{1}{L_m}\int_0^{L_m}(1/R)^2 dl\,, \tag{5.3}$$

then this loop must be such that the average rate of change (square) of its curvature is minimum:

$$\frac{1}{L_m}\int_0^{L_m}[(1/R)']^2 dl \rightarrow \min\,. \tag{5.4}$$

Introducing the dimensionless forms

$$\zeta = \frac{l}{L_m} \quad ; \quad \eta = \frac{L_m}{R} \quad ; \quad \eta_{av} = \frac{L_m}{R_{av}}\,, \tag{5.5}$$

we can express (5.3) and (5.4) as

$$\eta_{av}^2 = \int_0^1 \eta^2 d\zeta \quad \text{and} \quad \int_0^1 \eta'^2 d\zeta \rightarrow \min\,, \tag{5.6}$$

where η_{av} is to be treated as a given dimensionless constant. Thus the determination of $\eta = \eta(\zeta)$ is an isoperimetric variational problem. Accordingly, augmenting the functional on the right of (5.6) into the form

$$\int_0^1 (\eta'^2 - \lambda\eta^2)d\zeta \rightarrow \min\,, \tag{5.7}$$

where λ is the Lagrange multiplier, and substituting the integrand $F(\eta, \eta') = \eta'^2 - \lambda\eta^2$ into the Euler form

$$\frac{\partial F}{\partial \eta} - \frac{d}{d\zeta}\left(\frac{\partial F}{\partial \eta'}\right) = 0\,, \tag{5.8}$$

we arrive at the differential equation

$$\eta'' + \lambda\eta = 0\,. \tag{5.9}$$

The general solution of (5.9), viz

$$\eta = C_1\cos(\lambda^{1/2}\zeta) + C_2\sin(\lambda^{1/2}\zeta) \qquad (= - d\theta/d\zeta)\,, \tag{5.10}$$

yields (by integration)[4]

$$\theta = -\frac{C_1}{\lambda^{1/2}}\sin(\lambda^{1/2}\zeta) + \frac{C_2}{\lambda^{1/2}}\cos(\lambda^{1/2}\zeta)\,. \tag{5.11}$$

Using the three boundary conditions

$$[\eta(0) = 0], \quad [\theta = 0 \text{ when } \zeta = 1/4] \quad \text{and} \quad [\theta = \theta_0 \text{ when } \zeta = 0], \tag{5.12}$$

we determine for the three constants involved

$$C_1 = 0, \quad C_2 = 2\pi\theta_0 \quad \text{and} \quad \lambda^{1/2} = 2\pi. \tag{5.13}$$

Hence

$$\theta = \theta_0\cos\left(2\pi\frac{l}{L_m}\right), \tag{5.14}$$

which is the sine-generated curve (5.1).

iii- Note that the l-derivatives (of *any* order) of the sine-generated curve (5.14) and thus of the magnitude of its curvature η/L_m, viz

$$\frac{1}{R} = 2\pi\frac{\theta_0}{L_m}\sin\left(2\pi\frac{l}{L_m}\right), \tag{5.15}$$

vary continuously along l (they are sine or cosine functions of l). Figs. 5.4a and b show the plan view of railway tracks formed by tangentially joined circles. The continuity of these tracks alone, i.e. the absence of kinks in them, is not sufficient to ensure that the passengers will have a comfortable journey: they will experience a jolt whenever the train passes through a point such as O where the $1/R$-diagram is discontinuous, and thus where the suddenly arisen (or changing) centrifugal forces $\sim V^2/R$ are acting. Nature would never create a channel where moving fluid elements experience jolts – and the

---

4 The integration constant is obviously zero, for θ is distributed anti-symmetrically along l (or x).

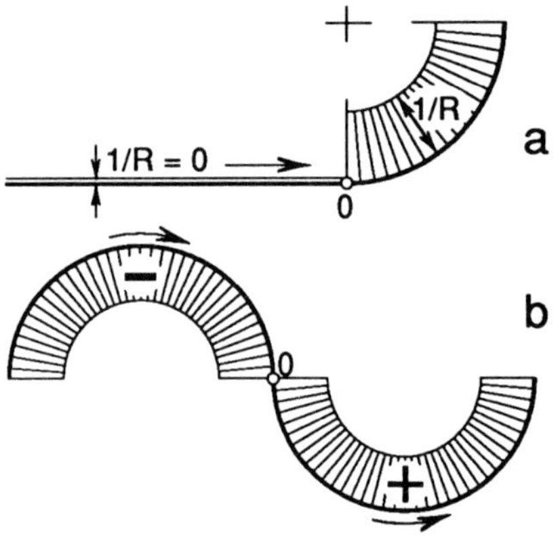

Fig. 5.4

sine-generated function is consistent with this fact.[5]

5.1.3 *Some consequences of sine-generated curve*

From (5.14), one obtains

$$\frac{l}{L_m} = \zeta = \frac{1}{2\pi} \operatorname{arc\,cos} \frac{\theta}{\theta_0} \quad \text{i.e.} \quad d\zeta = -\frac{1}{2\pi} \frac{d\theta}{(\theta_0^2 - \theta^2)^{1/2}}. \tag{5.16}$$

On the other hand, $dx = dl \cos\theta$ yields

$$d\left(\frac{x}{\Lambda_m}\right) = \frac{dl}{\Lambda_m} \cos\theta = \sigma\, d\zeta \cos\theta. \tag{5.17}$$

Eliminating $d\zeta$ between (5.16) and (5.17) and integrating, one determines

$$\int_0^{1/4} d\left(\frac{x}{\Lambda_m}\right) = \frac{1}{4} = -\frac{\sigma}{2\pi} \int_{\theta_0}^{0} \frac{\cos\theta}{(\theta_0^2 - \theta^2)^{1/2}}\, d\theta, \tag{5.18}$$

and thus

$$\frac{1}{\sigma} = \frac{2}{\pi} \int_0^1 \frac{\cos(\theta_0 \xi)}{(1 - \xi^2)^{1/2}}\, d\xi \tag{5.19}$$

(with $\xi = \theta/\theta_0$).

The expression on the right is the Bessel function of the first kind and zero-th order, $J_0(\theta_0)$: its graph is shown in Fig. 5.5a. Hence, in the case of a sine-generated curve, the sinuosity is completely determined by the (initial) *deflection angle* θ_0:

---

[5] The conditions associated with the schemes in Fig. 5.4 are, of course, well known to railway and highway engineers. And these schemes are never used in an important project: the circles are joined by special transitional curves. (Only some "meandering studies" were conducted in flumes having plan shapes as in Fig. 5.4.)

$$\frac{1}{\sigma} = J_0(\theta_0). \tag{5.20}$$

Observe, from Fig. 5.5a, that when $\theta_0 \approx 138°$ then $J_0(\theta_0) = 0$ and (5.20) yields $\sigma \to \infty$. This, however, can never be realized, for the loops begin to touch each other already when θ_0 has grown to $\approx 126°$ (Fig. 5.5b). If $\theta_0 = 126°$, then $\sigma \approx 8.5$ which is the maximum possible theoretical sinuosity (of a sine-generated curve). In natural rivers, the value of σ is seldom larger than ≈ 5, say. Fig. 5.5c shows the agreement of the function (5.20) with the data obtained in Russian rivers (data from Ref. [23]).

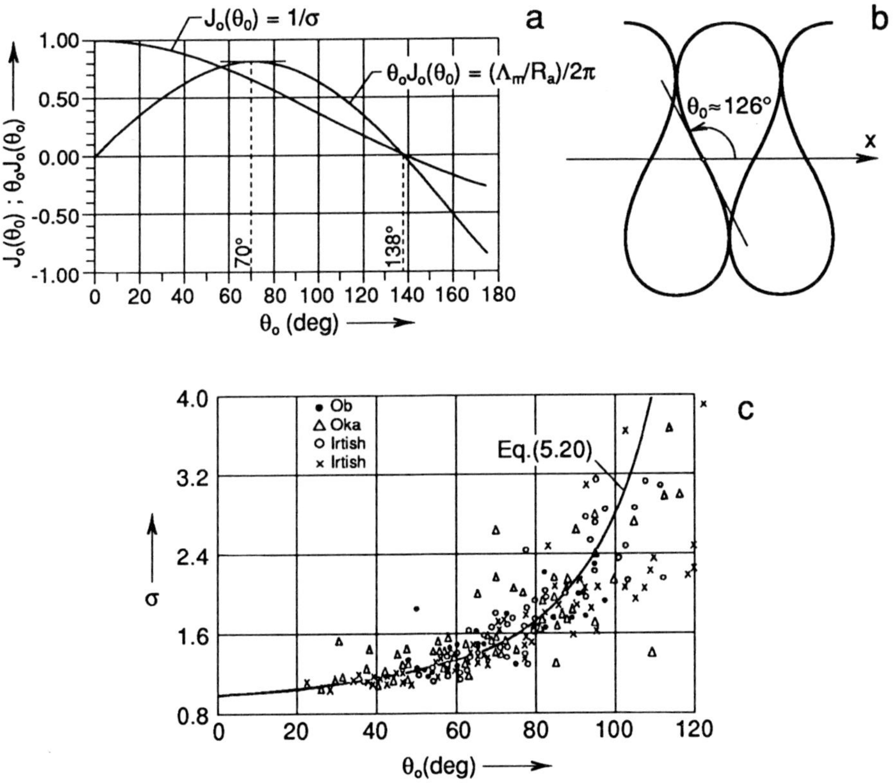

Fig. 5.5

Note from (5.15) that the largest curvature (smallest R) is at the apex (a) of the meander loop, where $l/L_m = \zeta = 1/4$. Denoting this R at a by R_a, one determines

$$\frac{1}{R_a} = 2\pi\frac{\theta_0}{L_m} = \frac{2\pi}{\Lambda_m}\frac{\theta_0}{\sigma} \quad \text{i.e.} \quad \frac{1}{2\pi}\frac{\Lambda_m}{R_a} = \theta_0 J_0(\theta_0). \tag{5.21}$$

The graph of $(\Lambda_m/R_a)/2\pi$ is also shown in Fig. 5.5a. Observe that, for a given (constant) Λ_m, the progressive increment of θ_0 (from zero onwards) first causes $1/R_a$ to increase, then to achieve its maximum at $\theta_0 \approx 70°$, and then to decrease again as to become zero when $\sigma \to \infty$ (at $\theta_0 \approx 138°$).

Since the shape of the sine-generated curve is specified by θ_0 alone, any dimensionless characteristic related to the shape of that curve, such as

$$\sigma, \quad L_m/\Lambda_m, \quad \Delta_m/\Lambda_m, \quad R_a/\Lambda_m \quad \ldots \text{ etc.,} \tag{5.22}$$

can be regarded as given if θ_0 is specified (and each of them can be used in lieu of θ_0).

5.2 Origin of Meanders

5.2.1 *Meandering and horizontal bursts*

i- Fig. 5.6 shows the meander length data plotted versus flow width B. As

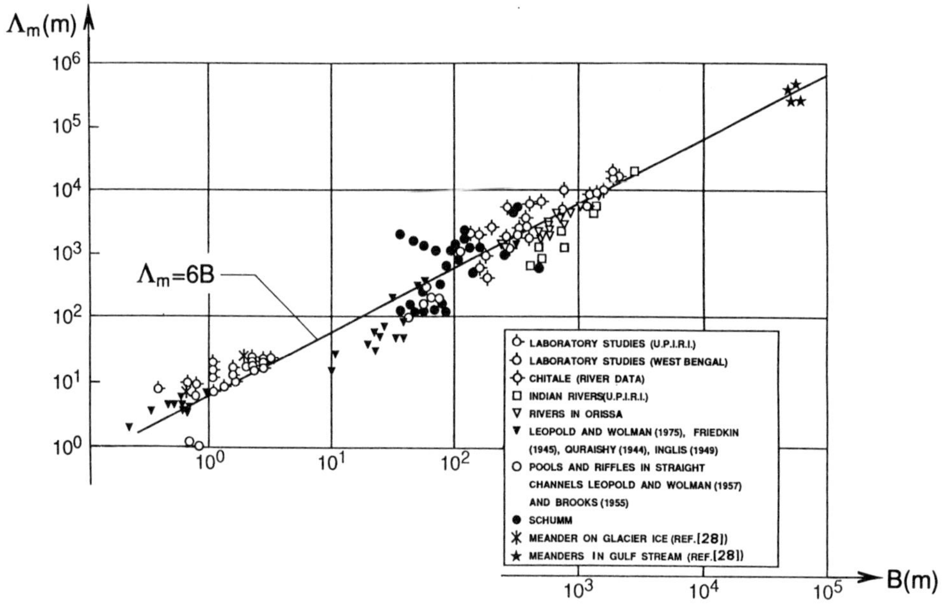

Fig. 5.6 (after Ref. [13])

seen from this graph, the length Λ_m of meanders can be given by

$$\Lambda_m \approx 6B, \tag{5.23}$$

which is the same as the expression of the length Λ_a of alternate bars and of the length $L_H = (L_H)_1$ of horizontal bursts having the basic arrangement $N = 1$ (Fig. 5.7):

$$\Lambda_m \equiv \Lambda_a \equiv L_H \quad (\approx 6B). \tag{5.24}$$

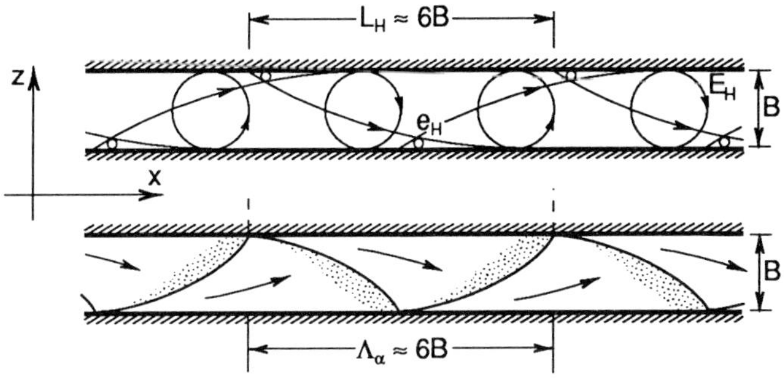

Fig. 5.7

Consider Fig. 5.8. It is the extended version of Fig. 3.16 where the available field and laboratory data of meandering and braiding streams are also plotted (points $M;m$ and $B;b$ respectively).[6] The scatter of the points is

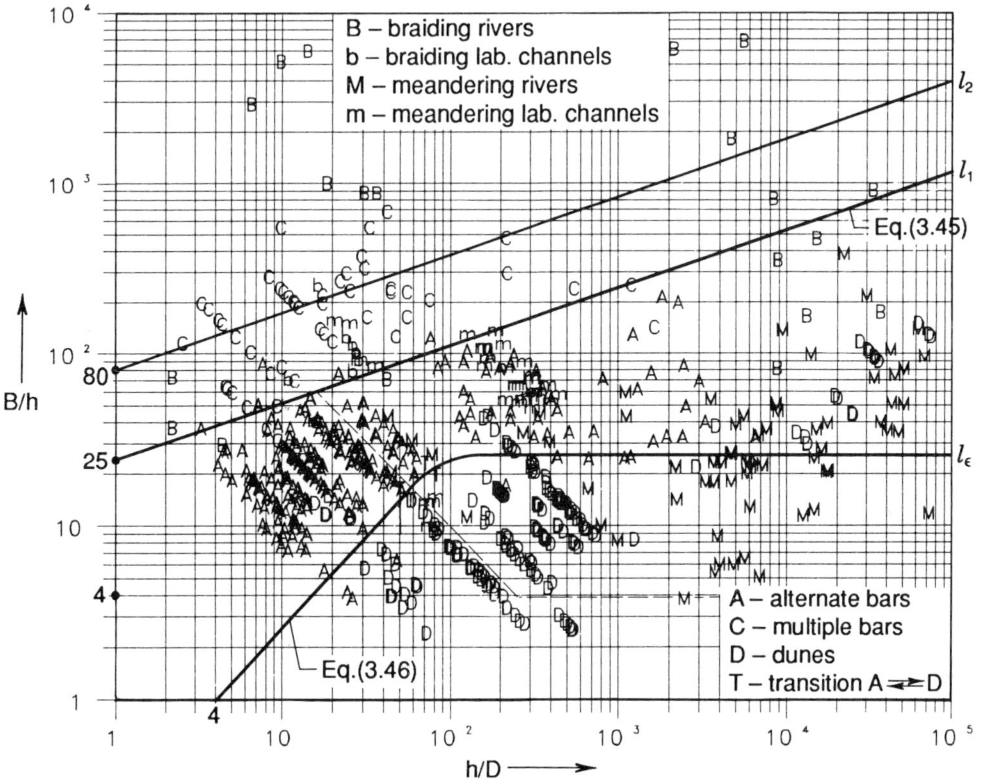

Fig. 5.8 (from Ref. [43])

---

6 The Refs. to the data sources are given in "References A".
River data:
Point-symbol M: [2a], [3a], [5a], [7a], [9a], [10a], [11a], [12a], [15a];

gross, and their "diffusion" from one region to another is substantial. Nonetheless, one can still realize from this graph that the regions of alternate bars (points A) and meanders, though overlapping, are not congruent: the points M extend more to the right and downwards than the points A, which, on the other hand, extend more to the left than M. The 1/3-inclined line l_1 appears as the common upper boundary for both A and M; m. The regions mentioned are sketched and labelled in Fig. 5.9.

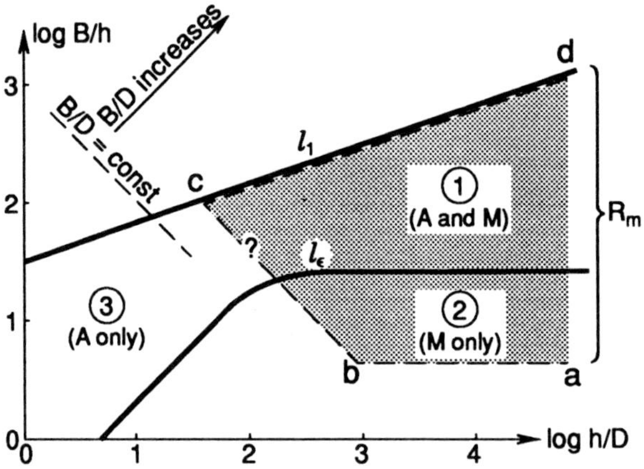

Fig. 5.9

The basic arrangement of horizontal bursts depicted in Fig. 5.7 is present at any location of the $(B/h;\, h/D)$-plane which is below the line l_1. If this location happens to be between the lines l_1 and l_ε (zones 3 and 1), then the horizontal burst-forming eddies e_H are rubbing the bed and in their final stage E_H they are also impinging on the banks (Fig. 5.10a); and if it happens to be below the line l_ε (zone 2), then the eddies e_H are not rubbing the bed – and yet at their final stage E_H they are still impinging on the banks (Fig. 5.10b).

In the case depicted in Fig. 5.10b, where no bed-rubbing horizontal eddies and thus no alternate bars are present, the meanders originate because of the direct action of horizontal bursts (i.e. of their eddies E_H) on the banks – whereby they imprint on the latter their length $L_H \approx 6B$ in the form of meander length $\Lambda_m \approx 6B$ (zone 2 in Fig. 5.9, which contains the points M, but not A). In the case depicted in Fig. 5.10a, where the bed-rubbing eddies e_H and thus the alternate bars are present, it is these bars which come into being first (i.e. before the banks can be deformed appreciably by the direct action of E_H). The (earlier) formation of alternate bars – whose wave length

Point-symbol B: [2a], [5a], [6a], [13a], [16a], [17a].
Laboratory data:
Point-symbol m: [1a], [4a], [14a];
Point-symbol b: [4a], [8a], [14a].
For data sources of point-symbols A, C and D, see footnote 19 in Chapter 3.

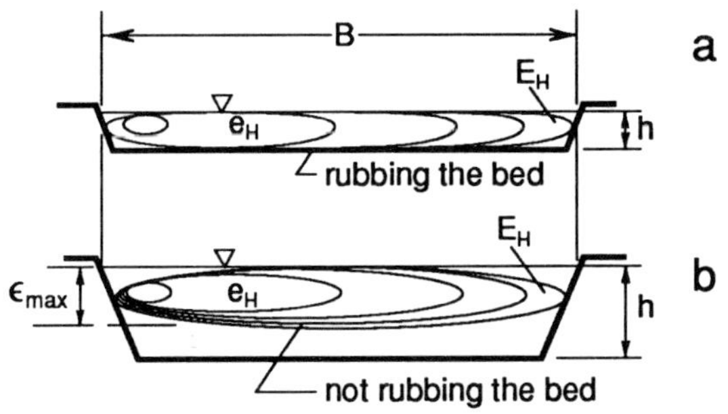

Fig. 5.10

is the same as that of meanders – can be viewed as the introduction of "guide-vanes" which direct and regularize the flow so as to initiate meandering in the most efficient manner (zone 1 in Fig. 5.9, which contains the points M, as well as A). The initiation of meandering with the aid of alternate bars is shown schematically in Fig. 5.11.

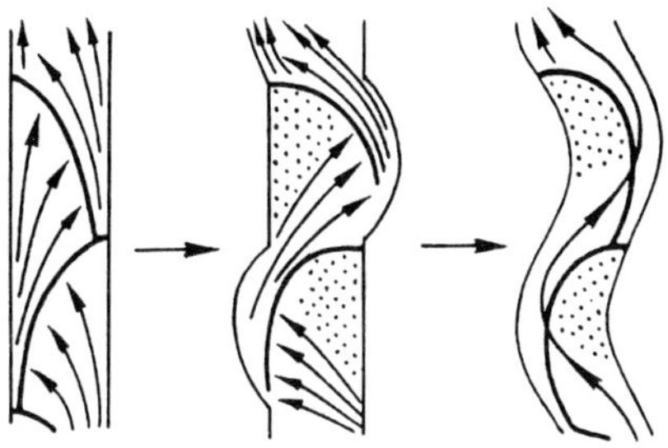

Fig. 5.11 (from Ref. [46])

The fact that meanders can be present when alternate bars are not present (all points M in Fig. 5.8 which are systematically and substantially below the lower limit l_ϵ of the alternate bar region) is a sufficient indication that alternate bars *cannot* be the cause of meanders. In contrast to the prevailing view (Refs. [1], [22], [23], [31], [46]), alternate bars are merely the "catalysts" which accelerate the formation of meanders, which would take place even without them: the "prime mover" of the periodic bed and bank deformation of the length $\approx 6B$ is the sequence of horizontal bursts.

ii- Let R_m be the "meander-region", i.e. the part of the $(B/h\,;\,h/D)$-plane that is within the (shaded) area *abcd* in Fig. 5.9.

1- If a point P of the $(B/h\,;\,h/D)$-plane is situated above the line l_1 forming the upper boundary of R_m, then the eddies E_H cannot reach the opposite bank. In this case, neither alternate bars nor meanders can form. And this is why the line l_1 is the commom upper boundary of them both.

2- If the point P is situated below the line l_ϵ, then the horizontal bursts are not rubbing the bed and the eddies E_H can affect only the upper parts of the banks. The height of the affected region is comparable with the thickness ϵ_{max} of the eddy E_H. Since $\epsilon_{max} \sim E_H \sim B$, the smaller is $\epsilon_{max}/h \sim B/h$, the less significant must be the bank deformation caused by the impingement of E_H. And if $\epsilon_{max}/h \sim B/h$ is sufficiently small, then the eddies E_H should not be able to deform the banks at all; hence the reason for the existence of the lower limit of R_m ($B/h \approx 4$ in Fig. 5.8, line *ab* in Fig. 5.9).

3- The occurrence of the point P within the meander region R_m, in short $P \in R_m$, is only a necessary condition for the occurrence of meandering (all meandering streams satisfy $P \in R_m$). However, there are many sediment transporting streams which satisfy $P \in R_m$, and yet which do not meander. The necessary and sufficient conditions for meandering will be introduced in the next section.

5.2.2 *Regime slope and valley slope*

i- Consider an experiment which starts at $t = 0$ in a straight initial channel excavated in an alluvial valley – the slope of the initial channel is the same as the valley slope:

$$S_0 = S_v \,. \tag{5.25}$$

We assume that the granular material and fluid are specified, $Q = const$ is given, and $\eta_0 \gg 1$. We also assume that the initial channel (B_0, h_0, S_0) is such that the formation of the regime channel (B_R, h_R, S_R) is possible (see 4.7.2).

The transition $B_0 \rightarrow B_R$ takes place much faster than $S_0 \rightarrow S_R$, and therefore the time-development of S occurs for the practically constant $B \approx B_R$ (4.7.1 (iii)).[7] Since for a given experiment Q and $B \approx B_R$ are constants, while h and c merely "adjust" themselves depending on Q, B and S (4.7.1 (i)), the development $S_v \rightarrow S_R$ unambiguously reflects the trend $Fr = c^2 S \rightarrow \min$ which motivates this development. If $S_v < S_R$, then the development of S can be only by aggradation. If, on the other hand, $S_v > S_R$, then two "options" are open to the stream: degradation and meandering (Figs. 5.12a and b respectively). And, in general, the stream "takes" a certain combination of both.

Hence, according to the present approach, the time growth of loops (or waves) of a meandering stream, and thus of its sinuosity σ, is because this stream "endeavours" to reduce its slope to S_R. [Essentially the same assertion

---

[7] In an actual meandering stream, B varies along l only because of accidental reasons [4], [23], [52]: no systematic relation $B \sim f(l/L_m, \theta_0)$ is present. Hence it is not an averaged-along L_m value of B, but theoretically the same-for-any l value of B ($\approx B_R$), which is referred to here.

was made previously by H. Chang [4], [5] – the difference is only in the terminology, and in the methods adopted for the determination of the "target slope" (which in this text is S_R)].

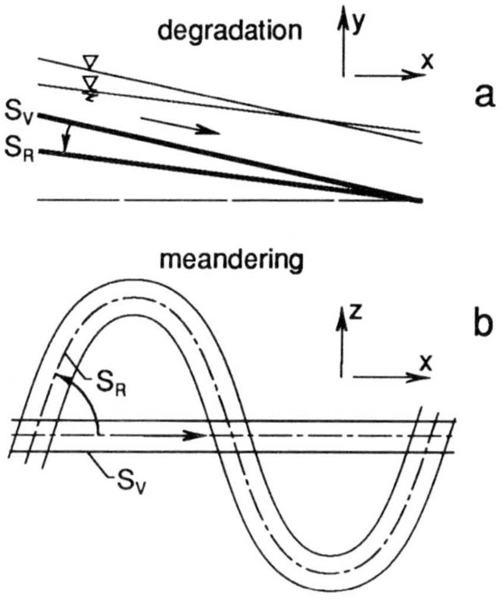

Fig. 5.12

ii- From the explanations above, it follows that the slope S of a meandering stream must, at any stage of its development, be within the interval

$$S_v > S > S_R \qquad \text{(for any } t\text{)}, \tag{5.26}$$

and since the valley slope S_v has to be regarded as arbitrary, the S-values of meandering rivers must be spread in the region $S > S_R$. The natural river data appear to confirm this (Fig. 5.13).[8]

Consider now the plot in Fig. 5.14. It indicates that the sinuosity σ tends to decrease with the increment of slope of meandering streams: or, which is the same, with the increment of their relative grain size D/h (for the probability to possess a certain $\eta \sim S(h/D)$ is comparable for all of them). The trend exhibited in Fig. 5.14 is not surprising. Indeed, let us interpret this trend in terms of the aforementioned two "options", viz degradation and meandering, by means of which a stream can reduce its slope from S_v to S_R. Since S_v is arbitrary, the same value of the ratio S_v/S_R can be present in a stream corresponding to any S or D/h. And the fact that streams having large S and D (gravel bed rivers) exhibit a small σ can only mean that they achieve their regime slopes S_R mainly by degradation. Conversely, streams

---

[8] The Refs. to the data sources are given in "References A":
Sand rivers: [2a], [3a], [5a], [7a], [9a], [10a], [15a].
Gravel rivers: [5a], [11a], [12a].
S_R values were computed according to the method in Section 4.6.

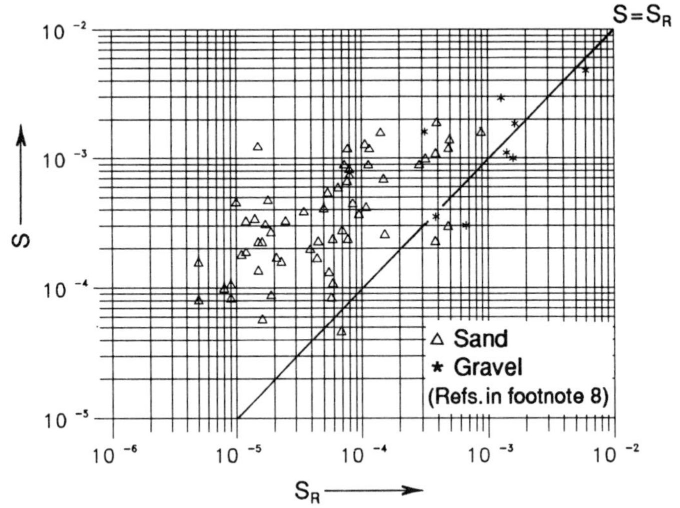

Fig. 5.13

which have small S and D (large fine sand bed rivers), and which exhibit large σ, acquire their S_R mainly by meandering.

Let $\mu_m[S_v/S_R]$ be that part of S_v/S_R which is "covered" by meandering (and $(1-\mu_m)[S_v/S_R]$ by degradation). Assuming that the same proportion applies to the development of S (and thus of σ) corresponding to any $t \in [0; T_m]$, we can write for a meandering experiment

$$1 \leq [\sigma = 1 + \mu_m(S_v/S - 1)] \leq [\sigma_R = 1 + \mu_m(S_v/S_R - 1)], \qquad (5.27)$$

where the lower and upper limits correspond to $t = 0$ and $t \geq T_m$ respectively. Clearly:

$$\left.\begin{array}{l} \text{if } \mu_m = 1 \text{ then } 1 \leq \sigma = S_v/S \leq \sigma_R = S_v/S_r \text{ (only meandering),} \\ \text{if } \mu_m = 0 \text{ then } 1 = \sigma = \sigma_R \text{ (only degradation).} \end{array}\right\} \qquad (5.28)$$

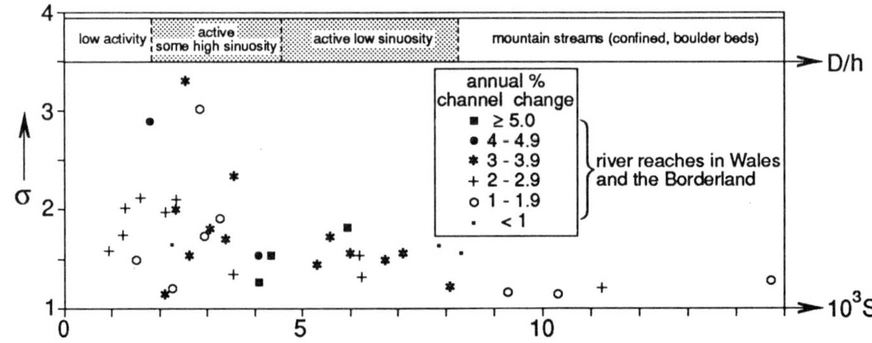

Fig. 5.14 (from Ref. [30])

It would certainly be worthwhile to reveal by future research how the factor μ_m is determined (by S_v, h/D, B/h, etc.) in its interval $0 \leq \mu_m \leq 1$.

The following should be noted:

1- When B/D (or h/D) decreases, then the prominence of alternate bars increases (Eq. (3.78)), whereas the prominence of meanders decreases (Fig. 5.14). It appears that gravel streams produce "good bars" and "bad meanders", whereas large rivers with fine sand do the opposite.

2- Although the plot in Fig. 5.14 shows that with the increment of S (and thus with the decrement of h/D) the sinuosity σ tends to become unity, this graph cannot supply the conditions corresponding to $\sigma \to 1$. Hence, the left-hand boundary (bc) of the meander region R_m, viz $B/D = (B/h)(h/D) \approx 10^3$, was determined on the basis of the population of M-points in Fig. 5.8.[9]

3- It has been often mentioned (but never clarified) that the small scale laboratory streams are not able to meander extensively (with "large" σ) [23], [41]. It should be noted that a certain grain size D can be possessed by all streams, large or small, and therefore the smaller the stream, the smaller its value of B/D is likely to be. And since σ decreases with the decrement of B/D, it is only natural that the meandering of small streams should only be limited.

5.2.3 *Necessary and sufficient conditions for meandering*

i- From the preceding parts of this section it follows that an initially straight tranquil stream in a cohesionless alluvium can meander only if it satisfies the following necessary conditions:

1- it transports the sediment (otherwise the flow boundaries cannot deform),
2- it is turbulent (otherwise there will be no bursts to initiate the periodic bank deformation of the wave length $\Lambda_m \approx L_H \approx 6B$),[10]
3- its slope is larger than the regime slope (otherwise the stream will not "endeavour" to reduce its slope by increasing its length, which brings the sinuosity σ into being),

---

[9] The upper and lower boundaries (cd and ab) of R_m are associated with certain physical changes; hence, they are (sharp) lines. In contrast to this, bc is but a center line of a transitional zone where the sinuosity σ is gradually "fading away" (with the decrement of $B/D = (B/h)(h/D)$).

[10] A series of laboratory experiments was conducted with sediment transporting laminar flows ($\eta > 2$; $Re < 200$ – various percentages of water-glycerine mixture. $3m$-long fine sand and polystyrene channels were used. Tests by H.A. Brown under the author's supervision – Grad. Hydr. Lab., Queen's Univ., 1987). No repetitive along x deformation of the channel bed or banks could have been obtained – despite the utilization of numerous combinations of v, h, S and B/h.

4- its initial values of B/h and h/D (at $t = 0$) must be such that the initial point P_0 is in a certain region, viz in the meander region R_m, of the $(B/h; h/D)$-plane (otherwise the horizontal burst sequences will not be of the type needed to deform the banks in the anti-symmetrical and periodic (along x) manner).

The totality of these necessary conditions forms the set of the *necessary and sufficient* conditions for an alluvial stream to meander. This set can be shown symbolically as

$$\eta > 1 \ ; \quad Re > \approx 500 \ ; \quad S_v > S_R \ ; \quad P_0 \in R_m . \tag{5.29}$$

ii- Since the meandering of an alluvial stream is but a part of its regime development (it is the development of its regime slope S_R for the already (nearly) formed B_R), the characteristic parameters determining meandering must be the same as those determining the regime development. If the stream is the channel R in Chapter 4, then its meandering is determined by the parameters (4.1), viz

$$Q, \rho, \nu, \gamma_s, D, g, \tag{5.30}$$

and by the fact that $S_v > S_R$, of course. [The nature of the regime channel (R or otherwise) has no bearing on the statements made hitherto with regard to meanders (as long as that regime channel does not correspond to a "given" S) – although it may affect the numerical value of the target slope (present S_R)].[11]

5.2.4 $(B/h; h/D)$-plane

i- From the resistance equation $Q/(Bh) = c\sqrt{gSh}$, one obtains

$$\frac{B}{h} = a\left(\frac{h}{D}\right)^{-2.5} \quad \text{(with } a = (Q/D^{2.5})/(c\sqrt{gS})\text{)}, \tag{5.31}$$

which indicates that each 2.5/1-declining straight line on the log-log $(B/h; h/D)$-plane is associated with a certain value of a (Fig. 5.15).

Consider a meander-development experiment which is specified by some values of its characteristic parameters (e.g. (5.30), if R), and we have also $S_v > S_R$. The value of a in (5.31) varies with the time $t \in [0 ; T_m]$, for both c and S vary with it. However, the variation of c is much weaker than that of S, and it is not worthwhile to encumber the present explanations by taking it into account.[12] Thus, we assume that the variation of a in the course

---

[11] e.g. the "equilibrium slope" in Ref. [5] is different from the present S_R.

[12] Eliminating h from $Q = Bhc\sqrt{gSh}$ and $c = 7.66(h/K_s)^{1/6}$, one obtains
$c \approx 1/[(QB)^{1/15} K_s^{1/6} (gS)^{1/30}]$ ($= c^{16/15}$),
which may help to infer how weakly c varies with S. Here, $K_s \sim D$ if ripples, and $K_s \sim h$

of a given experiment is due to the variation of S alone; consequently, each of the 2.5/1-declining lines in Fig. 5.15 "carries" a certain value of S – the more to the right is the line, the smaller is its S.

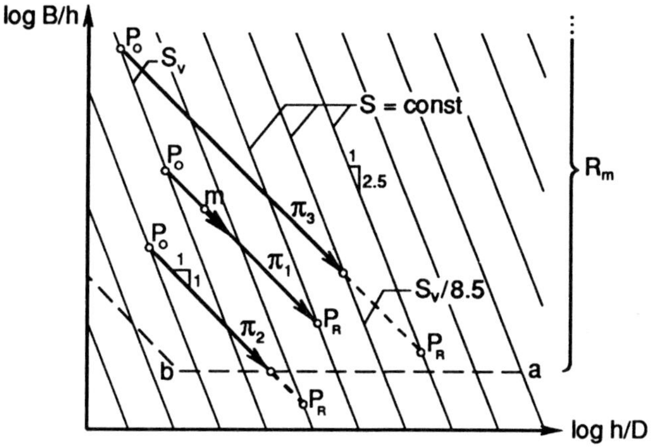

Fig. 5.15

In analogy to (4.4.4), the meander development can also be traced, in the $(B/h;\, h/D)$-plane, with the aid of the "moving point m".

Since $B \approx B_R$ and thus $B/D \approx B_R/D$ remain constant during the experiment (5.2.2 (i)), the point m is moving down the 1/1-declining line

$$\frac{B}{h} = \frac{[B_R/D]}{h/D} \tag{5.32}$$

(path π_1 in Fig. 5.15), starting from the initial point P_0 ($\in R_m$) which represents the initial channel ($B_0 \approx B_R$, h_0, S_v). During its motion, the point m crosses the 2.5/1-declining lines: thus, S progressively decreases (and h increases). Eventually, m reaches the regime-point P_R where $S = S_R$, and the channel development terminates.

Consider the case $\mu_m = 1$ (no degradation, meandering only):

a) If $P_0 \notin R_m$, then the meander development will not even start. However, it is possible that $P_0 \in R_m$ but $P_R \notin R_m$. In this case the meander development stops when m has reached the lower boundary (ba) of R_m (path π_2 in Fig. 5.15).

b) Similarly, S_R cannot be reached if S_v/S_R ($= \sigma_R$) happens to be larger than ≈ 8.5 – where the adjacent meander loops touch each other (Fig. 5.5b). In such cases the development will be interrupted when m has reached the 2.5/1-declining line which (theoretically) corresponds to $S \approx S_v/8.5$ (path π_3 in Fig. 5.15). [The meander develop-

if dunes. In the latter case, $c \sim (h/K_s)^{1/6}$ is practically unaffected by h; in the former, it varies with the 1/6-power of it. Ripples and/or dunes must be considered to be present for any $t \in [0, T_m]$ as they develop even (much) faster than B_R.

ment is "interrupted" – not accomplished. Having formed its ox-bow, the river, at that region, virtually returns to an earlier stage of its development and "tries" again to achieve its S_R, only to be interrupted again, etc. This type of meander formation never ends, for the river never achieves what it "wants" to achieve, viz S_R (Mississippi)].

If $\mu_m < 1$ (meandering and degradation), then the termination of meander development mentioned in a) does not necessarily mean the termination of regime development of the channel. Indeed, if $\mu_m < 1$, then the further reduction of S can be achieved by degradation (m continues to move along π_2 until P_R ($\notin R_m$) is reached). If $\mu_m < 1$, then all what has been said in b) in terms of S_v/S and S_v/S_R must be interpreted in terms of σ and σ_R respectively. Thus the loops begin to touch each other when σ (and not S_v/S) reaches ≈ 8.5, the value of S at this stage being $\approx S_v\mu_m/(7.5 + \mu_m)$ (use $\sigma \approx 8.5$ in (5.27)).

ii- *Additional remarks*

1- From the content of this section it should be clear that meandering is *initiated* by the large-scale turbulence (more precisely, by the sequence of horizontal bursts) which provides it also with the wave length $\Lambda_m \approx 6B$; its subsequent *development* (the time growth of its amplitudes Δ_m) is due to the regime trend ($S \rightarrow S_R$). This presence of two agents in one phenomenon is analogous to the situation depicted in Fig. 5.16. The ball at rest at O cannot

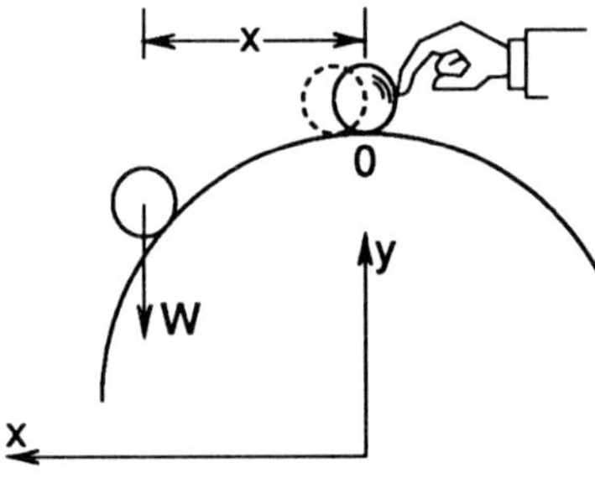

Fig. 5.16

start to move without the finger-push. Yet the finger-push merely *initiates* the ball motion; its motion as such, i.e. the time-*development* of its "amplitude" x, is due to a different agent: gravity. The amplitude x grows because the potential energy of the ball $E_p = Wy$ tends to acquire its minimum.

2- If meandering is not in alluvium, then the alluvium-related factors (S_v, S_R, η) are, of course, no longer meaningful. In such cases, the initiation of meanders (having the wave length $\Lambda_m \approx 6 \times$(flow size)) is still because of the large-scale turbulence (as e.g. the "internal meandering" in the region I of Fig. 5.17), their subsequent development (if any) being due to some other factors (e.g. due to the diffusion, as in the "free meandering" in the region

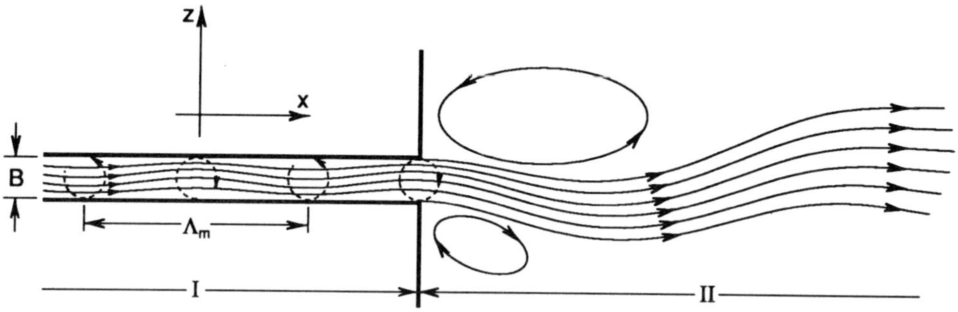

Fig. 5.17

II). The sketch in Fig. 5.18 shows a laboratory example of the stream depicted in II of Fig. 5.17.[13] Meanders can occur for any boundaries which can be deformed (eventually) by a turbulent flow (chalk, turf, ice, ... , etc. [32], [23], [18]), as well as for the liquid boundaries (which can be deformed instantly), as in the regions II of Figs. 5.17 and 5.18.

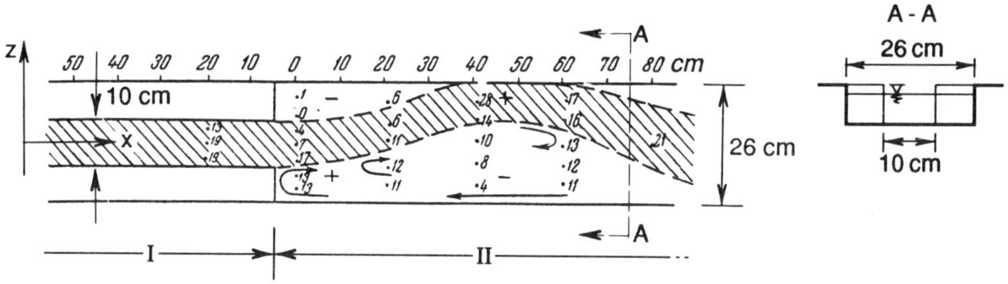

Fig. 5.18 (from Ref. [32])

3- Since for the present it is not known how the value of μ_m is to be determined, it will be assumed henceforward that no degradation takes place during the meander development (it will be assumed that $\mu_m = 1$).

5.3 Meander Kinematics (schematical outline)

5.3.1 *General*

i- Horizontal bursts begin to deform the banks in much the same way as vertical bursts begin to deform the bed. There is one difference, however. In the case of dunes, only one flow-boundary (viz the bed) is deforming,

---

[13] In Ref. [32] it is reported that the undulated stream pattern in Fig. 5.18 occurs only if the open-channel flow is tranquil and turbulent: at the beginning of the wider channel the stream turns to the left or to the right with the same probability.

whereas in the case of meanders, the deformation occurs in both boundaries. Consequently, in the case of dunes, the curvature of streamlines progressively decreases with the distance (y) from the bed, and therefore dunes can grow to a limited steepness only. No such restriction is present in the case of meanders: the simultaneous anti-symmetrical deformation of both banks, if anything, enhances the flow deformation which can go on until prevented by external reasons (S_R has been reached, loops are touching each other, etc.).

ii- The longitudinal bank-load rate, q_s' say, at any location on the bank b_1b_2 obviously satisfies $0 \leq q_s' \leq (q_s)_{b_2}$, where $(q_s)_{b_2}$ is the bed-load rate at the bank-toe b_2 (Fig. 5.19). Let $\bar{q}_s'$ be that (typical) value of q_s' wich is responsible

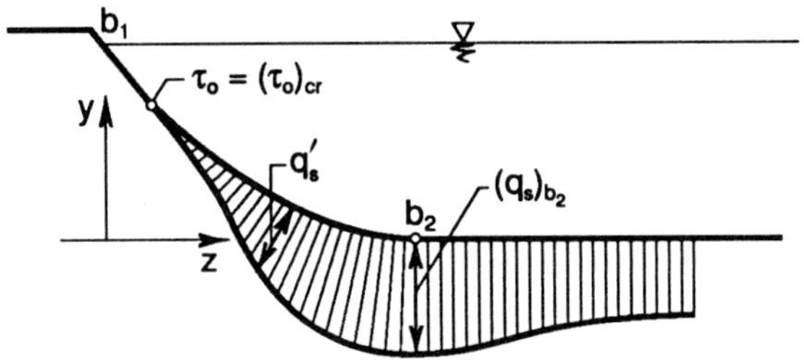

Fig. 5.19

for the bank deformation and displacement in plan. In this case, the horizontal counterpart of the transport continuity equation (1.76), with $p_s \equiv 0$, can be expressed as

$$\frac{\partial \bar{q}_s'}{\partial x} + \frac{\partial z_b}{\partial t} = 0 \qquad \left(\frac{\partial z_b}{\partial t} = \frac{dz_b}{dt} - U_m \frac{\partial z_b}{\partial x} \right) \tag{5.33}$$

where z_b is the horizontal coordinate of a point on the bank line.

At the early stages of meander development, i.e when θ_0 is small, the flow past the bank waves (in the $(x; z)$-plane) is similar to that past the bed forms (in the $(x; y)$-plane). Thus the flow adjacent e.g. to right bank in Fig. 5.20a is convergent along $a'Oa$, and divergent along $aO'a''$: at the apex a it is nearly parallel. This means that along $a'Oa$ we have $\partial\bar{q}_s'/\partial x > 0$ and thus $\partial z_b/\partial t < 0$; whereas along $aO'a''$, $\partial\bar{q}_s'/\partial x < 0$ and $\partial z_b/\partial t > 0$ (with $\partial z_b/\partial t \approx 0$ at the apex). Consequently, when θ_0 is small the meander waves are mainly migrating along x, rather than expanding along z. Note that the most intense deviation of streamlines from the "curvilinear parallelism" is at the inflection points O and O', where the *deviation angle* ω at the flow center line is maximum ($\omega = \omega_{max}$).

Nonetheless, the meander waves are also growing in their amplitude, however slowly at first; and eventually θ_0 will become as large as to deflect the flow away from the banks. Consequently, the flow which was convergent along the right bank $a'Oa$, will turn into a divergent one along its part Oa

(Fig. 5.20b), the most intense deviation from parallelism being roughly at the apex a ($\omega_a \approx \omega_{max}$). The maximum expansion of the diverging flow at the right bank is thus at a section, a_u say, downstream of a. At the section a_u we have, at the same time, the maximum contraction of flow at the left bank, and consequently the largest flow velocity (u_{max}) of the loop. Hence, the flow is not symmetrical with respect to the apex section a. And this is apparently due to the inclination of the flow plane in the general direction x: the forces gS_v deform the streamlines 1 into 2 (insert in Fig. 5.20b). Note that when θ_0 is large, then the flows at the left and right banks are convectively decelerated and accelerated, respectively, throughout the stream region $a_u a_u'$. Experiment shows that the meander length Λ_m remains practically constant during the meander development.

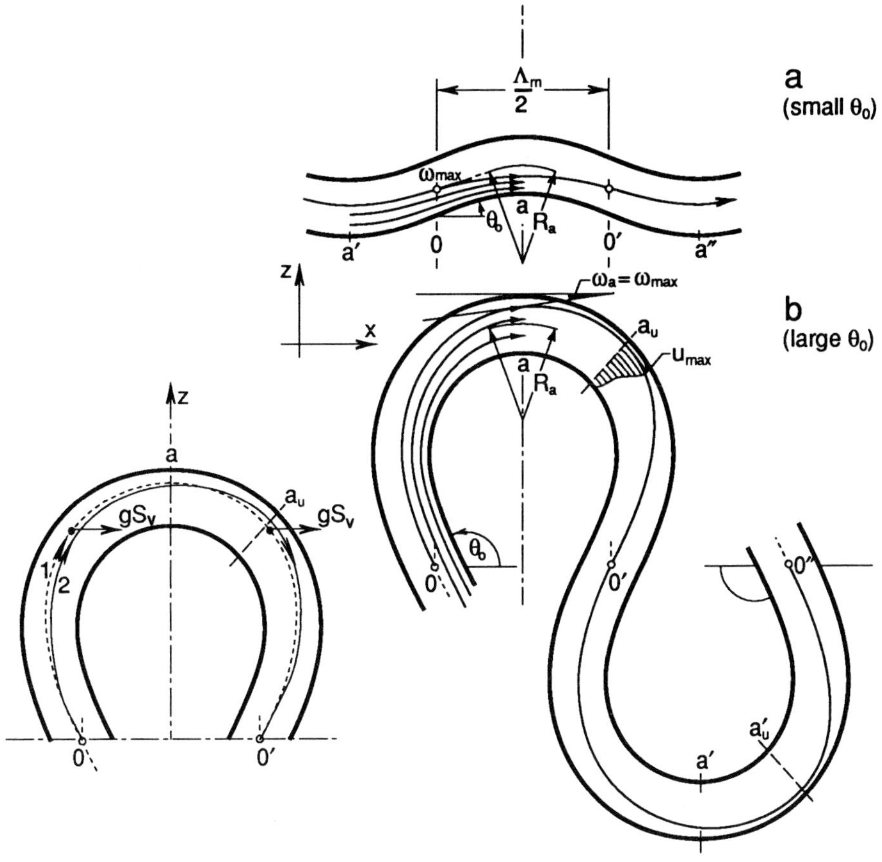

Fig. 5.20

For the sake of simplicity the present explanations are given on a two-dimensional (vertically-averaged) basis; any influence of a possible cross-circulation is ignored. The "small" and "large" θ_0, in the present context, cannot be separated from each other in a clear cut manner, for even in the case of a rectangular cross-section, the boundary-θ_0 separating them varies depending on B/h and c. [In P. Whiting and W.E. Dietrich's experiments [35], $\theta_0 = 10°$ was certainly "small"; in R.L. Hooke's experiments [16], $\theta_0 = 55°$ was "large"].

iii- The relation (5.33), with $p_s \equiv 0$, cannot reflect the expansion of meander loops adequately; at least because it yields $\partial z_b/\partial t = 0$ at the section a_u (where u is maximum and thus $\partial \bar{q}_s'/\partial x$ is zero). Introducing p_s ($\neq 0$) and using the natural bank coordinates $s = l$ and n, one can express (5.33) as

$$\frac{\partial n}{\partial t} = -\frac{\partial \bar{q}_s'}{\partial l} + p_s, \tag{5.34}$$

which is more suitable for large θ_0. Consider Fig. 5.20b. The material eroded from the left bank around a_u is conveyed – in suspension – downstream, and then deposited (by the decelerating flow $a_u a_u'$) on the left bank (and on its point bar) at a_u'. This transfer of material, which causes the expansion ($\partial n/\partial t$) of meander loops, must be considered as due to p_s (for the grains arriving at a_u' are from the "external source" a_u). The (local) bank-load rate derivative $\partial \bar{q}_s'/\partial l$ has different signs along Oa_u and a_uO', hence it contributes mainly to the deformation (skewness) of expanding loops. The larger h/D (large river; fine sand) the more effective is the material transfer due to p_s, and the more pronounced is thus the development of meander loops.[14] For the expansion velocity at the apex a of a regular meander loop we have

$$(V_m)_a = \left(\frac{\partial n}{\partial t}\right)_a = (p_s)_a. \tag{5.35}$$

5.3.2 *Migration and expansion of meanders*

i- From the observations and measurements carried out in freely meandering natural rivers it follows that "at the early stages (small θ_0), it is the downstream migration of meander waves which is mainly observable; and when $\theta_0 >$ [65° to 75°] (i.e. when $\sigma >$ [1.35 to 1.6]) it is their expansion which dominates" (Ref. [23], p.108). And also "From the observations and comparisons (of aerial photos taken at regular intervals over a long period of time) it has been found that for $\sigma <$ [1.4 to 1.6] ($\theta_0 <$ [65° to 75°]) the expansion speed of meander loops ($d\theta_0/dt$) increases, for $\sigma \approx$[1.4 to 1.6] it reaches its maximum value, and for $\sigma >$ [1.4 to 1.6] it progressively decreases". "It should be noted that the (same) magnitude $\sigma \approx$[1.4 to 1.6] is

---

14 If h/D is "small", then so is p_s. This explains why gravel rivers and laboratory streams cannot acquire large θ_0 and tend to minimize S by degradation (Fig. 5.14): the regime trend $S \rightarrow S_R$ may lead to the formation of extensive loops only if p_s is substantial. [The meanders in ice, turf, etc., correspond to $p_s \equiv 0$ and therefore their θ_0 can only be limited. The same applies to meandering of fluid in fluid (Fig. 5.18)]. The importance of the transfer by suspension on meander development has already been emphasized in [32], [34], [41].

exhibited by substantially different freely meandering rivers" (Ref. [23], p.126 and 128).

It can be shown, on a purely mathematical basis, that the angular expansion speed $d\theta_0/dt$ is related to the linear expansion speed $(V_m)_a$ at the apex a as

$$\frac{d\theta_0}{dt} = \frac{\theta_0}{R_a}(V_m)_a, \tag{5.36}$$

where R_a is the (smallest) radius of curvature of the meander loop at a (see the derivation of (5.36) in Ref. [43]).

From the explanations above, one infers that the (normalized) migration velocity U_m and the expansion speed $d\theta_0/dt$ ($\sim (V_m)_a$) should vary with θ_0 (or σ), as shown schematically in Fig. 5.21. Attention is drawn to the fact that

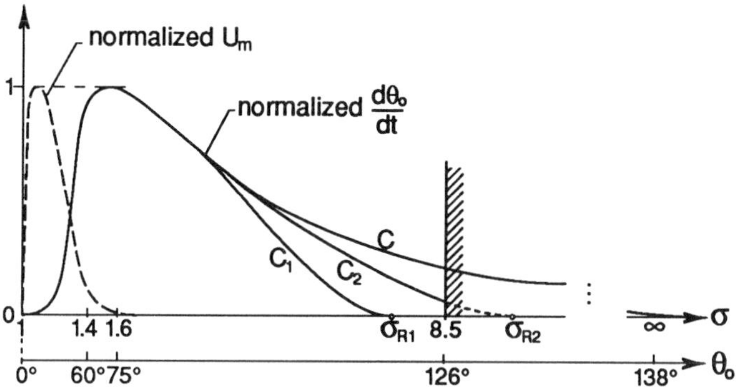

Fig. 5.21

the aforementioned empirical interval θ_0 = [65° to 75°] where $d\theta_0/dt$ achieves its maximum, includes in its middle the value $\theta_0 = 70°$ where the curvature $(1/R_a)$ of the sine-generated curve, corresponding to any given Λ_m, acquires its maximum (Fig. 5.5a). This suggests (as is clear from (5.36)) that $(V_m)_a$ must be proportional to $(1/R_a)^N$ (with $N \geq 0$); which, as will be shown presently, is indeed so. In Fig. 5.21, the angle $\theta_0 = 126°$ is in coincidence with $\sigma \approx 8.5$ (touching of meander loops); $\theta_0 = 138°$ implies $\sigma \to \infty$ (see 5.1.3). Since the regime value σ_R is finite, the $(d\theta_0/dt)$-curve corresponding to a concrete case must eventually deviate from the "envelope curve" C as to become zero at σ_R (curves C_1 and C_2 in Fig. 5.21), C itself being zero at $\sigma \to \infty$. If $S_v/S_R = \sigma_R < 8.5$, then the regime state can be achieved (curve C_1). If, on the other hand, $S_v/S_R > \approx 8.5$, then the regime state can never be achieved (without degradation), for the meander development will be interrupted at $\sigma \approx 8.5$ (broken curve C_2).

ii- Let q_s be the cross-sectional average specific transport rate, which can be assumed to be the same for any section l. Clearly, the right-hand side of (5.35) must be proportional to q_s/L_m, and we can express (5.35) as

$$(V_m)_a = A_a \frac{q_s}{L_m} . \tag{5.37}$$

In the next section it will be shown (Eq. (5.57) with $l/L_m = 0.25 = const$) that a dimensionless quantity such as A_a is determined by a functional relation of the form

$$A_a = \frac{h}{R_a} \phi\left(\frac{B}{h}, c, \theta_0\right),$$

which, in conjunction with (5.37), gives

$$(V_m)_a = \frac{h}{R_a} \frac{q_s}{L_m} \phi\left(\frac{B}{h}, c, \theta_0\right). \tag{5.38}$$

Here $L_m = \sigma \Lambda_m \approx \sigma(6B)$ (where B is, in fact, B_R), and therefore

$$(V_m)_a \sim \frac{h}{R_a} \frac{q_s}{\sigma B} \phi\left(\frac{B}{h}, c, \theta_0\right). \tag{5.39}$$

Note that $(V_m)_a$ is proportional to $1/R_a$ (i.e. $N = 1 > 0$, as required). Substituting (5.39) in (5.36) and eliminating σ and R_a with the aid of (5.20) and (5.21), one determines for the angular expansion speed

$$\frac{d\theta_0}{dt} \sim \frac{q_s}{B^2} \Phi\left(\frac{B}{h}, c, \theta_0\right)[\theta_0 J_0(\theta_0)]^3 \tag{5.40}$$

(where $\Phi = (h/B)\phi$),

which, with the aid of the dimensionless time $t_* = tq_s/B^2$, can be expressed in the dimensionless form

$$\frac{d\theta_0}{dt_*} \sim \Phi\left(\frac{B}{h}, c, \theta_0\right)[\theta_0 J_0(\theta_0)]^3. \tag{5.41}$$

Hence, if the secondary influence of θ_0 within Φ is ignored, then for a set of specified B, h, q_s and c, the angular expansion speed $d\theta_0/dt$ must vary in proportion to the third power of the $[\theta_0 J_0(\theta_0)]$-curve shown in Fig. 5.5a. The solid curve in Fig. 5.22 (where $\sigma = 1/J_0(\theta_0)$ is used as the ordinate, instead of θ_0) is the normalized graph of $d\theta_0/dt \sim [\theta_0 J_0(\theta_0)]^3$: it corresponds to the envelope curve C in Fig. 5.21. Note that the grossly scattered data of Russian Rivers plotted are "within the envelope". They form the curves C_i, which deviate from C as to approach "their" σ_R. The plot in Fig. 5.22 is the normalized version of the plot in Fig. 8.15 of Ref. [23].[15]

---

[15] A thorough and interesting study of the time deformation of meander loops can be found in the work of Hasegawa 1989 [15]. This study also rests on the transport continuity equation; however, it invokes both the longitudinal and radial transport rates q_s and q_r.

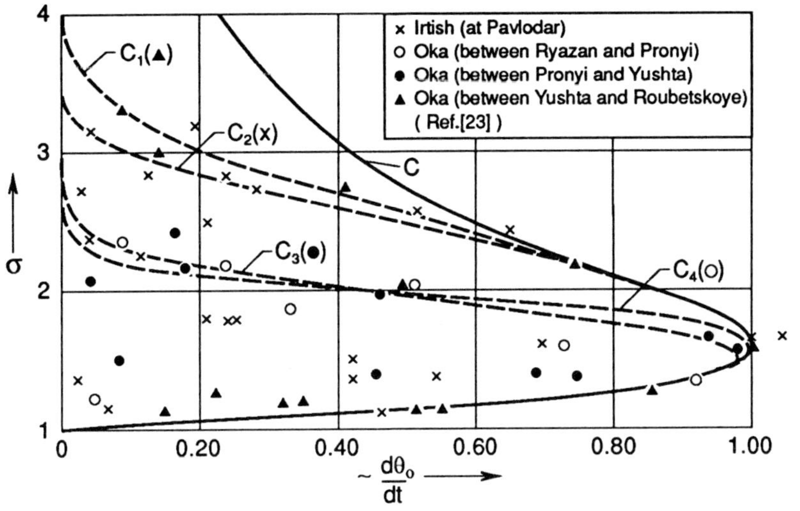

Fig. 5.22 (after Ref. [23])

5.4 Meander Dynamics (schematical outline)

5.4.1 *Dimensionless formulation of curved channel flows*

i- Much research (both theoretical and experimental) has been carried out on circular open-channel flows, with the hope of uncovering the mechanism of meandering streams.

Consider the steady state tranquil flow in a circular open-channel segment shown in Fig. 5.23a: $L > \approx B$, the non-deforming cross-section is rectangular, the flow is turbulent. Any cross-sectional property A of this flow is determined by the functional relation

$$A = f_A(\rho, B, h, R, c, U, \bar{\theta}_0, l/L), \tag{5.42}$$

where R is the radius of the circle forming the plan center line and U is a

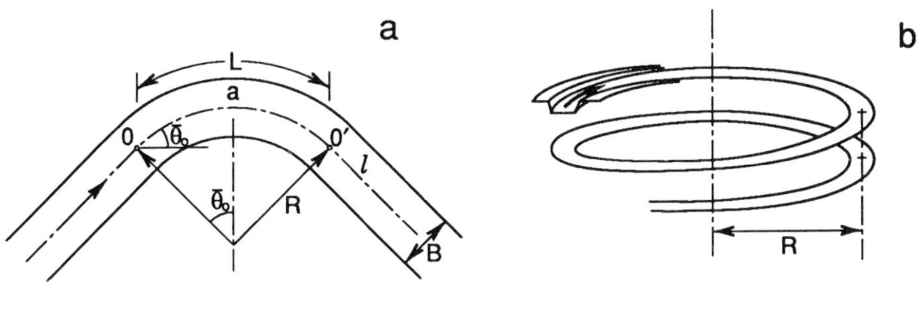

Fig. 5.23

typical flow velocity.[16] This velocity can be identified (depending on the nature of the quantity A under investigation) with v, v_*, u_{max}, etc.; for all longitudinal velocities are interrelated by c alone (which is present in the unspecified function f_A):

$$\text{e.g.} \quad v_* = \frac{v}{c} \;, \quad u_{max} = v\left(\frac{2.5}{c} + 1\right) \;, \quad \text{etc.} \tag{5.43}$$

Selecting ρ, h and U as basic quantities, one arrives at the following dimensionless form of (5.42):

$$\Pi_A = \rho^x h^y U^z A = \bar{\Phi}_A\left(\frac{h}{R}, \frac{B}{h}, c, \bar{\theta}_0, \frac{l}{L}\right). \tag{5.44}$$

Here we will focus our attention only on the flow region at the apex a (where $l/L \approx 1/2 = const$). Furthermore, we will assume, following the convention, that $\bar{\theta}_0$ and $R/B = (R/h)(h/B)$ are sufficiently large, and thus that the influence of $\bar{\theta}_0$ can be ignored (in the neighbourhood of a). This means that the flow in the a-region of the present circular channel, having a finite L, is identified with the so-called "developed flow" in the corresponding infinite circular channel $L \to \infty$ (spiral channel in Fig. 5.23b). Consequently, we will replace (5.44) by its reduced form

$$\Pi_A = \rho^x h^y U^z A = \Phi_A\left(\frac{h}{R}, \frac{B}{h}, c\right). \tag{5.45}$$

Let $\bar{A}$ be those A which are finite only when the channel has a finite curvature ($|\bar{A}| > 0$ only when $|1/R| > 0$). Clearly, for such $\bar{A}$ the relation (5.45) can more suitably be expressed as

$$\Pi_A = \rho^x h^y U^z \bar{A} = \left(\frac{h}{R}\right)^m \phi_{\bar{A}}\left(\frac{h}{R}, \frac{B}{h}, c\right) \tag{5.46}$$

(with $m > 0$).

We will use in the following the "channel fitted" system of orthogonal cylindrical coordinates l, r and y, shown in Fig. 5.24; and we will mark the characteristics associated with these coordinates by the corresponding subscripts. The origin O is at the bed level $y = 0$, and it is in coincidence with the center of the circular channel. The upper fluid layer adjacent to the free surface will be referred to as "layer 1"; the lower fluid layer adjacent to the bed, as "layer 2". The characteristics (A) of the layers 1 and 2 will be marked by the subscripts 1 and 2, respectively.

---

[16] Since the flow is tranquil ($Fr < 1$), g is not included as a characteristic parameter. In the present phenomenon, g can appear only in conjunction with the longitudinal and tranverse free surface slopes (S and S_r respectively) as gS and gS_r (see e.g. the equations of motion in 5.4.2).

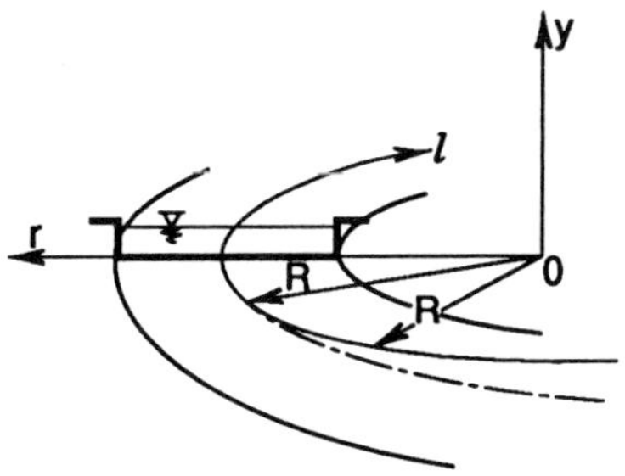

Fig. 5.24

ii- Each of the properties listed below is obviously of the type $\overline{A}$ (for which the form (5.46) is valid) – they correspond to the middle of the flow cross-section ($r \approx R$):

1- The transverse free surface slope S_r or, to be more precise, gS_r (see footnote 16).
2- The deviation angle ω_1 of the free surface streamlines at the center circle: $\tan \omega_1 = u_{r1} / u_{max}$.
3- The deviation angle ω_2 of the bed streamlines at the center circle: $\tan \omega_2 = u_{r2} / u_2$ (u_2 is at a specified level y near the bed).
4- Radial force F_{r1} acting per unit fluid volume at the free surface.
5- Radial shear stress $(\tau_0)_{r2}$ acting on the flow bed.

Substituting

$$\begin{aligned} \overline{A}_1 &= gS_r \quad (\text{with } U = u_{max}) \\ \overline{A}_2 &= \tan \omega_1 = u_{r1} / u_{max} \quad (\text{with } U = u_{max}) \\ \overline{A}_3 &= \tan \omega_2 = u_{r2}/u_2 \quad (\text{with } U = v_* = \sqrt{\tau_0/\rho}\,) \\ \overline{A}_4 &= F_{r1} \quad (\text{with } U = u_{max}) \\ \overline{A}_5 &= (\tau_0)_{r2} \quad (\text{with } U = v_* = \sqrt{\tau_0/\rho}\,) \end{aligned} \tag{5.47}$$

in (5.46), and taking $m = 1$, one determines

$$S_r = \frac{u_{max}^2}{gR} \lambda_S \tag{5.48}$$

$$\tan \omega_1 = \frac{u_{r1}}{u_{max}} = \frac{h}{R} \lambda_{\omega_1} \tag{5.49a}$$

$$\tan \omega_2 = \frac{u_{r2}}{u_2} = \frac{h}{R} \lambda_{\omega_2} \tag{5.49b}$$

$$F_{r1} = \frac{\rho u_{max}^2}{R} \lambda_F \tag{5.50}$$

$$(\tau_0)_r = \tau_0 \frac{h}{R} \lambda_c = \rho v^2 \frac{h}{R} (\lambda_c / c^2), \tag{5.51}$$

where each $\lambda_{\overline{A}}$ is determined as

$$\lambda_{\bar{A}} = \phi_{\bar{A}}\left(\frac{h}{R}, \frac{B}{h}, c\right). \tag{5.52}$$

The above derived relations are consistent with their counterparts encountered in the literature.[17] From the derivation above it should be clear that the multipliers $\lambda_{\bar{A}}$ are not just some "correction factors" (as they are often referred to), but certain functions of three dimensionless variables. And it is the knowledge of these functions (and not of the combinations $(h/R)^m(\rho^x h^y U^z)^{-1}$ which follow directly from the π-theorem) that signifies the knowledge on a $\bar{A}$-formula. In most cases the factors $\lambda_{\bar{A}}$ are given as functions of c only.

iii- Consider now a region Δl ($< \approx B$) of a sinuous channel whose plan center line is a sine-generated curve: the cross-section is rectangular $(B; h)$, the steady state tranquil flow is turbulent. In this case, R varies as a function of θ_0 and l/L_m as indicated by (5.15). The segment Δl of this sine-generated channel can always be identified with the segment of a circular channel whose center-circle radius is equal to the radius of curvature of the sine-generated center line (at the location l/L_m of Δl). The attempts to formulate the flow within Δl by using the "matching" circular flow expressions were not successful, for the flow structure within Δl is not determined by the (circular) geometry of Δl alone − it is determined by the *total* geometry of the sinuous channel. [Consider e.g. the sinuous channels in Figs. 5.20a and b. These channels have at their apexes a the same R_a (and B);[18] and, suppose, the same h and c. Consequently, their B/h, h/R and c (at $l/L_m = 0.25$) are identical. Yet the behaviour of their streamlines (at a) is essentially different. And this is because their total plan geometry (which in the case of a sine-generated curve is determined by θ_0 alone) is very different]. It follows that a cross-sectional property A of the flow in a sine-generated channel is determined by the relation

$$\Pi_A = \rho^x h^y U^z A = \Phi_A\left(\frac{h}{R}, \frac{B}{h}, c, \theta_0, \frac{l}{L_m}\right), \tag{5.53}$$

which is equivalent to (5.44). Using the sine-generated relations (5.15) and (5.20), viz

$$\frac{L_m}{R} \approx 6\theta_0 \sin\left(2\pi \frac{l}{L_m}\right) \quad \text{and} \quad \frac{\Lambda_m}{L_m} = J_0(\theta_0), \tag{5.54}$$

---

[17] Compare (5.48) with the S_r-formulae e.g. in [4], [14], [51]; (5.49a) and (5.49b) with the u_{r1}- and u_{r2}- formulae in [4], [21], [40]; (5.50) and (5.51) with the F_{r1}- and $(\tau_0)_r$-formulae in [4], [40] and [4], [9], [20], [40] respectively.

[18] Note from the (Λ_m/R_a)-curve in Fig. 5.5a that it is possible to have the same $1/R_a$ for two essentially different θ_0.

and considering that in the case of a real meandering channel $\Lambda_m \approx 6B$, one obtains

$$\frac{h}{R} \approx \frac{h}{B}\left[J_0(\theta_0)\,\theta_0 \sin\left(2\pi \frac{l}{L_m}\right)\right], \tag{5.55}$$

i.e.

$$\frac{h}{R} = \frac{h}{B}\phi\left(\theta_0, \frac{l}{L_m}\right). \tag{5.56}$$

Substituting (5.56) in (5.53), one arrives at the (reduced) form

$$\Pi_A = \rho^x h^y U^z A = \Phi_A\left(\frac{B}{h}, c, \theta_0, \frac{l}{L_m}\right), \tag{5.57}$$

which corresponds to a sine-generated channel having $\Lambda_m \approx 6B$. Clearly, the independent variables B/h, c and θ_0 in this relation are, in the case of an actual meandering stream, themselves the properties (Π_A) determined by the parameters (5.30), the channel width B being B_R.

5.4.2 *Fluid motion in sinuous channels*

i- Consider a sinuous channel whose plan center line is a sine-generated curve: the (non-deforming) cross-section is rectangular, the steady state flow is turbulent.

We introduce two extreme cases of flow in this channel. The first of them is the helicoidal flow, or "α-flow", in Fig. 5.25a; the second is the laterally oscillating flow, or "β-flow" in Fig. 5.25b. The α-flow is induced by the channel curvature $1/R$ and it consists of a parallel (to the curvilinear boundaries) longitudinal fluid motion (u_l), upon which a cross-circulation (Γ) is superimposed.[19] β-flow is caused by the *variation* of channel curvature, i.e. by $d(1/R)/dl$, and it is formed by fluid mass which shifts (in all its thickness h) periodically left and right as it moves along l (Fig. 5.25b): the lateral oscillation amplitude of streamlines decreases as the bed is approached. The keyword for β-flow, is convective acceleration.

Experiment shows that the behaviour of small-scale meandering streams, having small B/h, is similar to that of an α-flow (and the more so, the smoother are their boundaries), whereas the large-scale meandering rivers, having large B/h (and a rough bed), tend to behave as a β-flow ([40], [16],

---

[19] In the present text the circulation Γ is assumed to be in the (readily identifiable) radial $(r; y)$-plane, and it is identified with a u_r-diagram (along y) which consists of two negative and positive parts having equal areas (Fig. 5.25a). Some authors prefer to consider Γ as taking place in the surfaces which are orthogonal to streamlines. In the case of a (pure) α-flow, both definitions are congruent; in a combined (α and β)-flow, the shape of the (curved) orthogonal surfaces varies depending on θ_0 and l/L_m [a substantial difficulty in measurements].

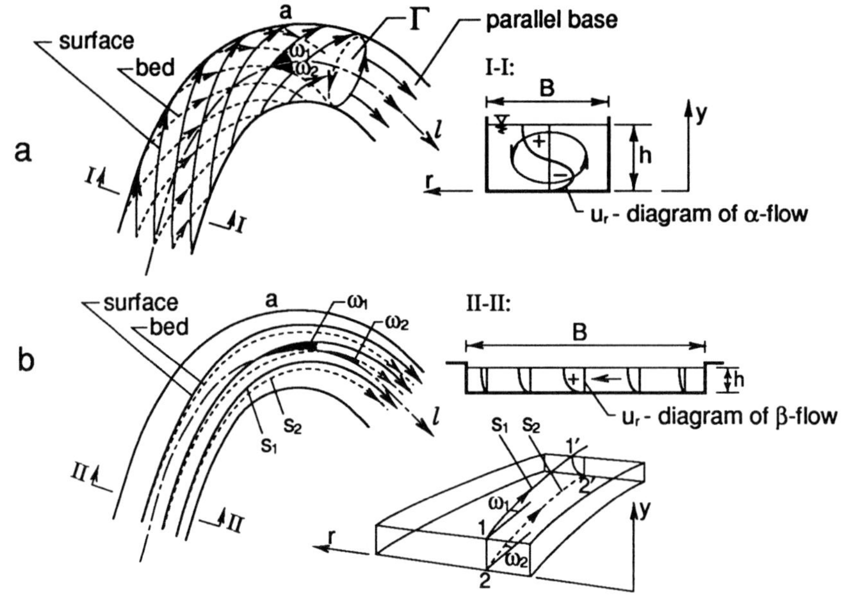

Fig. 5.25

[34], [26], etc.).[20] The radial velocities u_{r1} and u_{r2} of an α-flow are in the opposite r-directions, and therefore the angles ω_1 and ω_2 of its streamlines s_1 and s_2 are of the opposite sign (Fig. 5.25a). In contrast to this, the velocities u_{r1} and u_{r2} of a β-flow are in the same r-direction, and its angles ω_1 and ω_2 are thus of the same sign (Fig. 5.25b): the velocity gradient $\partial u_r/\partial y$ merely causes the vertical $1 - 2$ to deform into $1' - 2'$ as it is shifted sideways (Fig. 5.25b). Let q_r be the net specific cross-flow rate passing through a surface of unit length (along l) parallel to the banks. In the case of an α-flow, q_r is obviously zero; in the case of a β-flow, it is not. In the latter case, the elementary cross-flow rate $\delta q_r = q_r \delta l$ entering ABC through AB ($= \delta l$), leaves it as $q\delta r$ through BC ($= \delta r$) (Fig. 5.26). Thus, $q_r = q \tan \omega_{av}$.

ii- α- and β-flows are applicable to the limiting cases of small and large B/h respectively. In general (intermediate B/h), the u_r-diagram, at the center line, should be "somewhere in between": it should be as one of the normalized u_r-diagrams in Fig. 5.27. Let us denote the u_r-diagrams of (purely) α- and β-flows by $(u_r)_\alpha$ and $(u_r)_\beta$ respectively. The $(u_r)_\alpha$ and $(u_r)_\beta$ diagrams form the "envelope" containing the intermediate u_r-diagrams. Each

---

20 In a meandering river $B \approx B_R$, while h, at any stage of development, is comparable with h_R. Recall from Chapter 4 that $B_R \sim Q^{1/2}$ and $h_R \sim Q^{n_h}$ ($1/3 < n_h < 0.42$). Hence $B_R/h_R \sim Q^m$ (where $m > 0$), which indicates that it is indeed the large rivers, carrying large Q, which usually possess large B/h.

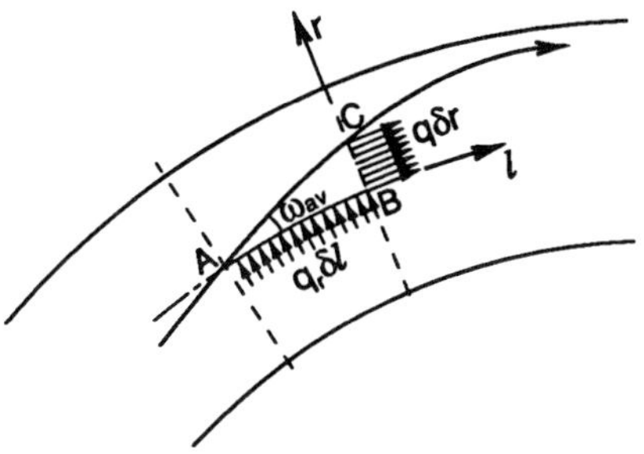

Fig. 5.26

u_r-diagram is the sum of two component diagrams: the circulatory $u_{r\alpha}$-diagram, and the translatory $u_{r\beta}$-diagram (Fig. 5.27). i.e.

$$u_r = u_{r\alpha} + u_{r\beta} \qquad (\text{for any } y \in [0,\, h]). \tag{5.58}$$

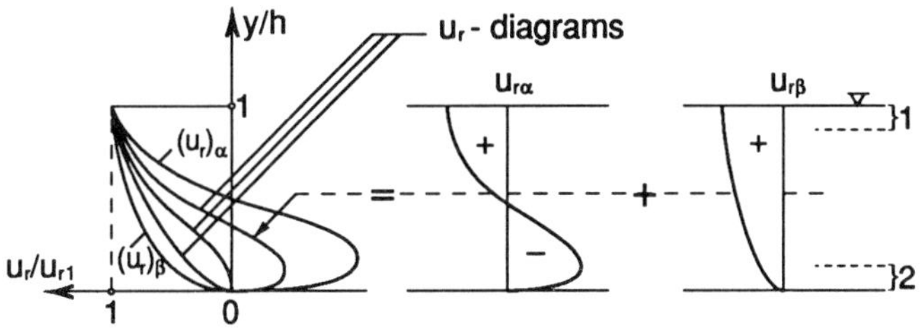

Fig. 5.27

The vertical average of radial velocities of an α-flow is zero, whereas that of a β-flow is not. The vertically-averaged streamlines of a β-flow are within the (narrow) areas between the streamlines s_1 and s_2 (see Fig. 5.25b), and therefore the plan behaviour of a β-flow can be reflected very adequately by its vertically-averaged counterpart. Clearly, no information can be gained from the vertically-averaged (or depth-averaged) equations on the internal structure of a meandering flow (on the velocity distributions along y, the comparative prominence of α- and β-components, etc.). On the other hand, they give a clear and adequate picture on the convective accelerations and decelerations in plan, and it is these characteristics which determine $\nabla \mathbf{q}_s$, and consequently the deformation of the bed surface (Eqs. (1.78) and (1.87)). Owing to this reason many recent theoretical works on bed deformation of meandering streams were conducted with the aid of the vertically-averaged equations ([8], [35], [45], etc.).

It may also be added that, since the vertical average of an α-component $u_{r\alpha}$-diagram is zero, the non-zero vertical average of the radial velocity field of a meandering flow is, in fact, the vertical average of its β-component ($(u_r)_{av} \equiv (u_{r\beta})_{av}$). Hence, if the vertically-averaged flow is convectively accelerated/decelerated, then that flow necessarily contains the β-component: the α-component may or may not be present − its vertical average consists of parallel sinuous streamlines. (e.g. the β-component is certainly present in the flows depicted in Figs. 5.20).

From the explanations above it should be clear that the detection of a finite radial velocity u_r is not necessarily an indication of the existence of cross-circulation; and even when the cross-circulation is present, it is only a part ($u_{r\alpha}$) of that u_r which "belongs" to it. It is difficult to visualize the existence of a cross-circulation in a meandering flow having $B/h = 100$ or

200, say, and it has been realized since a long time that the "transverse circulation only takes place in distorted experimental models, and in such natural channels whose width is small as compared with the depth" [34] (see also [33], [8]). The analogous cannot be said with regard to the β-component: a non-zero $u_{r\beta}$-diagram is present in any turbulent flow in a sinuous channel.

The fact that the laboratory streams having comparatively small B/h possess a β-component is evident e.g. from the classical experiments of J.F. Friedkin [10]. Indeed, he has observed that "the sand from the eroded concave bank almost entirely deposited on the convex bank downstream, close to the same side of the channel, and only a small part of it crossed the channel and deposited on the opposite convex bank" [10], [24]. In the case of a (pure) α-flow, the amount of material transported on the side of eroded bank, to the most, is comparable with that transported across the channel (as can be inferred e.g. from Fig. 5.29). Yet in the case of a (pure) β-flow, all of the eroded material is transported on the side of eroded bank. [If θ_0 is sufficiently large, then the material eroded from the apex zone of the outer (left) bank (Fig. 5.20b) is likely to be deposited along the decelerating flow region $a_u a_u'$ adjacent to that bank (5.3.1 (ii))]. Hence if in the experiments reported in Ref. [10], the eroded sediment was transported remaining "almost entirely" at the same bank, then β-component must have been present in the flows of those experiments. Similarly, the cross-circulation Γ is (nearly) symmetrical with respect to the center line of a symmetrical cross-section. Yet, in the laboratory experiments of R.L. Hooke [16][21] the location of maximum "helix strength" (near the apex section a) was substantially shifted toward the outer bank (Fig. 11.B in Ref. [16]); and the more so, the smaller was the flow rate Q and thus the ratio h/B.

However, in the past the β-component was usually ignored, and attention was focused almost exclusively on cross-circulation.

iii- No vertical velocities u_y are present in a laterally oscillating β-flow, and in the case of an α-flow they are concentrated in the neighbourhood of the banks (laboratory experiments in [40], see also [24]). Considering this, and taking into account that the variation of u_r with r (at any y) at the flow center line must be negligible, one can assert that the flow in a sinuous channel must possess a central region b_m (however narrow) where both u_y and $\partial u_r/\partial r$ can be identified with zero.

The radial component of the equation of motion in the channel-fitted cylindrical system of coordinates is given by

$$u_l \frac{\partial u_r}{\partial l} + u_r \frac{\partial u_r}{\partial r} + u_y \frac{\partial u_r}{\partial y} = \frac{u_l^2}{r} - gS_r + \frac{1}{\rho}\frac{\partial \tau_r}{\partial y} \tag{5.59}$$

[21] R.L. Hooke's [16] experiments: sine-generated channel, rigid banks, cohesionless mobile bed. $\theta_0 = 55°$, $\Lambda_m = 10.33m$, $B = 1m$, $S = 0.002$, $D \approx 0.3mm$. $Q = 10, 20, 35, 50\ l/s$ were used: $0.27 < Fr < 0.40$, $8 < B/h < 20$.

where the left-hand side is the radial convective acceleration. Substituting $u_y = 0$ and $\partial u_r / \partial r = 0$ in this equation, one arrives at its reduced form

$$u_l \frac{\partial u_r}{\partial l} = \frac{u_l^2}{r} - gS_r + \frac{1}{\rho}\frac{\partial \tau_r}{\partial y} , \tag{5.60}$$

which is valid for the central region b_m. Using the value of u_r given by (5.58), one can express (5.60) as

$$u_l \frac{\partial u_{r\alpha}}{\partial l} + u_l \frac{\partial u_{r\beta}}{\partial l} = \frac{u_l^2}{r} - gS_r + \frac{1}{\rho}\frac{\partial \tau_r}{\partial y} . \tag{5.61}$$

Most of the theoretical studies aiming to reveal the cross-sectional behaviour of a curved channel flow (i.e. to reveal the form of the function $u_r = f_r(y)$) differ from each other mainly because of the methods used for the evaluation of u_l, τ_r and S_r. Basically, however, they tend to identify the flow in a region Δl ($< \approx B$) of a sinuous channel with the developed flow in a circular channel (5.4.1 (i)). Thus more often than not they rest on the following three postulates (which were apparently first introduced by J.L. Rozovskii [40]):

1- In a comparatively short region Δl ($< \approx B$), a sinuous channel can be treated as a circular channel.
2- The variation of cross-circulation within Δl is negligible.
3- The sinuous channel flow within Δl can be identified with the developed circular flow.

The postulate (1) has no direct bearing on the diff. Eq. (5.61). The postulate (2) is consistent with reality and it implies

$$u_{r\alpha} = const \quad \text{i.e.} \quad \frac{\partial u_{r\alpha}}{\partial l} = 0 , \tag{5.62}$$

for any y and r within Δl and b_m. The postulate (3), however, imposes a parallel longitudinal flow (concentric circles of the longitudinal base of the developed circular α-flow) – and, consequently, it eliminates automatically the convective (non-parallel) β-component:

$$u_{r\beta} \equiv 0 \quad \text{i.e.} \quad \frac{\partial u_{r\beta}}{\partial l} \equiv 0 . \tag{5.63}$$

Substituting (5.62) and (5.63) in (5.61), one arrives at

$$0 = \frac{u_l^2}{r} - gS_r + \frac{1}{\rho}\frac{\partial \tau_r}{\partial y} , \tag{5.64}$$

which was used as basic equation in most of the theoretical contributions produced ([4], [9], [14], [21], [40], [51], etc.).

Since in the postulates leading to (5.64) α-component (cross-circulation) is mentioned, whereas β-component is not, it is only natural that the radial shear stress τ_r in (5.64) was interpreted and evaluated in terms of $u_r = u_{r\alpha}$ only (as, for example, $\tau_r = \tau_{r\alpha} \sim \rho u_{r\alpha}^2$ or $\tau_{r\alpha} \sim (du_{r\alpha}/dy)^2$). Most of the relations

($u_r = f(y)$) determined in this way were tested by the laboratory flows having comparatively small B/h; and their validity was thus "experimentally confirmed". Clearly if u_r is identified with $u_{r\alpha}$, then *anything* that has a radial component, or occurs along r (deviation of streamlines ($\omega_1; \omega_2$), expansion of loops, cross-sectional non-uniformity of transport rate, etc.) *must be* explained by cross-circulation.[22] It is thus not surprising that it was the field researchers dealing with large natural rivers (having large B/h), in particular G.H. Matthes [34], N.I. Makaveyev [32] and S. Leliavskii [26], who were first to realize that the role of cross-circulation was overestimated (see e.g. [52], [24], [23], [4]).

It appears that the longstanding debate on the relevance of cross-circulation is thus somewhat out of focus: it is not *whether* but *when* the cross-circulation is relevant (or prominent) — what is right for small B/h may turn out not to be so when B/h is large.

iv- We conclude this subsection by giving a schematical outline of the cross-sectional mechanics of α- and β-components.

α-component: Eq. (5.64); Fig. 5.28a.
Since $u_{l1} > u_{l2}$, we have for the same r (within b_m)

$$F_{r1} = u_{l1}^2/r > u_{l2}^2/r = F_{r2}\,.$$

We assume that the pressure is distributed according to the hydrostatic law, and thus that the pressure difference force can be taken as gS_r, for any y. Furthermore, $F_{r1} > gS_r > F_{r2}$. The leftward-directed active force $F_{r1} - gS_r$ moves the fluid of layer 1 to the left (toward the outer bank): the positive force $F_{r1} - gS_r$ is balanced by the negative $T_{\alpha 1} = (\partial\tau_r/\partial y)_1/\rho$. Having turned downwards at the outer bank and descended to layer 2, the fluid is now under the action of the rightward-directed force $gS_r - F_{r2}$, which moves it to the right: the negative $gS_r - F_{r2}$ is balanced by the positive $T_{\alpha 2} = (\partial\tau_r/\partial y)_2/\rho$. At the inner bank, the fluid turns upwards and returns to layer 1 — the cross-circulation is completed.

β-component: Eq. (5.60); Fig. 5.28b.
We consider the convectively accelerating region *Oa* at the outer bank (Figs. 5.20b or 5.30b), where the left-hand side of (5.60), viz $u_l(\partial u_r/\partial l)$ is positive for all y (and l). Since the fluid moves in radial direction only to the left, we have $F_{r1} > F_{r2} > gS_r$. The leftward-directed positive forces $F_{r1} - gS_r$ and $F_{r2} - gS_r$ are balanced by the negative $T_{\beta 1} = -u_{l1}(\partial u_r/\partial l)_1 + (\partial\tau_r/\partial y)_1/\rho$ and $T_{\beta 2} = -u_{l2}(\partial u_r/\partial l)_2 + (\partial\tau_r/\partial y)_2/\rho$ respectively. Since the flow converges toward the outer bank throughout its thickness h, the projections of τ_s and $\partial\tau_s/\partial y$ on the $(r; y)$-plane are positive. And because $\partial\tau_s/\partial y < 0$, for any y, so

---

[22] A tendency to explain any "radial phenomenon" by cross-circulation is still noticeable even in some recent works. e.g. if the left-hand side of (5.64) is known to be different from zero, then this non-zero value is, as a rule, attributed to $\partial u_{r\alpha}/\partial l \neq 0$, rather than to $\partial u_{r\beta}/\partial l \neq 0$ (irrespective of B/h and c).

must be its projection $\partial\tau_r/\partial y$. Hence the balancing forces $T_{\beta 1}$ and $T_{\beta 2}$ consist of the (rightward-directed) negative terms.

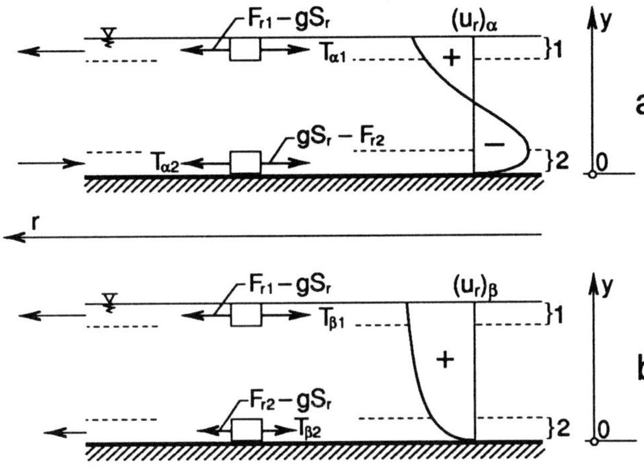

Fig. 5.28

5.4.3 *Flow patterns*

i- We assume, as before, that the sinuous channel is sine-generated, the (non-deforming) cross-section is rectangular, the steady state flow is turbulent.

1) *α-flow*

The increment of $1/R$ causes the increment of the magnitude of all ω_i and u_r (Eqs. (5.49a) and (5.49b)). Since the longitudinal flow, upon which the cross-circulation is superimposed, is parallel to the (sine-generated) channel boundaries, and since both u_l and u_r corresponding to any y are the same for all r of the central region b_m, the streamlines at a level y are also parallel to each other (within b_m). The angles ω_1 and ω_2 at a flow section are of opposite sign but same magnitude. This magnitude varies along l so as to become zero at the points of inflection (O) and acquire maximum at the apexes (a). The bed streamlines of an α-flow are sketched in Figs. 5.29a and b (which are due to J.L. Rozovskii [40]). Here the lines σ separate the bed streamlines which cross and do not cross the channel.

2) *β-flow*

In this case, the angles ω_1 and ω_2 are of the same sign, and the magnitude of ω_1 is larger than that of ω_2 at any location of flow. At a given cross-section, the magnitudes of ω_1 and ω_2 and thus of their average $\omega \approx \approx (\omega_1 + \omega_2)/2$ are largest at the center line; and they are zero at the banks. The streamlines of a β-flow do not cross the channel: they form periodically alternating convergence-divergence zones in plan (see Figs. 5.30, which are essentially the same as Figs. 5.20). When θ_0 is small, then the largest ω, viz

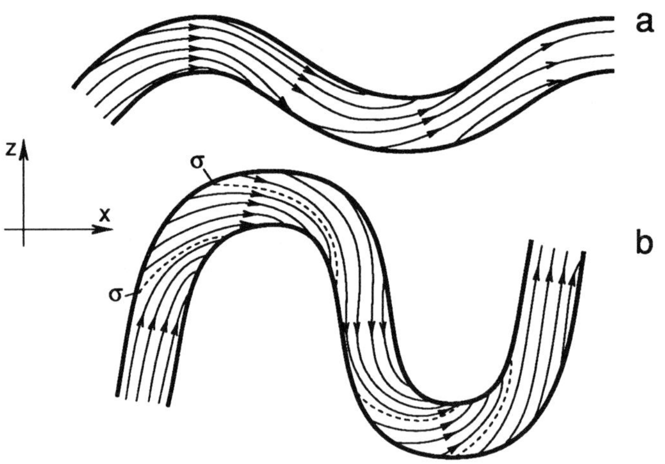

Fig. 5.29 (from Ref. [40])

ω_{max}, is at the inflection points O and O', the largest flow velocity (u_{max}) being at the apex section a (Fig. 5.30a). With the increment of θ_0, the flow picture

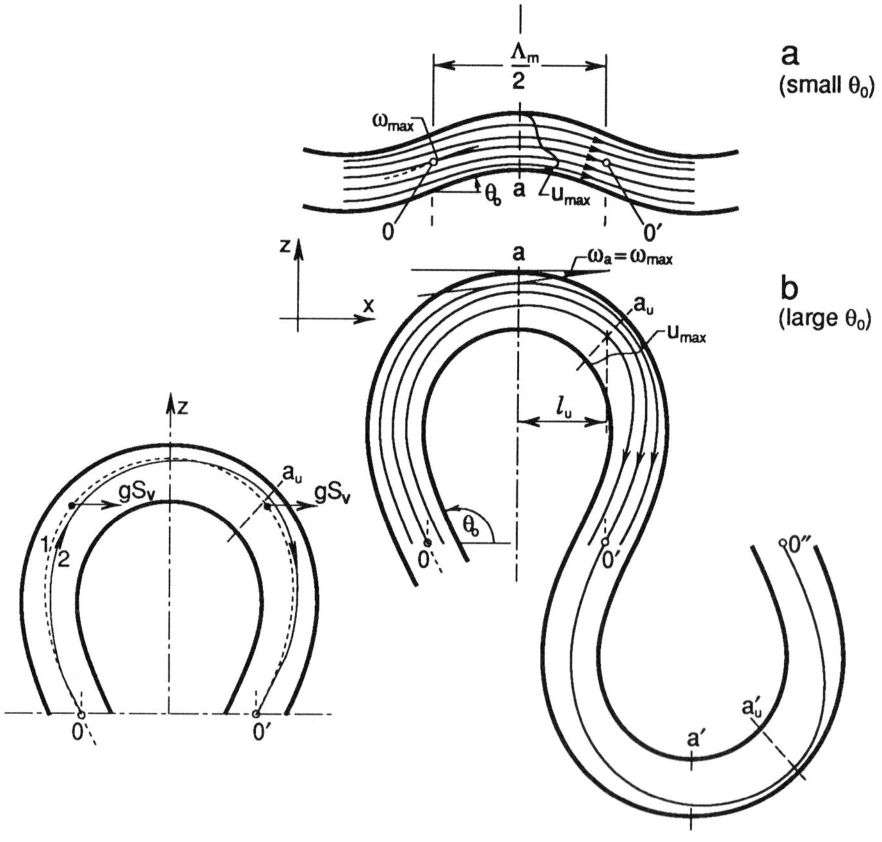

Fig. 5.30

progressively changes, so that when θ_0 is large, then ω_{max} is approximately at the apex section, while u_{max} is at the section a_u – downstream of a (Fig. 5.30b). The dimensionless location, l_u/Λ_m say, of the section a_u must be expected to vary depending on θ_0, B/h and c.

3) *General case*

The flow patterns formed by both, α- and β-components, are sketched in Fig. 5.31. The circulation is present, but the streamlines are not parallel.

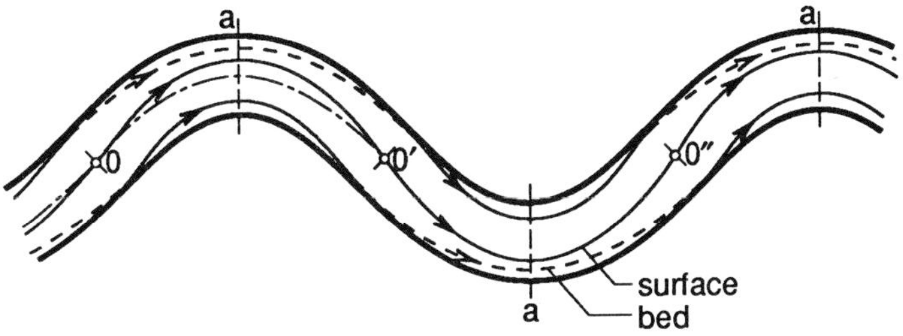

Fig. 5.31

While the streamlines at one bank rotate (incompletely, around longitudinal axes) as to come from the free surface to the bed, those at the other bank rotate as to come from the bed to the free surface. The streamlines far from the banks do not rotate: they merely oscillate laterally, in their horizontal planes. The percentage of rotating and oscillating lines depends on the comparative intensity of α- and β-components.

For the same remaining conditions, all ω_i's and u_{ri}'s (see (5.49a) and (5.49b)) corresponding to a flow section, increase with the channel curvature ($1/R$) of that section. On the other hand, $1/R$ is proportional to $1/R_a$, which acquires its maximum at $\theta_0 \approx 70°$ (Fig. 5.5a). Hence, for the same remaining conditions, the intensity of divergence of streamlines from the general direction l (in all flow patterns shown in Figs. 5.29 to 5.31) must reach its maximum when $\theta_0 \approx 70°$. To put it slightly differently, during the formation of a meandering stream, the intensity of radial phenomena and the expansion speed ($d\theta_0/dt$) of loops acquire their maximums when θ_0 passes through $\approx 70°$ (see Fig. 5.21).

ii- The increment of β-component intensifies the increment of energy losses due to the convergence and divergence of streamlines, while the increment of α-component leads to the increment of energy losses due to the rotation of flow at the banks and the consequent radial bed friction.[23] One would

[23] In the case of a β-flow, τ_{0_r} is but the projection of τ_{0_s} on the r-direction.

expect that the flow in a sinuous channel "organizes" its α- and β-components so as to pass the loops with minimum energy expenditure.

Consider the radial velocity u_r at a specified level y; e.g. at the level $y \approx h$, implying the layer 1. On the basis of (5.58), we have for $u_r = u_{r1}$

$$u_{r1} = u_{r\alpha 1} + u_{r\beta 1} \tag{5.65}$$

and thus

$$1 = n_\alpha + n_\beta . \tag{5.66}$$

The relative input of α- and β-components can reasonably well be reflected by the values of $n_\alpha = u_{r\alpha 1}/u_{r1}$ and $n_\beta = u_{r\beta 1}/u_{r1}$. It is not known yet how n_α and n_β ($= 1 - n_\alpha$) are determined by B/h, c, θ_0 and l/L_m; and all that can be asserted on the score at present is that for a given l/L_m (e.g. for $l/L_m = 0.25$ at the apex section) and a specified θ_0, they should vary with B/h and c as shown schematically in Fig. 5.32.

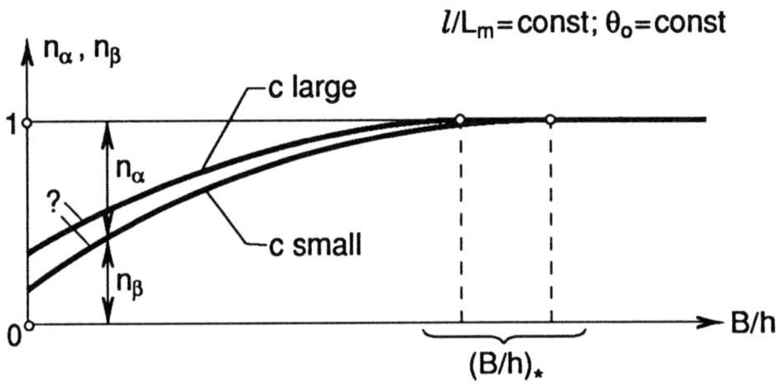

Fig. 5.32

Since β-component is present for all B/h whereas α-component progressively vanishes with the increment of B/h, there must exist such a $(B/h)_*$ that α-component is "practically zero" if

$$B/h > (B/h)_* = \phi_*(c, \theta_0, l/L_m)$$

(as indicated in Fig. 5.32). The graphs of n_α and n_β in Fig. 5.32 and thus the function $(B/h)_*$ can be readily determined from the family of radial velocity diagrams (in Fig. 5.27) which, in turn, can be determined either theoretically or from laboratory measurements: future research on the topic would certainly be worthwhile.

Suppose that the u_r- and u_l-distributions along y, e.g. at the center line of the apex section a, have been revealed from laboratory measurements (u_l-distribution should be nearly logarithmic, if B/h is sufficiently large). In this case, the cross-circulation diagram $u_{r\alpha}$ can be determined as follows. We have

$$u_{r\alpha} = u_r - u_{r\beta} = u_r - u_l \tan\omega ,$$

where the unknown ω reflects the (vertically averaged) β-deviation of flow. Now the upper and lower areas of the $u_{r\alpha}$-diagram must be equal, and therefore ω must be of that particular value which satisfies

$$\int_{k_s}^{h} (u_r - u_l \tan \omega)dy = 0$$

(where $u_r = f_r(y)$ and $u_l = f_l(y)$ are known from measurements).

From the content of this section it should be clear that the mechanical structure of a meandering stream (corresponding to a given θ_0 and c) varies fundamentally depending on its aspect ratio B/h. Utmost caution should thus be exercised when using the distorted models to predict the behaviour of a meandering flow (let alone the sediment transport or channel deformation caused by it). Much of the conflict in the field is due to the attempts to explain the meandering of large-B/h streams by using information gained from small-B/h streams.

5.4.4 *Bed deformation*

i- The presence of convectively accelerated and decelerated flow zones in a sinuous channel means the presence of non-uniformities (in plan) of the bed-load rate ($\nabla \mathbf{q}_s \neq 0$). Hence, as is clear from the transport continuity equation (1.78), viz

$$\nabla \mathbf{q}_s + \frac{\partial y_b}{\partial t} = 0 \, ,$$

in some areas the bed elevation y_b must increase, while in some others decrease with the passage of time, as to form eventually the "bed topography" (of a meandering stream): the circulation, if present, is a catalyst only.[24] Considering this, the bed deformation will be discussed in this subsection on the (convective) β-basis. (It may be noted in passing here, that the increment of β-input with the increment of B/h, and thus of Q (see footnote 20), forms an additional reason for large rivers to exhibit a "better meandering").

We will assume, in analogy to R.L. Hooke's experiments [16], that the banks of the (sine-generated) channel are virtually rigid: the transport is by the bed-load ($q_s = q_{sb}$). At $t = 0$, the bed is a (flat) plane $y_b = 0$; at $t = T_b$, it is a (curved) surface[25] $y_b = \phi_b(r, l)$.

---

[24] The cross-circulation is effective mainly in the formation of point bars ([32], [52], [24], etc.).

[25] When referring to the "bed" in this paragraph, we refer to that (smooth) geometric surface on which the bed forms (ripples and/or dunes) are superimposed. The bed forms are already developed when the deformation of this surface has just started. The geometry of the developed bed forms cannot vary significantly during the formation of the bed surface, and the roughness K_s caused by them is reflected by the practically non-varying $c \sim K_s^{-1/6}$.

ii- If the changes in the flow structure due to the bed deformation are ignored, then the general (qualitative) nature of the resulting bed topography can be predicted on the basis of transport continuity form (1.87), i.e.

$$\frac{\partial y_b}{\partial t} = -[\alpha u_b]\frac{\partial u_b}{\partial s} \quad (= -\nabla \mathbf{q}_s), \tag{5.67}$$

where $[\alpha u_b] = u_b\, \partial\phi(u_b)/\partial u_b$ is a monotonously increasing function of u $(> u_{cr})$.

1- Since $[\alpha u_b]$ is always positive, $\partial y_b/\partial t$ changes its sign when $\partial u_b/\partial s$ does so. Hence the zones of the downward and upward displacements (i.e. the erosion and deposition zones) must coincide with zones of convective acceleration and deceleration respectively. Consequently, the erosion and deposition zones due to the flow shown in Figs. 5.30a and b should be in the areas sketched in Figs. 5.33a and b. The scheme thus predicted is, in a broad sense, consistent with the bed topography of actual meandering streams (Figs. 5.34a and b).

2- The deepest erosion or pool should be at (or around) that flow section where the bed-load rate "convergence" $\nabla \mathbf{q}_s$, i.e. the product $[\alpha u_b]\partial u_b/\partial s$, is maximum. Since the multiplier $[\alpha u_b]$ is a monotonous function of u_b $(> u_{cr})$, its maximum is at the section of maximum u_b; the maximum $\partial u_b/\partial s$ is at the section of $\omega = \omega_{max}$ (where the deviation of streamlines from the curvilinear parallelism is the largest). But this means that the deepest erosion should occur at a section that is somewhere between the sections of ω_{max} and $(u_b)_{max}$. Hence if θ_0 is "small", then the deepest erosion should be at the inner bank of a section that is between the sections O and a (Fig. 5.33a); and if θ_0 is "large" then it should be at the outer bank of a section that is between the sections a and a_u (Fig. 5.33b) – "somewhat downstream of the apex" [23]. As is well known, this, as a rule, is indeed so in reality.

iii- The determination of the quantitative bed topography is a popular research topic since a long time. Yet it has been realized only recently that this topography cannot be determined on the basis of static equilibrium of forces acting on bed grains; at least because the bed grains of a meandering river in sand are not static (η is usually comparable with 10). Let $(Q_s)_{in}$ and $(Q_s)_{out}$ be the bed-load rates entering and leaving an area, A say, of the bed surface. The point bar grows because $(Q_s)_{in} > (Q_s)_{out}$ for its areas A. However, the larger the steepness of the bar becomes, the more it impedes the sediment-bringing flow to "climb" on it. One can say that the steeper the bar, the more effectively it steers away from itself the flow which tends to amplify its steepness further. Eventually the bar reaches such a final equilibrium steepness; and the flow bed as a whole, such an equilibrium topography, for which $(Q_s)_{in}$ and $(Q_s)_{out}$ are equal all over.

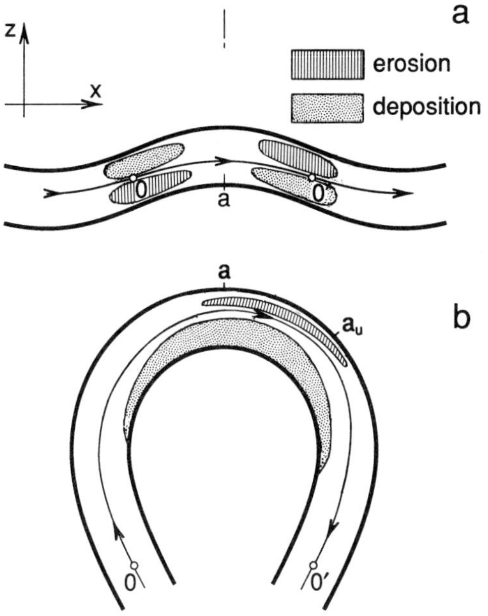

Fig. 5.33

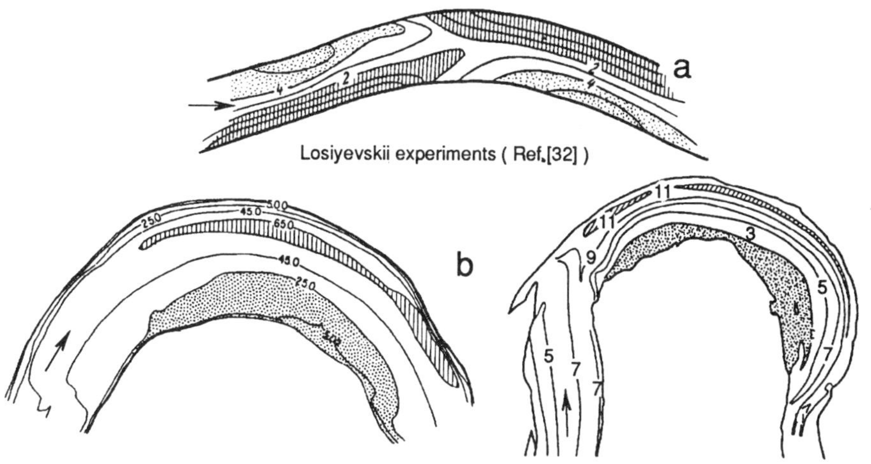

Fig. 5.34

iv- The prevalent current approach to the theoretical determination of bed topography of meandering streams rests on the dynamic equilibrium $((Q_s)_{in} = (Q_s)_{out} \neq 0)$ achieved by the above described "topographic steering". This approach is mainly due to the works of H.J. de Vriend, J.D. Smith, S.R. McLean, J.M. Nelson, W.E Dietrich, P. Whiting, N. Struiksma and others ([7], [35], [8], [45], [42]). In these works, the formation of bed topography is formulated by the convective accelerations of flow and thus of the bed-load rate: all equations are vertically averaged, the banks are treated as rigid. It is not possible to enter a detailed explanation and comparison of these works here, and the following is but an outline of their common principles.

It is assumed that the plan geometry of the sinuous channel is specified (θ_0 is specified, if sine-generated), and that Q, B and c are given.

The vertically-averaged local flow velocities $\mathbf{u}$ and the local shear velocities $\mathbf{v}_*$ satisfy the continuity and momentum equations which can be symbolized as

$$\phi_1(\mathbf{u}, h) = 0 \quad \text{and} \quad \phi_2(\mathbf{u}, \mathbf{v}_*, hy_b) = 0 \tag{5.68}$$

respectively. The velocities $\mathbf{u}$ and $\mathbf{v}_*$ are interrelated by the (known) "closures":

$$\phi_3(\mathbf{u}, \mathbf{v}_*) = 0 . \tag{5.69}$$

Substituting $y_b = 0$, one determines from the three equations above the three field functions: $\mathbf{u}$, $\mathbf{v}_*$ and h which correspond to the flat initial bed. Knowing thus the vector field $\mathbf{v}_*$, one determines the corresponding $\mathbf{q}_s$-field, with the aid of a (selected) bed-load relation,

$$\mathbf{q}_s = \phi_4(\mathbf{v}_*) , \tag{5.70}$$

and uses it in the transport continuity equation

$$\nabla \mathbf{q}_s + \frac{\partial y_b}{\partial t} = 0 . \tag{5.71}$$

Then, from this equation, one determines $(\delta y_b)_1$ corresponding to the time-step δt_1, alters the field functions according to $(\delta y_b)_1$, and repeats the procedure for δt_2, δt_3, ... etc. The computation ends (the equilibrium bed surface $y_b = \phi_b(r, l)$ is achieved) when $\partial y_b / \partial t \rightarrow 0$ is reached. The results thus obtained appear to be remarkably realistic: compare e.g. the equilibrium bed surfaces computed by J.M. Nelson and J.D. Smith [35] with those determined from experiment (see Figs. 5.35a and b).

PART II: Braiding

5.5 Origin of Braiding

5.5.1 *General*

i- Braiding, i.e. the splitting of an alluvial channel into a multitude of channels, is also, like meandering, a self-induced form of an alluvial stream motion (which is not forced upon the stream by its environment). Furthermore, it initiates also because of the large-scale horizontal turbulence, and develops because of the regime trend $Fr \rightarrow \min$.

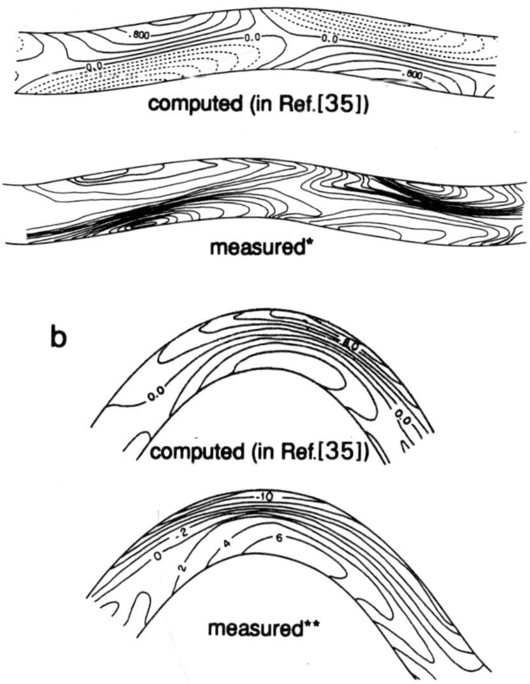

* by P. Whiting and W.E. Dietrich: Ref. [35]
** by R.L Hooke [16]

Fig. 5.35

Note from Fig. 5.8 that the braiding data (points B) are situated above the line l_1 forming the upper boundary of the meander region R_m, and therefore the line l_1 can be taken as the boundary between the meander (and alternate bar) region R_m and the braiding (and multiple bar) region R_b.

Numerous laboratory experiments have been conducted with the intention to explore the mechanism of braiding ([2], [11], [17], [29], [41], etc.). However, in some of them the essential characteristic of a natural braiding stream, viz its self-induced development (by successive splittings, i.e. subdivisions), appears to be overlooked. The fact that the region of braiding coincides with the region of multiple bars does not mean that a braiding stream is a "static image" of multiple bars. (A meandering stream is not a static image of alternate bars, although their regions also overlap on the $(B/h\,;\,h/D)$-plane). Most of laboratory experiments were conducted for a constant Q and erodible bed and banks. The bars occurred first: the channel width continued to increase and flow depth h to decrease. When h was sufficiently small, the bars were exposed. Does it mean that the flow was braiding? Take another example, $Q = const$, $B = const$ (flume experiments): again mutiple bars occured first. The slope S (which adjusts itself depending on sediment fed into the flume) continued to increase and h to decrease; eventually the bars were exposed, and the same question can be asked again. The answer is "no" in both cases, for each of the resulting flows around the emergent islands did not possess the dynamic quality of a self-induced further development. Indeed, a further decrement of h would merely cause the islands to become wider and the flow channels between them narrower: yet the *number* of these islands and channels would *not* increase. On the other hand, e.g. in the experiments of L.B. Leopold and M.G. Wolman [29] the laboratory stream was certainly braiding: the increment of number of islands and channels is clearly observable from Fig. 34 of Ref. [29]. (This valuable paper, though not "new" any longer, will often be referred to in the following). See also the experiments of S.A. Schumm and H.R. Khan [41].

Since braiding can initiate and develop only if: the sediment is moving, the flow is turbulent, $(Fr)_0 > (Fr)_R$ (i.e. $S_0 = S_\upsilon > S_R$) and the initial point P_0 is in the braiding region R_b, the necessary and sufficient conditions for the occurrence of braiding in a tranquil alluvial stream can thus be summarized by the set

$$\eta > 1 \ ; \quad Re > \approx 500 \ ; \quad S_\upsilon > S_R \ ; \quad P_0 \in R_b , \tag{5.72}$$

which differs from (5.29) only because R_b appears instead of R_m.

ii- As has been mentioned earlier, the regime development of flow width takes place much faster than that of the slope, and therefore one can assume, in analogy to meanders, that braiding also develops starting from a single initial channel whose width is nearly the same as the regime width:

$$B_0 \approx B_R \sim Q^{1/2}. \tag{5.73}$$

This means (according to the resistance equation $Q = B_0 h_0 c_0 \sqrt{g S_\upsilon h_0}$) that the initial depth and slope of both braiding and meandering experiments are given by

$$c_0^2 h_0^2 S_\upsilon \sim Q = const. \tag{5.74}$$

Consider now the braiding and meandering experiments which correspond to the same Q and D, and thus the same $B_0/D \sim Q^{1/2}/D$ (we will distinguish the characteristics of braiding and meandering flows by the subscripts b and m, respectively). As should be clear from 5.2.4 (i), the initial points $(P_0)_b$ and $(P_0)_m$ of these experiments must lie on the same 1/1-declining straight line, π say, in the log-log $(B/h; h/D)$-plane (Fig. 5.36). The 2.5/1-declining straight lines in this graph characterize the values of S: the more to the right is the line, the smaller is its S (5.2.4 (i)). Let P be the intersection of the line π with the line l_1 (boundary between R_m and R_b).

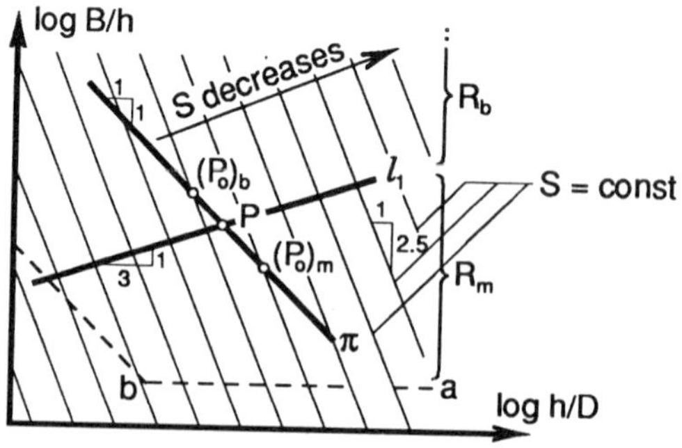

Fig. 5.36

Since each $(P_0)_b$ is higher than P and thus than $(P_0)_m$, its slope must be larger than that of $(P_0)_m$:

$$(S_v)_b > (S_v)_m . \tag{5.75}$$

As will be apparent in the next section, the overall slope of a braiding stream does not vary significantly as it develops with the passage of time. Yet, the slope of a meandering stream progressively decreases. Consequently the inequality (5.75), which is valid for $t = 0$, can only be intensified with the passage of time. Thus, (5.75) can be generalized into

$$(S)_b > (S)_m \tag{5.76}$$

which is valid for any t.

It follows that the slope of a braiding stream must, in general, be larger than that of a meandering stream corresponding to the same Q. This (derived) statement is in agreement with reality [recall the well known plot of L. Leopold and M. Wolman (Fig. 46 in Ref. [29]).

5.5.2 *Initial channel*

Although the braiding data (points B in Fig. 5.8) are scattered throughout the region R_b, the initial points $(P_0)_b$ are likely to be concentrated in the lower part of this region, i.e. in the zone of 2-row horizontal bursts ($N = 2$). This is invariably so in conventional laboratory braiding-experiments, where one can hardly have an initial channel, with a turbulent flow, whose B_0/h_0 is 300, say. In natural rivers, the initial channel problem is more ambiguous; for the braiding of a natural stream does not originate from an initial channel as we know it – the initial channel must be construed (inferred). A natural single-channel stream starts to braid, and remains braiding, in those stretches where the slope is steeper than usual. When the slope flattens, braiding disappears: separate channels merge again into a single-channel (Figs. 5.37 and 5.38). These conditions are shown schematically in Fig. 5.39. Here, the flatter stretch OA contains the single-channel stream (B', h', S'). The steeper stretch AB, whose slope is S'', is the braiding region: B'' and h'' are the characteristics of that section where the division of the single-channel begins. We assume that $B' \approx B'' \approx B_R \sim Q^{1/2}$, and that the single-channel characteristics h' ; S' and h'' ; S'' are interrelated, by the resistance formula, approximately as

$$\frac{h'}{h''} \approx \left(\frac{S''}{S'}\right)^{1/3} \quad \left(\approx \frac{B_R/h''}{B_R/h'}\right). \tag{5.77}$$

The ratios B_R/h'' and B_R/h' can be identified, respectively, with the ordinates of the initial braiding-point $(P_0)_b$ (above l_1) and a meander-point $(P_0)_m$ (below l_1), both points being on the same 1/1-declining straight line π (Fig. 5.36). Note from Fig. 5.8 that the line l_2 forming the upper limit of the zone of the

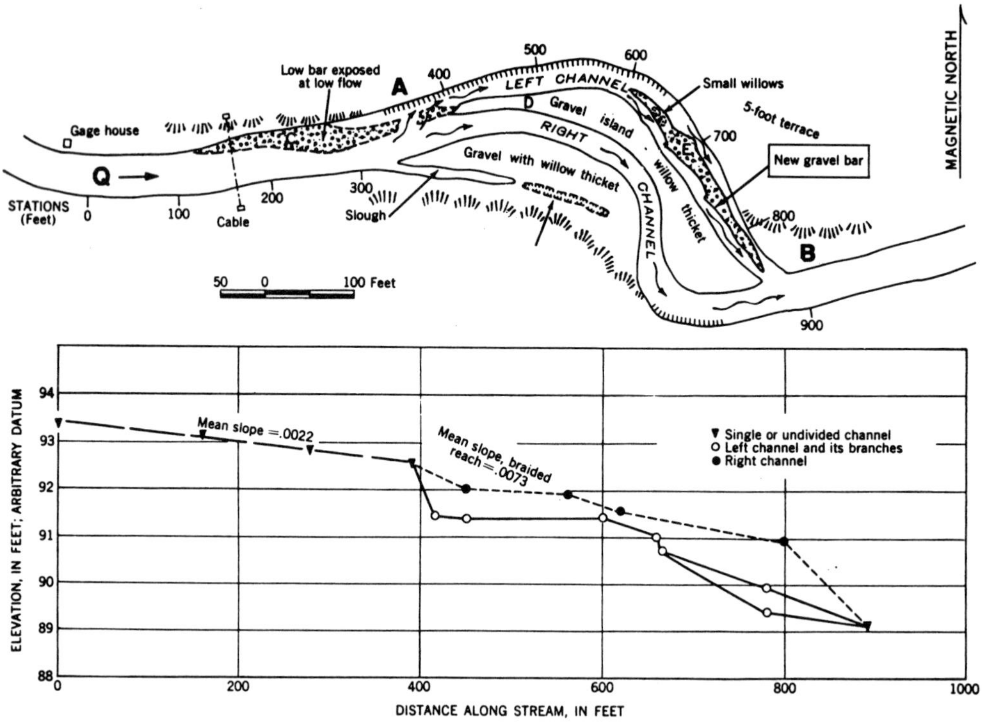

Fig. 5.37 (from Ref. [29])

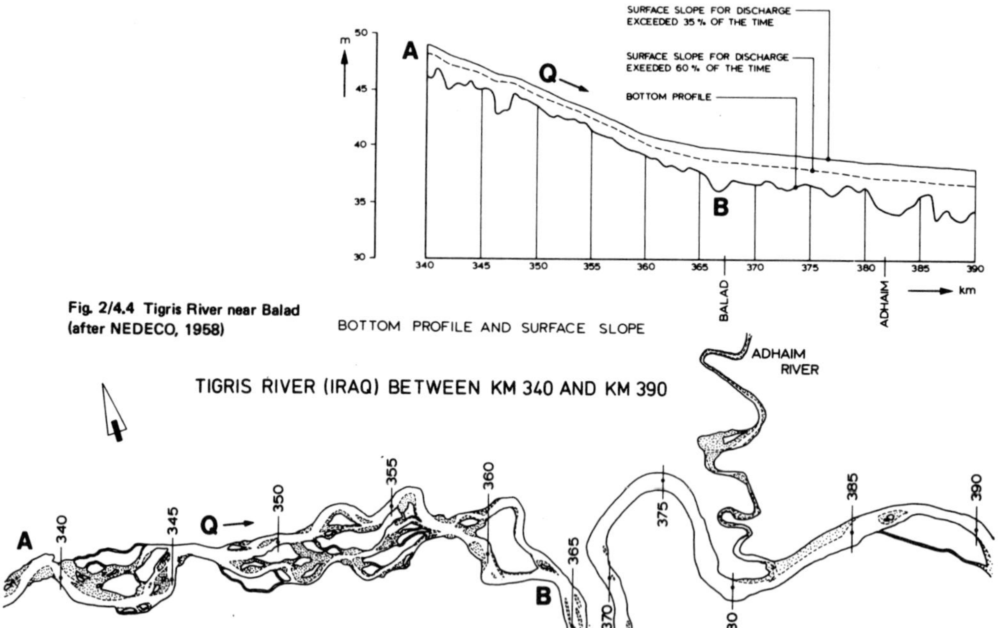

Fig. 5.38 (from Ref. [20])

2-row horizontal bursts and bars[26] is "higher" than the line l_1 by a factor of $80/25 \approx 3.2$. Hence even if the point $(P_0)_m$ were "just below" the line l_1, one would still need, roughly, $S''/S' > (3.2)^3 \approx 33$ to make the point $(P_0)_b$ to occur in the zone of 3-row bursts. Considering this, it will be assumed in the following that the initial channel of a braiding stream corresponds to $N = 2$.

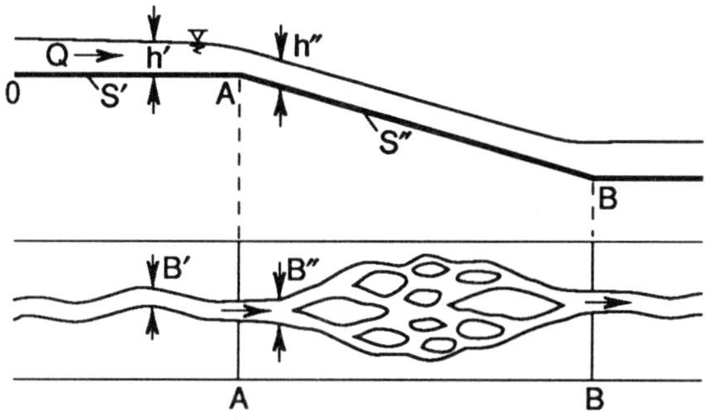

Fig. 5.39

5.5.3 *Minimization of the excess* $S_v - S_R$[27]

i- It should be recalled from Chapter 4 that the regime slope S_R is proportional to a negative power of the flow rate:

$$S_R = \alpha_S Q^{-n_S}. \tag{5.78}$$

Here $\approx 0.11 < n_S < \approx 0.42$ (lower limit for sand, upper limit for gravel) and $\alpha_S \approx$ *const* for a given experiment. Suppose that, owing to reasons which will be clarified in the next section, the initial channel (of $N = 2$) which has the width B_0 ($\approx B_R$) and which conveys the flow rate Q, splits into 2 channels, each of which has the width $B_2 = B_0/2$ and conveys $Q/2$. The regime slope of each of these channels, viz $(S_R)_2$, is larger than the regime slope S_R of the initial channel. Indeed

$$(S_R)_2 = \alpha_S (Q/2)^{-n_S} = S_R 2^{n_S} > S_R. \tag{5.79}$$

Hence, the original excess $S_v - S_R$ is reduced:

$$S_v - (S_R)_2 < S_v - S_R. \tag{5.80}$$

Now, each of these two channels is subdivided, in turn, into further two channels in the same way (all together four channels),[28] and then again ...,

---

[26] The line l_2 was located using the data in [12].

[27] See 5.2.2 (i) for the replacement of the trend $(Fr)_0 \rightarrow (Fr)_R$ by $S_v \rightarrow S_R$.

[28] Apparently, it is the authors of Ref. [29] who were first to recognize that the "new channels in the divided reach may become subdivided in the *same* manner".

etc. These successive subdivisions yield the decreasing sequence of deficits $S_\upsilon - (S_R)_k$, viz

$$(S_\upsilon - S_R) > (S_\upsilon - 2^{n_s} S_R) > (S_\upsilon - 2^{2n_s} S_R) > \cdots > (S_\upsilon - 2^{kn_s} S_R)\,. \quad (5.81)$$

Here the integer k is the number of successive subdivisions (splits), 2^k being the total number of channels which come into being at the end of the k'th split. The initial (single) channel corresponds to $k = 0$ or $2^0 = 1$, and it is referred to as the *first order channel* [3]. The $2^1 = 2$ channels corresponding to $k = 1$ are *second order channels*, the $2^2 = 4$ channels corresponding to $k = 2$ are *third order channels*, and so on ("order" = $k + 1$). The field studies usually do not go beyond the third order channels. It is thus tacitly assumed that the slopes of the divided channels are (nearly) the same: they are all equal to $\approx S_\upsilon$. This assumption will be clarified in 5.6.2.

ii- It follows that meandering and braiding are two means used by an alluvial stream to minimize its deficit $S_\upsilon - S_R$. In the case of meandering, the positive difference $(S_\upsilon - S_R)$ is minimized by the decrement of S_υ (S_R remaining constant); in the case of braiding, by the increment of S_R (S_υ remaining constant). The "choice" of a particular means depends on the geometry and roughness of the initial channel, and thus on the turbulent structure of flow conveyed by it (if $N = 1$, meandering; if $N \geq 2$, braiding). The decrement of S_υ in a meandering stream is by the growth of sinuosity, the increment of S_R in a braiding stream is by the increment of successive channel subdivisions: σ and k are the counterparts of each other. In the case of meandering, the growth of σ is not a guarantee that the regime state will be achieved, particularly if the excess $S_\upsilon - S_R$ is "large" (loops may touch each other, the "moving point" m may reach the lower boundary of R_m, etc. (see 5.2.4)). Similarly, the increment of k is not a guarantee that the stable regime state will ever be reached by a braiding stream (usually because of branching channels touching each other) – and the "restlessness", or instability, of braiding rivers is, in fact, more common.

5.6 Development of Braiding

5.6.1 *Channel division mechanism*

i- Consider an initial channel with a flat mobile bed at $t = 0$. The turbulent flow contains a sequence of 2-row horizontal bursts (Fig. 5.40a) caused by a discontinuity d (upstream end of mobile bed in laboratory, change of the bed slope at A (Fig. 5.39) in the field, etc.). If the channel banks were rigid, this sequence of bursts would generate a sequence of double-row bars (Fig. 5.40b). However, the banks are not rigid, and the alluvial formation is thus not quite the same.

The first bursts $(\beta_H)_1$ and $(\beta_H)_1'$ convectively accelerate the flow at the center line, and thus compell the transported sediment (which is assumed to be distributed uniformly along the flow width at d) to converge toward this line – along the distance $(\Lambda_H)_1 \approx 6(B_0/2) = 3B_0$, as indicated by the "arrows" in Fig. 5.40b.[29] Further downstream, along $(\Lambda_H)_2$, the flow at the

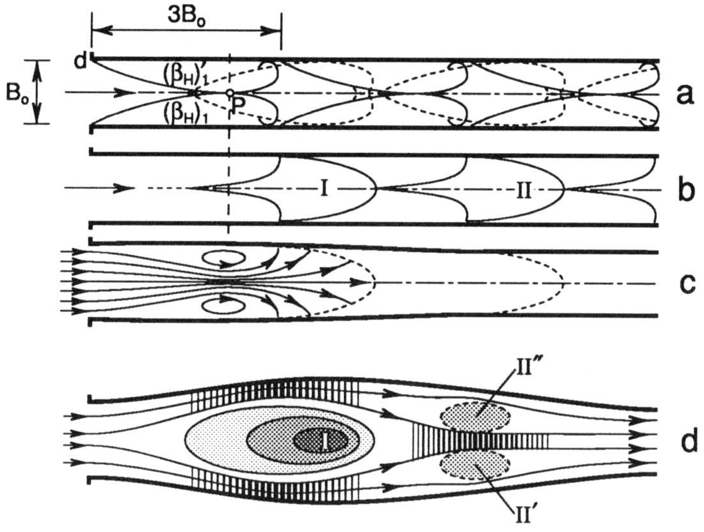

Fig. 5.40

center line decelerates, and the sediment is deposited so as to initiate the formation of the central bar I. With the increment of bar steepness its "topographic steering" (in the sense of 5.4.4) becomes more and more effective: the bar "forces the water into the flanking channels, which, to carry the flow, deepen and cut laterally into the original banks. Such deepening locally lowers the water surface and the central bar emerges as an island" (p. 39, Ref. [29]). In short, "when bank erosion occurs (and the channel widens), the water level (and thus the flow depth) decreases, and the bars emerge" (p. 1761, Ref. [40]).[30] The conditions described are shown schematically in Figs. 5.40c and d. The widening of the channel around the central bar I alters the plan geometry of the flow, and consequently the "intended" alluvial formation. The widened (and thus "weakened") flanking flows cannot transport the material they erode (from the side channels of I) as far downstream as the location of the potential central bar II. The material transported by them is deposited earlier: at II′ and II″. Thus the flanking streams create "their own" depositions (II′ and II″) which, in turn, will cause

[29] This (predicted) convergence of transport toward the center line is consistent with experimental observations: at the early stages of formation "the band of principal bed transport lies on top of the submerged central bar" (p. 47 in [29]).

[30] The original past tense has been altered to the present tense by the author; brackets added for clarification.

each of these two second order streams to be divided into two third order streams, etc.

It follows that it is indeed not appropriate to consider braiding as the "image", or "imprint", of multiple bars; at least because the initiation of braiding is associated with the *elimination* of the potential bar pattern (the bar II is replaced by II′ and II″) – in spite of the fact that it is one of the bars, viz the bar I, which, if present, triggers its initiation. This is analogous to the destruction of alternate bars by the initiation of meanders [12] (although it is the alternate bars which facilitate the initiation of meanders).

ii- The "New gravel bar" in Fig. 5.37 is a field example of the deposition caused by the second order flanking stream. Figs. 5.41a and b show the channel- (or island-) hierarchies in the braiding Donjek [49] and Brahmaputra [3] Rivers; "hierarchy numbers in circles refer to islands or bars,

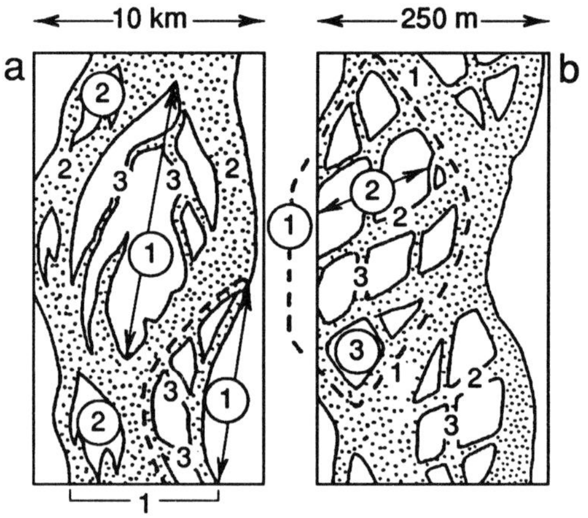

Fig. 5.41 (from Ref. [3])

other numbers to channels. The third-order channels still have bars within them which cannot be shown in this scale" [3]. Some of the third order channels occur on the top of the first- and second-order islands, some others on the bed of the first- and second-order channels. The former third-order channels are produced, during the decrement of Q (falling stage), by water level differences in the neighbouring lower-order channels: thus they cannot be formed in experiments with constant Q (a detailed description of hierarchy channels can be found in [3]).

5.6.2 *Slope and transport rate*

From the content of preceding sections, it should be clear that, for a given Q and thus $B_R \sim Q^{1/2}$, the initial single-channel stream can braid only if its slope $S_0 = S_\upsilon$ is as sufficiently large, and thus its flow depth h_0 is as sufficiently small as to make the point $P_0(B_R/h_0;\, h_0/D)$ to occur in the braiding region R_b (above the line l_1). Two types of $Q = const$ "experiments" are conceivable with a stream of a limited length AB:

1) The initial slope S_0 which ensures $P_0 \in R_b$ is provided, and the equality of the "in" and "out" transport rates $(Q_s)_A = (Q_s)_B$ is maintained for any t (as in the regime channel R). In this case the braiding development starts soon after $t = 0$ (S.A. Schumm and H.R. Khan experiments [41]).

2) The initial slope S_0 cannot yield $P_0 \in R_b$. A sufficiently large constant $(Q_s)_A$, larger than that which corresponds to S_0, is introduced at the channel entrance A (as in the regime channel R_1). In this case, the braiding development starts only after the slope has increased (because of the aggradation caused by $(Q_s)_A - (Q_s)_B > 0$) to a level which ensures $P_0 \in R_b$ (L.B. Leopold and M.G. Wolman experiments [29]).

Apparently, it was the encounter with conditions analogous to experiment 2, in both field and laboratory, which gave rise to the widespread view that it is the excess of the transport rate $((Q_s)_A - (Q_s)_B)$ which is the *cause* of braiding [23], [32], [52]. This is not so; "braiding does not necessarily indicate an excess of total load" (p. 39, Ref. [29]). The role of the transport excess is indirect: it directly affects S (by elevating it), which directly affects h (by reducing it) so that B/h becomes as large as to make the point P_0 to occur in R_b – and thus as to initiate braiding.

Consider Fig. 5.42 (from Ref. [29]). This graph shows that the braiding region AB coincides indeed with the elevated S region, and that the elevated S was achieved roughly at $t = 9h$. During $0 < t < 7$ to $9h$, the slope was increasing, but no braiding activity as such was present (Fig. 34 in Ref. [29]). Conversely in the later times, viz during $9h < t < 20h$, a substantial braiding development was present, but the slope was not changing in the process. (Note from Fig 5.42 that the bed-elevation lines corresponding to $9h$ and $20h$ are parallel to each other, the shift being due to the non-uniformity

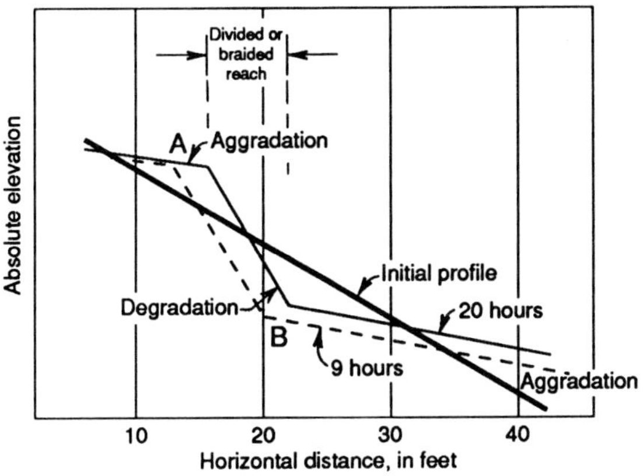

Fig. 5.42 (from Ref. [29])

$\partial Q_s / \partial x = const$ (< 0)). Hence, the overall slope of a braiding system either does not vary at all, or its variation is negligible; the observations reported in other sources are consistent with this view [3], [6], [44].[31] One can say that all branches of a braiding system which evolves on the "platform" *AB* in Fig. 5.39, remain parallel to this platform.[32]

5.6.3 $(B/h\,;\,h/D)$-plane

i- We assume that each of the $(k+1)$'th order channels (resulting after k successive divisions) first widens until it acquires its regime width $(B_R)_{k+1}$, and then divides so as to produce two $(k+2)$'th order channels. The regime width of the $(k+1)$'th order channel is given by

$$(B_R)_{k+1} = const(Q/2^k)^{1/2} = (const\, Q^{1/2})2^{-k/2} = (B_R)_1 2^{-k/2}, \quad (5.82)$$

where $(B_R)_1$ is the regime width of the initial first order channel (which was so far denoted as $B_0 = B_R$), the regime width of the $(k+2)$'th order channel being $(B_R)_{k+2} = (B_R)_1 2^{-(k+1)/2}$. Hence

$$\frac{(B_R)_{k+1}}{(B_R)_{k+2}} = \sqrt{2}. \quad (5.83)$$

This relation indicates that the ratio of the regime widths of successive channels does not depend on k: channels of all orders contract, because of their division, in a geometrically similar manner, that is, in the same proportion, viz $\sqrt{2} \approx 1.4$.

The above analysis is, of course, a highly idealized reflection of reality, and therefore one should be satisfied if the width ratio of successive real channels is (to the most) comparable with ≈ 1.4. Consider again Ref. [29], where it is reported that "two channels divided by the willow-covered islands are about equal in width, 30 to 40 feet. The sum of these widths is 20 to 30 percent greater than the 50-feet width of the undivided reach" (p. 41 in [29]). Thus $(B_R)_1 = 50ft$ and $(B_R)_2 \approx 35ft$, which yield $(B_R)_1/(B_R)_2 \approx 50/35 = 1.43$. Such a proximity to ≈ 1.4 is, of course, a pure coincidence. Nonetheless, it shows that the present approach is in line with the (typical) braiding pattern of Horse Creek (Fig. 5.37). Similarly, with reference to the Brahmaputra River, it is stated in Ref. [3] that "The first-order channel comprises the entire river. This has an average width of $10km$ ... ", and that "Second-order channels have maximum widths of $5km$ ..." (p. 73 in [3]). In this case we have $(B_R)_1/(B_R)_2 = 2$, which is larger than ≈ 1.4, but still comparable with it.

ii- Consider now the depth ratio h_{k+1}/h_{k+2}. The value of h_{k+1} corresponding to given B_{k+1} and S follows from the resistance equation $Q_{k+1} =$

---

[31] The statement concerns the overall (or average) slopes only: locally the slopes fluctuate.

[32] It should thus be clear that it is the elevated slope which causes braiding, and not the other way around (as it is assumed in some works).

$= c_{k+1}\sqrt{gS}\,B_{k+1}h_{k+1}^{3/2}$; and the analogous is valid for h_{k+2}. Assuming (in accordance with 5.6.2) that $S = S_\upsilon$, ignoring the difference between c_{k+1} and c_{k+2}, and taking into account (5.83), together with $Q_{k+1} = 2Q_{k+2}$, one determines

$$\frac{h_{k+1}}{h_{k+2}} = 2^{1/3}. \tag{5.84}$$

Hence the depth-ratio is also independent of k. Using (5.83) and (5.84), one obtains for the ratio of aspect ratios of successive channels

$$\frac{(B/h)_{k+1}}{(B/h)_{k+2}} = 2^{1/6}, \tag{5.85}$$

which indicates that the B/h-ratio decreases as the divided channels become smaller and smaller.

Each successive division of channels can be represented in the log-log $(B/h;\, h/D)$-plane (in Fig. 5.43) by a discrete (stepwise) displacement of the

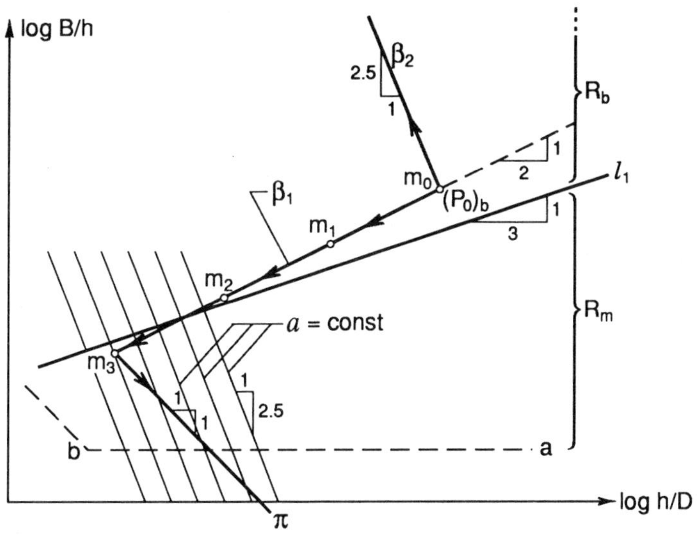

Fig. 5.43

moving point m. Before the division of the single-channel initial stream, $m = m_0$ is at the initial point $P_0 = (P_0)_b \in R_b$. After the first division, m is at m_1, after the second at m_2, ... after the k'th at m_k. From (5.84) and (5.85), it is clear that the displacements $m_0 \to m_1 \to \cdots \to m_k$ must take place (in the downward-to the left direction) along the 1/2-inclined straight line β_1. The line β_1 intersects the 1/3-inclined line l_1, and therefore eventually, i.e. after a certain k, the point m will enter the meander region R_m (after $k = 3$ in Fig. 5.43). And from then on each of the divided channels will continue to minimize its excess $S_\upsilon - 2^{kn_S}S_R$ by meandering: the point m will be moving along a 1/1-declining line π.

Each of the 2.5/1-declining straight lines in Fig. 5.43 signifies the variation of B/h with h/D for a constant value of the dimensionless complex

$$a = (Q/D^{2.5})/(c\sqrt{gS}),$$

which was introduced in 5.2.4 (Eq. 5.31).

In the case of a meandering channel (given Q, D and c), each of these lines corresponds to a particular S. Thus, they indicate how S decreases as the point m moves along π – after it has entered the region R_m. However, when m is still in the braiding region R_b (m_0, m_1 in Fig. 5.43), then the 2.5/1-declining lines do not signify particular S values; rather, they signify particular Q values. Indeed, the divided channels have nearly the same S ($\approx S_v$), and it is their flow rate which varies (decreases) in each division (m_0, m_1, m_2, ... have Q, $Q/2$, $Q/4$ ...). In other words, when the stream is meandering, then the variation of the complex a is due to the variation of S (Q remaining constant); when the stream is braiding, then its variation is due to the variation of Q (S remaining constant). The 2.5/1-declining lines in Fig 5.43 are designated accordingly.

ii- Consider now the braiding stream as a whole. With the increment of k (of divisions) its average depth decreases (see (5.84)), while its total width (the sum of widths of channels and islands across a section) increases; i.e. B/h increases and h/D decreases. On the other hand, the total Q and S remain constant, and consequently the complex a also remains constant (the variation of c is ignored). Hence, the development of a braiding stream as a whole can be reflected by the (upward-to the left) motion of the point m along the 2.5/1-declining straight line, β_2 say, which originates from $(P_0)_b$ and which represents a certain $a = const.$

5.6.4 *Delta formation*

When a single-channel stream enters a stationary fluid mass (when a river enters the sea), it convectively decelerates; its cross-section area Bh progressively increases, while its transport capacity decreases in the flow direction l or x. This leads to the increment of the aspect ratio B/h along x, and to the occurrence of deposition. Let x_e be the (end) section where the stream enters the sea, and x ($< x_e$) any other stream section in the decelerated region.

1- If the point P (on the (B/h; h/D)-plane) corresponding to any x remains below the line l_1 (i.e. if $P \in R_m$), then the single-channel stream enters the sea in the form of a (widening) "river mouth" (Thames, Seine, St. Lawrence, etc.).

2- If, on the other hand, there exists such a region, $\overline{x_b x_e}$ say, that the point P corresponding to any x in that region is above the line l_1 (i.e. if $P \in R_b$), then the stream braids in $\overline{x_b x_e}$ and it enters the sea in the form of a "delta" (Nile, Mississippi, Mackenzie, etc.).

Thus, in the case of a usual braiding, the increment of B/h (as to yield $P \in R_b$) is mainly due to the decrement of $h-$ caused by the increment of S; in the case of a delta, it is due to the increment of $B-$ caused by convective deceleration.

References

1. Ackers, P., Charlton, F.G.: *The geometry of small meandering streams.* Proc. Instn. Civ. Engrs., Paper 73286S, London, 1970.
2. Ashmore, P.E.: *Laboratory modelling of gravel braided stream morphology.* Earth Surface Processes and Landforms, Vol. 7, 1982.
3. Bristow, C.S.: *Brahmaputra River: channel migration and deposition.* in *Recent Developments in Fluvial Sedimentology,* Soc. Economic Paleontologists and Mineralogists, Special Publication No. 39, 1987.
4. Chang, H.H.: *Fluvial processes in river engineering.* John Wiley and Sons, 1988.
5. Chang, H.H.: *On the cause of river meandering.* Int. Conf. on River Regime, W.R. White ed., Wallingford, 1988.
6. Coleman, J.M.: *Brahmaputra River: channel processes and sedimentation.* Sediment. Geol., Vol. 3, 1969.
7. De Vriend, H.J.: *A mathematical model of steady flow in curved shallow channels.* J. Hydr. Res., Vol. 15, No. 1, 1977.
8. Dietrich, W.E., Whiting, P.: *Boundary shear stress and sediment transport in river meanders of sand and gravel.* in *River Meandering,* S. Ikeda and G. Parker eds., American Geophysical Union, Water Resources Monograph, 12, 1989.
9. Falcon-Ascanio, M., Kennedy, J.F.: *Flow in alluvial-river curves.* J. Fluid Mech., Vol. 133, 1983.
10. Friedkin, J.F.: *A laboratory study of the meandering of alluvial rivers.* U.S. Waterways Exp. Sta., Vicksburg, Mississippi, 1945.
11. Fujita, Y.: *Bar and channel formation in braided streams.* in *River Meandering,* S. Ikeda and G. Parker eds., American Geophysical Union, Water Resources Monograph, 12, 1989.
12. Fujita, Y., Muramoto, Y.: *Multiple bars and stream braiding.* Int. Conf. on River Regime, W.R. White ed., Wallingford, 1988.
13. Garde, R.J., Raju, K.G.R.: *Mechanics of sediment transportation and alluvial stream problems.* Wiley Eastern, New Dehli, 1977.
14. Grishanin, K.V.: *Dynamics of alluvial streams.* Gidrometeoizdat, Leningrad, 1979.
15. Hasegawa, K.: *Universal bank erosion coefficient for meandering rivers.* J. Hydr. Engrg., ASCE, Vol. 115, No. 6, June 1989.
16. Hooke, R.L.: *Distribution of sediment transport and shear stress in a meander bend.* Rept. 30, Uppsala Univ. Naturgeografiska Inst., 58, 1974.
17. Ikeda, H.: *A study on the formation of sand bars in an experimental flume.* (in Japanese) Geographi. Rev. Japan, 46-7, 1973.
18. Ikeda, S., Parker, G., Sawai, K.: *Bend theory of river meanders. Part 1. Linear development.* J. Fluid Mech., Vol. 112, 1981.
19. Jackson, R.J.: *Velocity-bed-form-texture patterns of meander bends in the lower Wabash River of Illinois and Indiana.* Geol. Soc. Am. Bull., Vol. 86, 1975.
20. Jansen, P. Ph., van Bendegom, L., van den Berg, J., De Vries, M., Zanen, A.: *Principles of river engineering.* Pitman, London, 1979.
21. Kikkawa, H., Ikeda, S., Kitagawa, A.: *Flow and bed topography in curved open channels.* J. Hydr. Div., ASCE, Vol. 102, No. HY9, Sept. 1976.
22. Kinoshita, R.: *Investigation of the channel deformation of the Ishikari River.* (In Japanese) Science and Technology Agency, Bureau of Resources, Memorandum No. 36, 1961.

23. Kondratiev, N., Popov, I., Snishchenko, B.: *Foundations of hydromorphological theory of fluvial processes.* (In Russian) Gidrometeoizdat, Leningrad, 1982.
24. Kondratiev, N.E., Lyapin, A.N., Popov, I.V., Pinkovskii, S.I., Fedorov, N.N., Yakunin, I.N.: *River flow and river channel formation.* Gidrometeoizdat, Leningrad, 1959. Translated from Russian by the Israel Program for Scientific Translations, Jerusalem, 1962.
25. Lapointe, M.F., Carson, M.A.: *Migration pattern of an asymmetric meandering river: the Rouge River, Quebec.* Water Resour. Res., Vol. 22, No. 5, May 1986.
26. Leliavskii, S: *An introduction to fluvial hydraulics.* Constable and Company, 1959.
27. Leopold, L.B., Langbein, W.B.: *River meanders.* Sci. Am., 214, 1966.
28. Leopold, L.B., Wolman, M.G., Miller, J.P.: *Fluvial processes in geomorphology.* W.H. Freeman, San Francisco, 1964.
29. Leopold, L.B., Wolman, M.G.: *River channel patterns: braided, meandering and straight.* U.S. Geol. Survey Professional Paper 282-B, 1957.
30. Lewin, J.: *British meandering rivers: the human impact.* in *River Meandering,* Proc. Conf. Rivers's 83, ASCE, 1983.
31. Lewin, J.: *Initiation of bed forms and meanders in coarse-grained sediment.* Geol. Soc. Am. Bull., Vol. 87, 1976.
32. Makaveyvev, N.I.: *River bed and erosion in its basin.* Press of the Academy of Sciences of the USSR, Moscow, 1975.
33. Matthes, G.H.: *Mississippi River cutoffs.* Trans. ASCE, Vol. 113, 1948.
34. Matthes, G.H.: *Basic aspects of stream meanders.* Amer. Geophys. Union, 1941.
35. Nelson, J.M., Smith, J.D.: *Evolution and stability of erodible channel beds.* in *River Meandering,* S. Ikeda and G. Parker eds., American Geophysical Union, Water Resources Monograph, 12, 1989.
36. Nelson, J.M., Smith, J.D.: *Flow in meandering channels with natural topography.* in *River Meandering,* S. Ikeda and G. Parker eds., American Geophysical Union, Water Resources Monograph, 12, 1989.
37. Parker, G., Andrews, E.D.: *On the time development of meander bends.* J. Fluid Mech., Vol. 162, 1986.
38. Parker, G. Diplas, P., Akiyama, J.: *Meander bends of high amplitude.* J. Hydr. Engrg., ASCE, Vol. 109, No. HY10, Oct. 1983.
39. Parker, G., Sawai, K., Ikeda, S.: *Bend theory of river meanders. Part 2. Nonlinear deformation of finite-amplitude bends.* J. Fluid Mech., Vol. 115, 1982.
40. Rozovskii, J.L.: *Flow of water in bends of open channels.* The Academy of Sciences of the Ukrainian SSR, 1957, Translated from Russian by the Israel Program for Scientific Translations, Jerusalem, 1962.
41. Schumm, S.A., Khan, H.R.: *Experimental study of channel patterns.* Geol. Soc. Am. Bull., 83, June 1972.
42. Shimizu, Y., Itakura, T.: *Practical computation of two-dimensional flow and bed deformation in alluvial streams.* Civil Engrg. Research Inst. Rept., Hokkaido Development Bureau, Sapporo, 1985.
43. Silva, A.M.F.: *Alternate bars and related alluvial processes.* M.Sc. Thesis, Dept. of Civil Engrg., Queen's Univ., Kingston, Canada, 1991.
44. Smith, D.G., Smith, N.D.: *Sedimentation in anastomosed river systems: examples from alluvial valleys near Banff, Alberta.* J. Sedimentary Petrology, Vol. 50, 1980.
45. Struiksma, N., Crosato, A.: *Analysis of a 2-D bed topography model for rivers.* in *River Meandering,* S. Ikeda and G. Parker eds., American Geophysical Union, Water Resources Monograph, 12, 1989.
46. Sukegawa, N.: *Condition for the occurrence of river meanders.* J. Faculty Engrg., Tokyo Univ., Vol. 30, 1970.
47. Von Schelling, H.: *Most frequent random walks.* Gen. Elec. Co., Rept. 64 GL92, Schenectady, N.Y., 1964.
48. Von Schelling, H.: *Most frequent particle paths in a plane.* Am. Geophysical Union Trans., Vol. 32, 1951.

49. Williams, P.F., Rust, B.R.: *The sedimentology of a braided river.* J. Sedimentary Petrology, Vol. 39, 1969.
50. Yen, B.C.: *Spiral motion of developed flow in wide curved open channels.* in *Sedimentation (Einstein),* H.W. Shen, ed., Fort Collins, Colorado, 1972.
51. Yen, C.L., Yen, B.C.: *Water surface configuration in channel bends.* J. Hydr. Div., ASCE, Vol. 97, No. HY2, Feb. 1971.
52. Znamenskaya, N.S.: *Sediment transport and alluvial processes.* Hydrometeoizdat, Leningrad, 1976.

References A
Sources of Meandering and Braiding River Data

1a. Ackers, P., Charlton, F.G.: *The geometry of small meandering streams.* Proc. Instn. Civ. Engrs., Paper 7328S, London, 1970.
2a. Chitale, S.V.: *River channel patterns.* J. Hydr. Div., ASCE, Vol. 96, No. HY1, 1970.
3a. Dietrich, W.E., Smith, J.D.: *Influence of the point bar on flow through curved channels.* Water Resour. Res., Vol. 19, No. 5, Oct. 1983.
4a. Fujita, Y., Muramoto, Y.: *Experimental study on stream channel processes in alluvial rivers.* Bull. Disaster Prevention Res. Inst., Kyoto Univ., Vol. 32, part I, No. 288, 1982.
5a Kellerhals, R., Neill, C.R., Bray, D.I.: *Hydraulic and geometric characteristics of rivers in Alberta.* Alberta Cooperative Research Program in Highway and River Engineering, 1972.
6a. Kinoshita, R.: *Model experiments based on the dynamic similarity of alternate bars.* (In Japanese) Research Report, Ministry of Construction of Japan, Aug. 1980.
7a. Lapointe, M.F., Carson, M.A.: *Migration pattern of an asymmetric meandering river: the Rouge River, Quebec.* Water Resour. Res., Vol. 22, No. 5, May 1986.
8a. Leopold, L.B., Wolman, M.G.: *River channel patterns: braided, meandering and straight.* U.S. Geol. Survey Prof. Paper 282-B, 1957.
9a. Neill, C.R.: *Hydraulic geometry of sand rivers in Alberta.* Proc. Hydrology Symposium, Alberta, May 1973.
10a. Odgaard, A.J.: *Streambank erosion along two rivers in Iowa.* Water Resour. Res., Vol. 23, No. 7, July 1987.
11a. Odgaard, A.J.: *Transverse bed slope in alluvial channel bends.* J. Hydr. Div., ASCE, Vol. 107, No. HY12, Dec. 1981.
12a. Pizzuto, J.E., Meckelnburg, T.S.: *Evaluation of a linear bank erosion equation.* Water Resour. Res., Vol. 25, No. 5, May 1989.
13a. Prestegaard, K.L.: *Bar resistance in gravel bed streams at bankfull stage.* Water Resour. Res., Vol. 19, No. 2, April 1983.
14a. Schumm, S.A., Khan, H.R.: *Experimental study of channel patterns.* Geol. Soc. Am. Bull., 83, June 1972.
15a. Schumm, S.A.: *River adjustements to altered hydrologic regime – Murrumbidgee river and paleochannels, Australia.* U.S. Geol. Survey Prof. Paper 598, 1968.
16a. Smith, N.D.: *Transverse bars and braiding in the Lower Platte River, Nebraska.* Geol. Soc. Am. Bull., Vol. 82, 1971.
17a. Struiksma, N., Klaasen, G.J.: *On the threshold between meandering and braiding.* Int. Conf. on River Regime, W.R. White ed., Wallingford, 1988.

SUBJECT INDEX